GEOMETRY AND TOPOLOGY

GEOMETRY AND TOPOLOGY

Editors
G. Stratopoulos
G.M. Rassias
The American College of Greece

World Scientific
Singapore • New Jersey • London • Hong Kong

Published by

World Scientific Publishing Co. Pte. Ltd.
P O Box 128, Farrer Road, Singapore 9128

USA office: World Scientific Publishing Co., Inc.
687 Hartwell Street, Teaneck, NJ 07666, USA

UK office: World Scientific Publishing Co. Pte. Ltd.
73 Lynton Mead, Totteridge, London N20 8DH, England

Library of Congress Cataloging-in-Publication data is available.

GEOMETRY AND TOPOLOGY

ISBN: 9971-50-678-5

Printed in Singapore by Utopia Press.

PREFACE

The GEOMETRY AND TOPOLOGY volume contains a series of specially invited papers written by eminent mathematicians of the international community.

These papers deepen our understanding of some of the longstanding research problems of Geometry and Topology and provide the opportunity for publication of new aspects of mathematical research problems that will be of the greatest value to all scientists currently working within these areas of research.

This volume hopes to provide a source of the latest developments in these important and growing fields of mathematics. The book's presentation of concepts and methods makes it an invaluable reference for advanced graduate students and research mathematicians interested in introducing themselves to these research areas as well as for teachers and other professionals in mathematics interested in these fields.

Our warmest thanks are due to all scientists who so willingly responded to our invitation and enthusiastically participated to the preparation of this work. We also express our sincere appreciation and gratitude to the Publishers for including this volume to their series of publications and presenting this work to the scientists and students of the International Mathematical Community.

October 31, 1988 The Editors
Athens, Greece

CONTENTS

GEOMETRY AND TOPOLOGY

GEOMETRY AND TOPOLOGY (pp. 1–14)
edited by G. M. Rassias & G. M. Stratopoulos

CAUSAL CURVE EXTENSIONS

John K. Beem*

Department of Mathematics

University of Missouri

Columbia, MO 65211

ABSTRACT

Let γ be a curve associated with a local extension of a space-time. If γ is causal, then the local extension is said to be a causal curve extension. Minkowski space-time and open Robertson-Walker big bang models have causal curve extensions. Complete analytic space-times fail to have endpoint convergent curve extensions. This work yields a new argument for closed cosmological models.

1. INTRODUCTION

An extension of a space-time (M, g) is a larger space-time (M', g') which contains an isometric copy of (M, g) as a proper open subset. If (M, g) is extendible, then it contains at least some incomplete

*Supported in part by the National Science Foundation grant number DMS-8601342

timelike, null and spacelike geodesics. Furthermore, in this case (M, g) is also bounded acceleration incomplete and b-incomplete [2]. In General Relativity one usually defines a space-time to be **singular** if it contains an incomplete causal (i.e., timelike or null) geodesic. [2,13] Since extendible space-times are always singular in this mathematical sense, the search for physically significant singularities is usually undertaken in **maximal** (i.e., inextendible) space-times. Every extendible space-time has at least one maximal extension [6], but maximal extensions need not be unique.

The study of extensions has led to consideration of local extensions. A first approach to local extensions is to use open subsets with noncompact closure and extend these sets to space-times where they have compact closure [1,2,3,13]. An alternative is to use open subsets which contain some curve defining a point of the b-boundary of Schmidt [4,8,15,16]. In the present paper we consider local extensions about curves and are primarily interested in the case in which the curve is causal. A curve extension will be an extension of an open set U containing an inextendible (i. e., endless) curve $\gamma : [0, b) \to U$. We will require that the image of γ have noncompact closure in the original space-time (M, g) but that the corresponding image have compact closure in the extension of U. We will say the curve extension is **endpoint convergent** if the image of γ in the extension of U converges to some point as $t \to b$. This is the requirement that the image of γ be C^0 extendible as a curve. We give analytic examples of causal curve extensions of Minkowski space-time. Furthermore, we consider open Robertson-Walker cosmological models with pressure p and cosmological constant Λ equal to zero. We indicate, using the Einstein-de Sitter model, how class C^2 causal curve extensions may be constructed.

We also prove that no analytic endpoint convergent curve extensions exist for complete analytic space-times.

This paper was motivated by two considerations. The first of these was a desire to better understand the nature of the local extensions of Minkowski space-time and how these might be generalized. The first local extension of L^n was given by the author [1] for a curve which diverged to spacial infinity. The second consideration was more philosophical. A causal curve represents the world line of an observer. Thus an inextensible future directed causal curve represents a theoretically possible path which some (accelerated) individual might take in traveling to the edge of the universe. If there is a causal curve extension for this curve, then in theory one might use this local extension to escape from our universe to another. The observer and a small part of our universe would be involved. Since this seems to be implausible, one obtains a new argument for a closed cosmological model. The (usual) closed big bang cosmological models have no curve extensions.

2. PRELIMINARIES

Let (M, g) be an n-dimensional **space-time**. This means M is a connected, C^∞, Hausdorff manifold with a countable base, a Lorentzian metric of signature $(-, +, \cdots, +)$ and a time orientation. A class C^1 curve $\gamma : (a, b) \to M$ is a **causal curve** if $g(\gamma', \gamma') \leq 0$ for all $a < t < b$. It is future directed if the tangent vector γ' is always future directed.

The space-time (M, g) has an **extension** (M', g') if there is a diffeomorphism $f : M \to M'$ of M onto a proper open subset $f(M)$ of M' such that f is an isometry. Since space-times are connected, the set $f(M)$ nust have a nonempty boundary in M' and this yields the existence of incomplete geodesics in the original (M, g). A space-time that is not extendible is said to be **maximal or inextendible.** The n-dimensional Minkowski space-time L^n is geodesically complete and hence maximal. On the other hand, let U be a connected open set in L^n with $\bar{U}$ compact. Then U has L^n as one maximal extension, but also one may obtain maximal extensions of U using flat metrics on the n-dimensional torus T^n. Thus maximal extensions need not be unique even when one requires real analytic extensions.

A local extension [2] of the space-time (M, g) is a connected open subset U of M which has noncompact closure $\bar{U}$ in M and an extension (U', g') of $(U, g|_U)$ with an isometry f mapping U into U' such that the image $f(U)$ has compact closure in U'.

A second definition of local extension used in the literature [4,8] involves the b-boundary. In this paper, we will say (M, g) has a **local b-boundary extension about** γ is there is an open set U in M, an isometric extension $f : U \to U'$ of U into a space-time (U', g') and a curve $\gamma : [0, b) \to U$ such that

(i) the curve γ is inextendible in M to $t = b$ and γ is of finite generalized affine parameter length,

(ii) the curve $(f \circ \gamma)(t)$ is C^0 extendible in (U', g').

We now define curve extensions and causal curve extensions. These (local) extensions have an important subcase in which the curve is end-point convergent and hence the curve is C^0 extendible in the image space.

Definition 2.1. *Let $\gamma : [0, b) \to M$ be an inextendible (to $t = b$) class C^1 curve with an image $\gamma[0, b)$ which does not have compact closure. A* **curve extension about** *γ is an open set U containing $\gamma[0, b)$,* *a space-time (U', g') and an isometry $f : U \to U'$ such that the image $f \circ \gamma[0, b)$ has compact closure in U'. If $f \circ \gamma(t) \to p' \in U'$ as $t \to b$, then the curve extension is said to be* **endpoint convergent.**

A curve extension about a causal curve γ is said to be a **causal curve extension.**

Notice that if one is given a space-time (M, g) and local extension $f : U \to U'$, then one may construct many curves γ such that the extension is a curve extension about γ. Conversely, given a curve extension about γ one may, if necessary, redefine U to obtain a local extension. The essential idea in the above Definition 2.1 is that a unique curve γ has been distinguished. It then makes sense to consider the causal nature of γ and possible extendibility of the corresponding curve in the image space. It should be noted that the curve γ does not necessarily approach a point of the b-boundary because in Definition 2.1 the curve γ is allowed to have infinite generalized affine parameter length.

Clarke [6] has defined primordial singularities as certain parts of the b-boundary accessible along causal curves. Our approach has certain similarities but is different. In particular, we are interested in some space-times that are b-complete and hence have no b-boundary.

Most of our notational conventions are standard [2,13,14] Discussions of singularities and space-time boundaries may be found in a number of articles. [5,7,9,10,11]

3. CAUSAL EXTENSIONS

Let (M, g) have a curve extension corresponding to a causal curve $\gamma : [0, b) \to M$. Let f, U, and (U', g') be as in Definition 2.1. The curve $f \circ \gamma : [0, b) \to U'$ is a causal curve of

(U',g') which is imprisoned in the compact set $\overline{f \circ \gamma[0,b)}$. Hence either $f \circ \gamma(t)$ approaches an endpoint $p' \in U'$ as $t \to b$ or else (U',g') has compactly imprisoned and inextendible (i.e., endless) causal curves. In the latter case one may conclude certain restrictions on the causal nature of (U',g'). In particular, (U',g') cannot be strongly causal.

Lemma 3.1. *Let (M,g) have a causal curve extension about $\gamma : [0,b) \to M$ given by $f : U \to U'$. If (U',g') is strongly causal, then the extension is endpoint convergent and there must be some $p' \in U'$ such that $f \circ \gamma(t) \to p'$ as $t \to b$.*

Remark 3.2. If (U',g') is only causal and not strongly causal, then the result of Lemma 3.1 may fail. One may obtain examples of this behavior based on a space-time constructed by Carter [13].

We now give a causal curve extension of Minkowski space-time L^n into a chronological space-time. This local extension is across the ideal point i^+ (i.e., future timelike infinity).

Example 3.3. Let $M = L^n$ have the standard metric $\eta = -dx_1^2 + dx_2^2 + \cdots + dx_n^2$. Let $U' = S^1 \times \mathbb{R}^{n-1}$ be given the flat metric $g' = -d\theta^2 + dx_2^2 + \cdots + dx_n^2$ where $S^1 = \{\theta | 0 \le \theta \le 1\}$ is $\mathbb{R}^1$ (mod 1). Set $\gamma(t) = ([1-t]^{-1}, 1-t, 0, \ldots, 0)$ for $0 \le t < 1$. Let $D(r) = \{(x_1, x_2, \ldots, x_n) | \, x_1 = r$ and $(x_2 - r^{-1})^2 + x_3^2 + \ldots + x_n^2 < (100r^4)^{-1}\}$ and let U be the union of all of the sets $D(r)$ for $r > .5$. The open set U is like a tube about γ of variable radius $.1x_1^{-2}$ which thus gets thinner as $x_1 \to \infty$. Define $f : U \to U'$ by $f(x_1, x_2, \ldots, x_n) = (x_1(\text{mod } 1), x_2, \ldots, x_n)$. The map f is an isometry which "wraps" U around the set U' and takes the curve γ to a spiral which approaches the copy of S^1 given by $x_2 = x_3 = \cdots = x_n = 0$ in U'.

Notice that in the above example the curve γ represents a traveler in the $(n-1)$-dimensional space $(x_2, \ldots, x_n)$ moving towards the origin along the x_2 axis.

This traveler has bounded acceleration which goes to zero as $t \to \infty$. Since the length $L(\gamma)$ of γ is infinite, it takes this individual an infinite amount of time to reach the origin and (under f) leave the original space.

In the next example an observer leaves Minkowski space-time after a finite length of time. Minkowski space-time is b. a. complete and hence the observer has unbounded acceleration.

Example 3.4. Let $M = \mathrm{R}^n$ have the flat matric $g = -dx_1 dx_2 + dx_3^2 + \cdots + dx_n^2$. Let $\gamma(t) = (t, \tan(t), 0, \ldots, 0)$ for $0 \leq t < \pi/2$. Set $U' = S^1 \times \mathrm{R}^{n-1}$ where S^1 is R^1 (mod 1) and $g' = -d\theta dx_2 + dx_3^2 + \ldots + dx_n^2$. As in Example 3.3 one may construct an appropriate open set U about γ and an isometric map $f : U \to U'$ of the form $f(x_1, x_2, \ldots, x_n) = (x_1 \,(\mathrm{mod}\ 1), x_2, \ldots, x_n)$.

Notice that both Examples 3.3 and 3.4 are real analytic causal curve extensions and that neither one is endpoint convergent. In both cases the image space-time is chronological but not causal.

A four dimensional **generalized Robertson-Walker space-time** is a manifold $M = (a, b) \times H$ for $-\infty \leq a < b \leq \infty$ with a metric of the form $g = -dt^2 + S^2(t)\, d\sigma^2$ where $(H, d\sigma^2)$ is a three dimensional isotropic (and homogeneous) positive definite Riemannian manifold [2,13]. Wang [17] has shown that $(H, d\sigma^2)$ must be one of the Euclidian, hyperbolic, spherical or elliptic spaces. The corresponding topological structures for $M = (a, b) \times H$ are R^4, R^4, $\mathrm{R}^1 \times S^3$ and $\mathrm{R}^1 \times \mathrm{R}P^3$. In cosmology, the first two cases are said to be open since H is not compact. In these cases one usually takes $a = 0$ and $b = \infty$.

The last two cases are said to be closed models because H is compact. In these cases one takes $a = 0$ and $0 < b < \infty$. Under reasonable physical assumptions [13] one has an "essential" singularity corresponding to the original big bang at $t = 0$ for all four cases.
In the open models the matter spreads out as $t \to \infty$ and the spacial universe tends towards flatness. In the closed models there is a second singularity as $t \to b$ corresponding to a second big bang.

In the present paper we will consider the case with pressure p and cosmological constant Λ both equal to zero. As usual we normalize the sectional curve K of $(H, d\sigma^2)$ to be -1 if $(H, d\sigma^2)$ is hyperbolic. The open model solutions for $S(t)$ using a parameter τ satisfying $d\tau/dt = S^{-1}(t)$ can be expressed as [12,13]

$$S = \left(\tfrac{E}{3}\right)(\cosh\tau - 1), \quad t = \left(\tfrac{E}{3}\right)(\sinh\tau - \tau) \quad \text{for } K = -1$$

$$S = \tau^2 \qquad , \quad t = \tfrac{\tau^3}{3} \qquad \text{for } K = 0$$

The case $K = 0$ is the **Einstein-de Sitter** universe. We will use this case to illustrate how one may obtain future directed causal curve extensions. The method is similar to that of Example 3.3 but due to differentiability considerations we have only been able to establish the existence of class C^2 extensions.

Let $M = (0, \infty) \times \mathbb{R}^3 = \{(t, x, y, z) | \, t > 0\}$ be given the Einstien-de Sitter metric

$$g = -dt^2 + At^{\frac{4}{3}}(dx^2 + dy^2 + dz^2). \tag{1}$$

One may assume without loss of generality that the constant A is equal to one since a simple dilation in x, y, z allows a rescaling of A. Let $M' = \{(s, u, v, w) | \, s > 0\}$ and define the diffeomorphism $F : M' \to M$ by

$$t = s \qquad y = s^{-\frac{2}{3}}v$$

$$x = s^{-\frac{2}{3}}u \qquad z = s^{-\frac{2}{3}}w.$$

The pull back metric $g' = F^*g$ on M' is given by

$$g' = \begin{cases} -\,ds^2 + du^2 + dv^2 + dw^2 + \left(\dfrac{4}{9}\right) s^{-\frac{10}{3}}(u^2+v^2+w^2)\,ds^2 \\ -\left(\dfrac{4}{3}\right) s^{-\frac{7}{3}}(u\,duds + v\,dvds + w\,dwds). \end{cases} \tag{2}$$

Restricting our attention to the set $u^2 + v^2 + w^2 < 2$ one finds that the partial derivatives of the components of g' are of the order $\mathcal{O}(s^{-\frac{7}{3}})$. Define $h_{ij} = h_{ij}(s,u,v,w)$ by

$$g' = -ds^2 + du^2 + dv^2 + dw^2 + h_{ij}du_idu_j \tag{3}$$

where $u_1 = s$, $u_2 = u$, $u_3 = v$ and $u_4 = w$. Define $\gamma : [-1,0) \to M'$ by $\gamma(t) = (t^{-4}, -t, 0, 0)$, set $D_0(r) = \{(s,u,v,w) \in M' | \, s = r \text{ and } (u - r^{-1/4})^2 + v^2 + w^2 < 10^{-6}r^{-5/2}\}$ and let $D_1(r) = \{(s,u,v,w) \in M' | \, s = r \text{ and } (u - r^{-1/4}) + v^2 + w^2 < 10^{-4}r^{-5/2}\}$. Let U_0 and U_1 be respectively the unions of all $D_0(r)$ and $D_1(r)$ for $r > .6$ and $r > .5$. The set U_0 is an open tube about γ corresponding to radius $10^{-3}s^{-5/4}$ and U_1 is a larger tube corresponding to radius $10^{-2}s^{-5/4}$. Now let $M'' = S^1 \times \mathbb{R}^3$ have coordinates (θ, u, v, w) where $S^1 = \mathbb{R}^1 \pmod 1$. Define $f : U_1 \to M''$ by $f(s,u,v,w) = (s \pmod 1, u, v, w)$. We wish to define a metric g'' on M'' such that $f|_{U_0}$ is a causal curve extention. On $f(U_0)$ define $g'' = (f^{-1})^*(g')$ and on the set $M'' - f(U_1)$ use the flat metric $g'' = -d\theta^2 + du^2 + dv^2 + dw^2$. To define g'' on $f(U_1 - U_0)$ first construct a smooth real valued function $\phi : M' \to [0,1]$ with $\phi \equiv 0$ on $M' - U_1$ and $\phi \equiv 1$ on U_0. The metric g'' on $f(U_1 - U_0)$ is then given by

$$g'' = -d\theta^2 + du^2 + dv^2 + dw^2 + (f^{-1})^*(\phi h_{ij}\, du_idu_j). \tag{4}$$

Clearly, we have obtained at least a class C^0 causal curve extension. Using $u \sim s^{-1/4}$ and the $s^{-7/3}$ factors in the coefficients h_{ij}, one has h_{ij} of order $\mathcal{O}(s^{-31/12})$ on $U_1 - U_0$. One may show that ϕ can be chosen in such a way that $f : U_0 \to M''$ is at least a class C^2 causal curve extension. The difficulty is that as $s \to \infty$ the set $U_1 - U_0$ becomes thin at a rate proportional to $s^{-5/4}$. Consequently, one expects the second partials of ϕ to increase at the rate $s^{5/2}$ and the third partials at the rate $s^{15/4}$. We do not know if there are class C^k local extensions for $k \geq 3$. We have indicated the construction of an example which shows the following.

Lemma 3.5. *The Einstein-de Sitter space-time has class C^2 causal curve extensions.*

The previous examples have all failed to be endpoint convergent. We now give an example of a (noncausal) endpoint convergent curve extension. It is not a local *b*-boundary extension because the curve γ has infinite generalized affine parameter length. This example is an illustration of the difference between local *b*-boundary extensions and endpoint convergent curve extensions. This example also shows that if one cuts a closed set A out of a space-time, then one does *not* necessarily get one or more points of the *b*-boundary for every point on the boundary of A which is the endpoint of a curve in $M - A$.

Example 3.6. Let $M = \mathbb{R}^2$ have the usual Minkowski metric $\eta = -dx_1^2 + dx_2^2$. Let U_0 be the open rectangle $\{(x_1, x_2) | -1 < x_1 < 0, -1 < x_2 < +1\}$. We will delete a countable number of closed line segements from U_0. Let T_n be the line given by $x_1 = -2^{-n}$. From T_n delete an open segment S_n of length 2^{-n} centered at $(-2^{-n}, (-1)^n/n)$. The sets S_n are slits that alternate between the third and fourth quadrants. Delete each set $(T_n - S_n)$ from U_0 to obtain the open set U. The set U

is connected and the point (0,0) is a boundary point of U. Construct a curve $\gamma : [0,1) \to U$ with $\gamma(0) = (-.75, 0)$ such that γ passes through each slit S_n and satisfies $\gamma(t) \to (0,0)$ as $t \to 1$. The Euclidean length of γ is infinite and since the Euclidean and Minkowskian connexions agree on $\mathbb{R}^2$, it follows that γ has infinite generalized affine parameter length. Clearly, U is extendible to M and γ is endpoint convergent in this extension. On the other hand, there is no local b-boundary extension for γ because $(0,0)$ is not a b-boundary point.

4. ANALYTIC EXTENSIONS

Let (M, g) be a real analytic space-time with an inextendible class C^1 curve $\gamma : [0, b) \to M$ with image $\gamma[0, b)$ which fails to have compact closure in M. Assume U, (U', g') and $f : U \to U'$ are as in Definition 2.1. If (U', g') and f are real analytic, then the (local) extension is said to be an analytic curve extension. We now generalize a result of Beem and Ehrlich [3] to the class of analytic endpoint convergent curve extensions.

Theorem 4.1. *Let (M, g) be an analytic space-time. If (M, g) has an analytic endpoint convergent curve extension about $\gamma : [0, b) \to M$, then (M, g) contains an incomplete geodesic.*

Proof: Assume (M, g) is geodesically complete and let U, (U', g') and f be as in Definition 2.1. Then $f \circ \gamma : [0, b) \to U'$ and $f \circ \gamma(t) \to p \in U'$ as $t \to b$. Let $W' \subset U'$ be a neighborhood of p such that W' is a convex normal neighborhood of each of its points. Assume without loss of generality that U is connected and $f(U) \subset W'$. Set $q = \gamma(0)$, $r = f(q)$ and $F = \exp_q \circ (f_{*q})^{-1} \circ \exp_r^{-1}$. Then $F : W' \to M$ is analytic and is at least defined near q.

Since f is an isometry the linear maps $f_{*q} : T_qM \to T_rU'$ and $(f_{*q})^{-1} : T_rU' \to T_qM$ are nonsingular, analytic and defined on all of T_qM and T_rU', respectively. Since (M, g) is assumed to be geodesically complete and analytic, $\exp_q : T_qM \to M$ is an analytic map defined on all of T_qM. The map $\exp_r^{-1} : W' \to T_rU'$ is analytic, defined on all of W' and maps W' to an open star shaped region set about the origin of T_rU'. It follows that F is an analytic map defined on all of W'. Using the definition of the exponential map, one finds that F maps geodesic segments in W' starting at r to geodesic segments in M starting at q. Furthermore, the $(f_{*q})^{-1}$ factor in F implies that F preserves the lengths of geodesic segments starting at r. It easily follows that F and the analytic isometry $f^{-1} : f(U) \to U$ must agree on at least a small neighborhood V' about $r \in W'$. If $W'' = f(U)$, then W'' is an open connected subset of W' and is thus an analytic manifold. The maps $F|_{W''}$ and f^{-1} are both analytic maps from W'' into M. Since they agree on the open set V', they must agree on all of W''. This yields $F \circ f \circ \gamma(t) = \gamma(t)$ for all $0 \leq t < b$ and hence we must have $\lim \gamma(t) = \lim F \circ f \circ \gamma(t) = F(p)$ for $t \to b$. Thus γ is extendible in M in contradiction to Definition 2.1. We conclude that (M, g) cannot be geodesically complete.

Corollary 4.2. *Minkowski space-time L^n has no analytic endpoint convergent curve extensions.*

The above corollary now makes clear that the analytic extensions of L^n in Example 3.3 and 3.4 were of necessity not endpoint convergent.

If the curve γ in Theorem 4.1 is a future directed causal curve, then one can prove the existence of an incomplete future directed causal geodesic. Thus the open Robertson-Walker models we have considered have no analytic future directed causal curve extensions which are endpoint convergent.

We now end with some open questions. We believe that Questions 1 and 2B have no for their answers.

Question 1. Does Minkowski space-time L^n have any class C^k endpoint convergent curve extensions for $0 \leq k \leq \infty$? In particular, are there any endpoint convergent causal curve extensions?

Question 2A. Do the open Robertson-Walker cosmological models have curve extensions of class C^k for $k \geq 3$?

If the answer to Question 2A is yes, then one may ask if the techniques of Examples 3.3 and 3.4 may be used to obtain analytic extensions of the open Robertson-Walker models.

Question 2B. Do the open Robertson-Walker cosmological models have analytic causal curve extensions?

Acknowledgement: The author wishes to thank P. E. Ehrlich for helpful discussions.

References

1. Beem, J. K. *Minkowski space-time is locally extendible*, Commun. Math. Phys. 72, 273–275, (1980).
2. Beem, J. K. and Ehrlich, P. E. **Global Lorentzian Geometry**, Pure and Applied Math. vol 67, Marcel Dekker, New York, 1981.
3. Beem, J. K. and Ehrlich, P. E. *Local extensions and the existence of singularities in General Relativity*, Einstein Centenary Symposium in General Relativity, Nagpur, India, edited by T. M. Karade, Duhita, 107–121, (1982).
4. Clarke, C. J. S. *Local extensions in singular space-times*, Commun. Math. Phys. 32, 205–214, (1973).

5. Clarke, C. J. S. *The classification of singularities*, Gen. Rel. and Grav. 6, 35–40, (1975).

6. Clarke, C. J. S. *Space-time singularities*, Commun. Math. Phys. 49, 17–23, (1976).

7. Clarke, C. J. S. *Boundary definitions*, Gen. Rel. and Grav. 10, 977–980, (1979).

8. Clarke, C. J. S. *Local extensions in singular space-time II*, Commun. Math. Phys. 84, 329–331, (1982).

9. Clarke, C. J. S. and Schmidt, B. G. *Singularities: the state of the art*, Gen. Rel. and Grav. 8, 129–137, (1977).

10. Dodson, C. T. J. *Space-time edge geometry*, Int. J. Theo. Phys. 17, 389–504, (1978).

11. Ellis, G. F. R. and Schmidt, B. G. *Singular space-times*, Gen. Rel. and Grav. 8, 915–953, (1977).

12. Frankel, T. **Gravitational Curvature**, W. H. Freeman, San Francisco, 1979.

13. Hawking, S. W. and Ellis, G. F. R. **The Large Scale Structure of Space-time**, Cambridge Univ. Press, Cambridge, 1973.

14. O'Neill, B. **Semi-Riemannian Geometry**, Academic Press, New York, 1983.

15. Schmidt, B. G. *A new definition of singular points in General Relativity*, Gen. Rel. and Grav. 1, 269–280, (1971).

16. Schmidt, B. G. *The local b-completeness of space-times*, Commun. Math. Phys. 29, 49–54, (1973).

17. Wang, H. C. *Two-point homogeneous spaces*, Pacific J. Math. 1, 473–480, (1951).

GEOMETRY AND TOPOLOGY (pp. 15–30)
edited by G. M. Rassias & G. M. Stratopoulos

WHEN IS THE TANGENT SPHERE BUNDLE LOCALLY SYMMETRIC?

David E. Blair

ABSTRACT. It is shown that the natural Riemannian structure of the tangent sphere bundle of a Riemannian manifold is locally symmetric if and only if either the base manifold is flat or is of dimension 2 and constant curvature 1. The motivation for this result is the study of the tangent sphere bundle as a contact manifold and the general open question of the classification of locally symmetric contact metric manifolds.

1. INTRODUCTION.

Given a Riemannian manifold (M,G), its tangent bundle TM admits a natural Riemannian metric $\bar{g}$, called the Sasaki metric; it was shown by Kowalski [5] that $(TM,\bar{g})$ is locally symmetric if and only if (M,G) is flat. Now the tangent sphere bundle, T_1M, considered as a hypersurface of TM, inherits a Riemannian metric g (for reasons of normalization we will make a homothetic change of the induced metric to obtain our metric g) and we will prove the following result.

<u>Theorem</u>. (T_1M,g) is locally symmetric if and only if either (M,G) is flat or M is 2-dimensional and of constant curvature 1.

Now in view of Kowalski's result this is not surprising but our motivation here is quite different. Classically there are two large classes of examples of contact manifolds.

I. The odd-dimensional spheres and more generally the principal circle bundles over symplectic manifolds of integral class.

II. The tangent sphere bundles.

The contact manifolds in the first class admit an associated Riemannian metric which is K-contact, i.e. the characteristic vector field of the contact structure is a Killing field with respect to the metric. However, despite the example of $T_1S^2 \cong P\mathbb{R}^3$, the natural contact metric structure on T_1M is not in general K-contact.

<u>Theorem (Tashiro [11])</u>. The natural contact metric structure on T_1M is K-contact if and only if (M,G) has constant curvature 1, in which case the structure on T_1M is Sasakian.*

On the other hand we have the following result of Tanno which generalizes the corresponding result for Sasakian manifolds due to Okumura [6].

<u>Theorem (Tanno [9])</u>. A locally symmetric K-contact manifold is of constant curvature.

The results of Okumura and Tanno show why Takahashi [8] introduced for Sasakian manifolds the notions of φ-symmetric and locally φ-symmetric spaces.

---

*There should be no confusion between the <u>Sasaki metric</u> on TM and a <u>Sasakian structure</u> on a contact manifold, the latter being a normal contact metric structure (see e.g. [1] for details) and will not be needed in any substantive way in this paper.

Now some questions arise. In particular are there contact metric manifolds which are locally symmetric but not of constant curvature? Yes, $E^{n+1} \times S^n$ as the tangent sphere bundle of Euclidean space, and a general problem is to classify all locally symmetric contact metric manifolds. In a different direction one might try to develop a theory of φ-symmetric spaces for tangent sphere bundles.

We now turn to a discussion of tangent sphere bundles and contact metric manifolds and then in section 5 prove the theorem.

2. GEOMETRY OF TM AND T_1M.

Let M be an $(n+1)$-dimensional C^∞ manifold and $\bar{\pi}:TM \to M$ its tangent bundle. If $(x^1,\dots,x^{n+1})$ are local coordinates on M, set $q^i = x^i \circ \bar{\pi}$; then $(q^1,\dots,q^{n+1})$ together with the fibre coordinates $(v^1,\dots,v^{n+1})$ form local coordinates on TM. If X is a vector field on M, its vertical lift X^V on TM is defined by $X^V\omega = \omega(X)\circ\bar{\pi}$ where ω is a 1-form on M, which on the left side of this equation is regarded as a function on TM. For an affine connection D on M, the horizontal lift X^H of X is defined by $X^H\omega = D_X\omega$. The local expression for X^H in terms of the connection coefficients of D is

$$X^H = X^i \frac{\partial}{\partial q^i} - X^i v^j \Gamma^k_{ij} \frac{\partial}{\partial v^k} \tag{2.1}$$

The connection map $K: TTM \to TM$ is given by

$$KX^H = 0, \quad K(X^V{}_t) = X_{\bar{\pi}(t)} ,$$

$t \in TM$. TM admits an almost complex structure J defined by

$$JX^H = X^V, \qquad JX^V = -X^H \quad .$$

Dombrowski [4] showed that J is integrable if and only if D has vanishing curvature and torsion.

If now g is a Riemannian metric on M and D its Riemannian connection, we define a Riemannian metric $\bar{g}$ on TM, called the Sasaki metric, by

$$\bar{g}(X,Y) = G(\bar{\pi}_* X, \bar{\pi}_* Y) + G(KX,KY) \tag{2.2}$$

where X and Y are vector fields on TM. Since $\bar{\pi}_* \circ J = -K$ and $K \circ J = \bar{\pi}_*$, $\bar{g}$ is Hermitian for the almost complex structure J.

Remark. Already we can anticipate what will be happening and why the constant curvature 1 case of the theorem occuring only in dimension 2 is not surprising. The metric on T_1M to be studied will be, to within homothety, the one induced from $\bar{g}$, but note the role played by the Riemannian connection D of (M,G) in (2.1) and in turn by means of K in (2.2). Thus, for example, even though for the sphere S^3, T_1S^3 is topologically $S^3 \times S^2$, the induced metric is not the product metric. (The geometry of $S^3 \times S^2$ has been studied by Tanno in [10].)

On TM define a 1-form β by $\beta(X)_t = G(t, \bar{\pi}_* X)$, $t \in TM$ or equivalently by the local expression $\beta = \sum G_{ij} v^j dq^i$. $d\beta$ is then a symplectic structure on TM and in particular $2d\beta$ is the fundamental 2-form of the almost Hermitian structure $(J,\bar{g})$. Thus TM has an almost Kähler structure which is Kählerian if and only if (M,G) is flat (Dombrowski [4], Tachibana and Okumura [7]).

Let $\underline{R}$ denote the curvature tensor of G, then the Riemannian connection $\bar{\nabla}$ of $\bar{g}$ at $t \in TM$ is given by

$$(\bar{\nabla}_{X^H} Y^H)_t = (D_X Y)^H_t - \frac{1}{2} (\underline{R}_{XY} t)^V ,$$

$$(\bar{\nabla}_{X^H} Y^V)_t = - \frac{1}{2} (\underline{R}_{Yt} X)^H + (D_X Y)^V_t , \tag{2.3}$$

$$(\bar{\nabla}_{X^V}Y^H)_t = -\frac{1}{2}(\underline{R}_{Xt}Y)^H \ ,$$

$$\bar{\nabla}_{X^V}Y^V = 0 \ .$$

The curvature tensor $\bar{R}$ of $\bar{\nabla}$ was computed by Kowalski [5] and is given by

$$\bar{R}_{X^V Y^V}Z^V = 0 \ ,$$

$$(\bar{R}_{X^V Y^V}Z^H)_t = (\underline{R}_{XY}Z + \frac{1}{4}\underline{R}_{tX}\underline{R}_{tY}Z - \frac{1}{4}\underline{R}_{tY}\underline{R}_{tX}Z)^H_t \ ,$$

$$(\bar{R}_{X^H Y^V}Z^V)_t = -(\frac{1}{2}\underline{R}_{YZ}X + \frac{1}{4}\underline{R}_{tY}\underline{R}_{tZ}X)^H_t \ ,$$

$$(\bar{R}_{X^H Y^V}Z^H)_t = (\frac{1}{4}\underline{R}_{\underline{R}_{tY}Z,X}t + \frac{1}{2}\underline{R}_{XZ}Y)^V_t + \frac{1}{2}((D_X\underline{R})_{tY}Z)^H_t \ , \qquad (2.4)$$

$$(\bar{R}_{X^H Y^H}Z^V)_t = (\underline{R}_{XY}Z + \frac{1}{4}\underline{R}_{\underline{R}_{tZ}Y,X}t - \frac{1}{4}\underline{R}_{\underline{R}_{tZ}X,Y}t)^V_t + \frac{1}{2}((D_X\underline{R})_{tZ}Y - (D_Y\underline{R})_{tZ}X)^H_t \ ,$$

$$(\bar{R}_{X^H Y^H}Z^H)_t = \frac{1}{2}((D_Z\underline{R})_{XY}t)^V_t + (\underline{R}_{XY}Z + \frac{1}{4}\underline{R}_{t,\underline{R}_{ZY}t}X + \frac{1}{4}\underline{R}_{t,\underline{R}_{XZ}t}Y + \frac{1}{2}\underline{R}_{t,\underline{R}_{XY}t}Z)^H_t \ .$$

The tangent sphere bundle $\pi : T_1M \to M$ is the hypersurface of TM given by $\sum G_{ij}v^iv^j = 1$. The vector field $N = v^i \frac{\partial}{\partial v^i}$ is a unit normal as well as the position vector for a point $t \in TM$. For a vertical vector field U tangent to T_1M

$$\bar{\nabla}_U N = (Uv^i) \frac{\partial}{\partial v^i} + v^i \bar{\nabla}_U \left(\frac{\partial}{\partial x^i}\right)^V = U \quad .$$

For a horizontal tangent vector field X, we may suppose that X is (the restriction of) a horizontal lift, say $X = \left(\frac{\partial}{\partial x^j}\right)^H$; then at $t \in T_1M$

$$(\bar{\nabla}_X N)_t = \left(\left(\frac{\partial}{\partial x^j}\right)^H v^i\right)\left(\frac{\partial}{\partial v^i}\right)_t - \frac{1}{2} v^i \left(\underline{R}_{\frac{\partial}{\partial x^i}, t} \frac{\partial}{\partial x^j}\right)^H$$

$$+ v^i \left(D_{\frac{\partial}{\partial x^j}} \frac{\partial}{\partial x^i}\right)_t^V$$

$$= 0$$

by equation (2.1). Thus the Weingarten map A of T_1M is given by $AU = -U$ for any vertical vector U and $AX = 0$ for any horizontal vector X.

Let g' denote the induced metric, ∇ its Riemannian connection, R' its curvature tensor and α the second fundamental form. Recall the Gauss equation

$$R'(X,Y,Z,W) = \bar{R}(X,Y,Z,W) + \alpha(Y,Z)\alpha(X,W) - \alpha(X,Z)\alpha(Y,W).$$

Then

$$\begin{aligned}(\nabla_V R')(X,Y,Z,W) &= V(\bar{R}(X,Y,Z,W) + \alpha(Y,Z)\alpha(X,W) - \alpha(X,Z)\alpha(Y,W)) \\ &\quad - \bar{R}(\nabla_V X,Y,Z,W) - \alpha(Y,Z)\alpha(\nabla_V X,W) - \alpha(\nabla_V X,Z)\alpha(Y,W) \\ &\quad - \cdots \\ &= (\bar{\nabla}_V \bar{R})(X,Y,Z,W) + \bar{R}(\bar{\nabla}_V X - \nabla_V X,Y,Z,W) + \cdots \\ &\quad + ((\nabla_V \alpha)(Y,Z))\alpha(X,W) + \alpha(Y,Z)((\nabla_V \alpha)(X,W)) \\ &\quad - ((\nabla_V \alpha)(X,Z))\alpha(Y,W) - \alpha(X,Z)((\nabla_V \alpha)(Y,W)) .\end{aligned}$$

But $\bar{\nabla}_V X - \nabla_V X = \alpha(V,X)N$ and hence

$$(\nabla_V R')(X,Y,Z,W) = (\bar{\nabla}_V \bar{R})(X,Y,Z,W) + \bar{R}(N,Y,Z,W)\alpha(V,X)$$
$$+ \bar{R}(X,N,Z,W)\alpha(V,Y) + \bar{R}(X,Y,N,W)\alpha(V,Z) + \bar{R}(X,Y,Z,N)\alpha(V,W)$$
$$+ ((\nabla_V\alpha)(Y,Z))\alpha(X,W) + \alpha(Y,Z)((\nabla_V\alpha)(X,W))$$
$$- ((\nabla_V\alpha)(X,Z))\alpha(Y,W) - \alpha(X,Z)((\nabla_V\alpha)(Y,W)) . \qquad (2.5)$$

3. CONTACT MANIFOLDS.

By a contact manifold we mean a C^∞ manifold M^{2n+1} together with a global 1-form η such that $\eta \wedge (d\eta)^n \neq 0$. Given a contact form η it is well known that there exists a unique vector field ξ on M satisfying $\eta(\xi) = 1$ and $d\eta(\xi,X) = 0$; ξ is called the characteristic vector field of the contact structure. A Riemannian metric g is said to be an associated metric if there exists a tensor field φ of type $(1,1)$ such that $d\eta(X,Y) = g(X,\varphi Y)$, $\eta(X) = g(X,\xi)$ and $\varphi^2 = -I + \eta \otimes \xi$. These metrics are constructed by the polarization of $d\eta$ evaluated on a local orthonormal basis of an arbitrary metric on the contact distribution (subbundle) $\{\eta = 0\}$; see [1] as a general reference. We refer to (η,g) or (φ,ξ,η,g) as a contact metric structure.

Given a contact metric structure (φ,ξ,η,g) we define a tensor field h by $h = (1/2)\,\mathcal{L}_\xi\varphi$ where $\mathcal{L}$ denotes Lie differentiation and a tensor field ℓ by $\ell X = R_{X\xi}\xi$ where R denotes the curvature tensor of g. h and ℓ are both symmetric operators and it is easy to see that (φ,ξ,η,g) is K-contact if and only if $h \equiv 0$. Clearly $h\xi = 0$. We have the following formulas, the first two of which can be found in [1] (p. 66) and the third in [3].

$$\nabla_X\xi = -\varphi X - \varphi hX \qquad (3.1)$$

$$\frac{1}{2}(\varphi \ell \varphi - \ell) = h^2 + \varphi^2 \tag{3.2}$$

$$\nabla_\xi h = \varphi - \varphi h^2 - \varphi \ell \tag{3.3}$$

where ∇ denotes the Riemannian connection of g.

4. CONTACT METRIC STRUCTURE OF T_1M.

Let g' denote the metric of T_1M induced from $\bar{g}$ on TM. Define φ, ξ' and η' on T_1M by

$$\xi' = -JN, \quad JX = \varphi'X + \eta'(X)N .$$

η' is the contact form on T_1M induced from the 1-form β on TM as one can easily check. However $g'(X,\varphi'Y) = 2d\eta'(X,Y)$, so strictly speaking (φ',ξ',η',g') is not a contact metric structure. Of course the difficulty is easily rectified and we shall take

$$\eta = \frac{1}{2}\eta', \quad \xi = 2\xi', \quad \varphi = \varphi', \quad g = \frac{1}{4}g'$$

as our contact metric structure on T_1M. As $g = \frac{1}{4}g'$ is a homothetic change, the question of g or g' being locally symmetric is the same. In local coordinates $\xi = 2v^i(\frac{\partial}{\partial x^i})^H$; on TM $v^i(\frac{\partial}{\partial x^i})^H$ is the so-called geodesic flow.

Computing $\nabla_X\xi$ by both (3.1) and (2.3) we have for X horizontal at $t \in T_1M$

$$-\varphi X_t - \varphi h X_t = (\nabla_X \xi)_t = 2\bar{\nabla}_X v^i \left(\frac{\partial}{\partial x^i}\right)^H$$

$$= 2X^j\left(\left(\frac{\partial}{\partial x^j}\right)^H v^i\right)\left(\frac{\partial}{\partial x^i}\right)^H_t + 2v^i\left(D_{\pi_* X}\frac{\partial}{\partial x^i}\right)^H_t - v^i\left(\underline{R}_{\pi_* X, \frac{\partial}{\partial x^i}} t\right)^V$$

$$= -(\underline{R}_{\pi_* X, t} t)^V .$$

Thus for a horizontal vector X tangent to T_1M and orthogonal to ξ, we have, by applying φ on the left,

$$hX_t = -X_t + (\underline{R}_{\pi_* X, t} t)^H .$$

Similarly for a vertical vector U tangent to T_1M

$$hU_t = U_t - (\underline{R}_{KU, t} t)^V . \qquad (4.1)$$

Now a direct computation yields $\bar{g}(\bar{R}_{X\xi}\xi, N) = 0$ and so from the Gauss equation we have that $\ell X = R_{X\xi}\xi = \bar{R}_{X\xi}\xi$, since ξ is horizontal. Thus from the curvature formulas (2.4) we can find ℓ. For a horizontal vector X we have

$$\ell X_t = 2((D_t \underline{R})_{\pi_* X, t} t)^V + 4(\underline{R}_{\pi_* X, t} t + \frac{3}{4}\underline{R}_{t, \underline{R}_{\pi X, t} t} t)^H$$

and for a vertical vector U we have

$$\ell U_t = (\underline{R}_{\underline{R}_{KU, t} t, t} t)^V + 2((D_t \underline{R})_{KU, t} t)^H . \qquad (4.2)$$

5. PROOF OF THE THEOREM.

We begin with a few lemmas.

Lemma 1. If a contact metric manifold is locally symmetric, $\nabla_\xi h = 0$.

Proof. Since the metric is locally symmetric and the integral curves of ξ on a contact metric manifold are geodesics (e.g. set $X = \xi$ in (3.1)), we see that the operator ℓ is parallel along ξ. Since $\nabla_\xi \varphi = 0$ for any contact metric structure (see e.g. [1] p. 54), equation (3.2) then gives

$$\nabla_\xi h^2 = 0 . \tag{5.1}$$

Thus equations (3.3) gives

$$\nabla_\xi \nabla_\xi h = 0 . \tag{5.2}$$

Now writing (5.1) as

$$(\nabla_\xi h)h + h(\nabla_\xi h) = 0 ,$$

differentiating and using (5.2) we have

$$(\nabla_\xi h)^2 = 0 .$$

Since $\nabla_\xi h$ is a symmetric operator we have the result.

Lemma 2. If the contact metric structure on T_1M is locally symmetric, then for any orthonormal pair $\{X,t\}$ of tangent vectors on the base manifold (M,G),

$$\underline{R}_{\underline{R}_{Xt}t,t}t = \underline{R}_{Xt}t .$$

Proof. Applying φ to equation (3.3) and using Lemma 1 we see that

$$\varphi^2 + h^2 + \ell = 0 . \tag{5.3}$$

Now for a point $t \in T_1M$ and a vertical tangent vector $U \in T_t(T_1M)$, (5.3) becomes by virtue of equations (4.1) and (4.2)

$$0 = -U_t + (U_t - 2(\underline{R}_{KU,t}t)^V + (\underline{R}_{\underline{R}_{KU,t}t,t}t)^V)$$

$$+ ((\underline{R}_{\underline{R}_{KU,t}t,t}t)^V + 2((D_t\underline{R})_{KU,t}t)^H) .$$

Simplifying and taking the vertical part we have

$$\underline{R}_{\underline{R}_{KU,t}t,t}t = \underline{R}_{KU,t}t . \qquad (5.4)$$

Finally, note that since U is any vertical vector in $T_t(T_1M)$, U as a vertical vector in $T_t(TM)$ is orthogonal to t regarded as its position vector. Thus KU represents an arbitrary vector X orthogonal to t on the base manifold and hence equation (5.4) gives the result.

__Lemma 3.__ If the contact metric structure on T_1M is locally symmetric, then at each point of the base manifold (M,G) we have for any unit tangent vector t, arbitrary vectors X,Y and arbitrary vectors U,V,W orthogonal to t

$$2G(U,W)G(\underline{R}_{XY}t,V) - 2G(U,V)G(\underline{R}_{XY}t,W)$$

$$= G(\underline{R}_{X,\underline{R}_{UV}Y}t,W) - G(\underline{R}_{Y,\underline{R}_{tV}X}U,W) - 2G(\underline{R}_{Y,\underline{R}_{tU}X}V,W)$$

$$+ \frac{1}{2}\{G(\underline{R}_{X,\underline{R}_{tU}\underline{R}_{tV}Y}t,W) + G(\underline{R}_{\underline{R}_{tV}Y,\underline{R}_{tU}X}t,W)$$

$$- G(\underline{R}_{\underline{R}_{tV}\underline{R}_{tU}X,Y}t,W) - G(\underline{R}_{\underline{R}_{tU}\underline{R}_{tV}X,Y}t,W)\} .$$

Proof. As we will work on the manifold $(TM,\bar{g})$ in this proof we adopt for simplicity the following notations: $t \in T_1M \subset TM$; X and Y will be horizontal vectors at t; U,V,W will be vertical vectors at t tangent to T_1M. Then the vectors π_*X, π_*Y, KU, KV, KW are the vectors in the desired formula. Kowalski in his proof that $(TM,\bar{g})$ is locally symmetric if and only if M is flat, computed $4(\bar{\nabla}_X\bar{R})(Y,U,V,W)_t$ ([5] p. 128) by differentiation of (2.4); the result of his computation is the right hand side of our formula. On the other hand T_1M being locally symmetric implies $\nabla R' = 0$ in equation (2.5) and hence (2.5) gives

$$(\bar{\nabla}_X\bar{R})(Y,U,V,W) = g'(U,V)(\nabla_X\alpha)(Y,W) - g'(U,W)(\nabla_X\alpha)(Y,V) \ . \qquad (5.5)$$

Now

$$\begin{aligned}(\nabla_X\alpha)(Y,W)_t &= g'(\nabla_X Y,W)_t = \bar{g}(\bar{\nabla}_X Y,W)_t \\ &= -\frac{1}{2}\,G(\underline{R}_{\pi_*X,\pi_*Y}t,W)\end{aligned}$$

and similarly for $(\nabla_X\alpha)(Y,V)$. Thus (5.5) yields for $4(\bar{\nabla}_X\bar{R})Y,U,V,W)_t$ the left hand side of our formula.

Proof of the Theorem. Supposing (T_1M,g) to be locally symmetric we have the formulas of Lemmas 2 and 3 which we use to study the base manifold (M,G). We first show that (M,G) must be of constant curvature 0 or 1.

For each unit tangent vector $t \in T_mM$ let $[t]^\perp$ denote the subspace of T_mM orthogonal to t and define a symmetric linear transformation $L_t:[t]^\perp \longrightarrow [t]^\perp$ by $L_tX = \underline{R}_{Xt}t$. Then by Lemma 2, $L_t^2 = L_t$. Thus the eigenvalues of L_t are 0 or 1. If now (M,G) is not of constant curvature 0 or 1, there must be some $m \in M$ and $t \in T_mM$ such that there exist unit vectors $Y,t' \in [t]^\perp$ satisfying $L_tY = Y$ and $L_tt' = 0$. In particular, note that $\dim M \geq 3$.

We first observe that $L_Y t = t$. For on the one hand

$$1 = G(L_t Y, Y) = G(\underline{R}_{Yt} t, Y) = G(\underline{R}_{tY} Y, t) = G(L_Y t, t)$$

and on the other hand, by Lemma 2

$$1 = G(L_Y t, t) = G(L_Y^2 t, t) = G(L_Y t, L_Y t).$$

Similarly $L_{t'} t = 0$.

We now consider

$$L_{\frac{t+t'}{\sqrt{2}}} Y = \underline{R}_{Y,\frac{t+t'}{\sqrt{2}}} \frac{t+t'}{\sqrt{2}} = \frac{1}{2} (Y + \underline{R}_{Yt} t' + \underline{R}_{Yt'} t + \underline{R}_{Yt'} t') \tag{5.6}$$

and shall show that each of the last three terms on the right side of (5.6) vanish.

First in Lemma 3 set $X = Y = t$, then

$$0 = G(\underline{R}_{t,\underline{R}_{UV}t} t, W) - G(\underline{R}_{t,\underline{R}_{tV}t} U, W) - 2G(\underline{R}_{t,\underline{R}_{tU}t} V, W)$$

$$+ \frac{1}{2} \{2G(\underline{R}_{t,\underline{R}_{tU}\underline{R}_{tV}t} t, W) + G(\underline{R}_{\underline{R}_{tV}t,\underline{R}_{tU}t} t, W) - G(\underline{R}_{\underline{R}_{tV}\underline{R}_{tU}t,t} t, W)\}. \tag{5.7}$$

Now recalling the above vectors Y and t' (i.e. $L_t Y = Y$, $L_t t' = 0$), in (5.7) set $U = Y$ and $V = t'$, then

$$0 = G(R_{tW} t, R_{Yt'} t) + 2G(R_{tY} t', W) - \frac{1}{2} G(R_{tW} t, R_{tt'} Y) .$$

Now denote the 0-eigenspace of L_t by $[0]_t$ and the 1-eigenspace by $[1]_t$. Then taking $W \in [0]_t$, we see that $\underline{R}_{tY} t' \in [1]_t$. Taking $W \in [1]_t$ gives by the Bianchi identity

$$0 = \frac{1}{2} G(W, R_{t't} Y) + 3G(W, R_{tY} t') . \tag{5.8}$$

On the other hand in (5.7) set $U = t'$ and $V = Y$, then

$$0 = G(R_{tW}t, R_{t'Y}t) + G(R_{tY}t', W) - G(R_{tW}t, R_{tt'}Y) .$$

Again taking $W \in [1]_t$ and using the Bianchi identity we have

$$0 = 2G(W, R_{t't}Y) .$$

Consequently (5.8) gives

$$\underline{R}_{tY}t' = 0 . \qquad (5.9)$$

Now in the formula of Lemma 3 set $X = t'$, $U = V = Y$ and W orthogonal to U, then using (5.9) and $L_Y t = t$,

$$-2G(\underline{R}_{t'Y}t, W) = \frac{1}{2} G(\underline{R}_{t'}, \underline{R}_{tY}\underline{R}_{tY}Y^{t,W})$$

$$= -\frac{1}{2} G(\underline{R}_{t'Y}t, W)$$

and hence $G(\underline{R}_{t'Y}t, W) = 0$ for W orthogonal to Y; also (5.9) yields $G(\underline{R}_{t'Y}t, Y) = 0$ and hence

$$\underline{R}_{t'Y}t = 0 . \qquad (5.10)$$

Finally, in the formula of Lemma 3, set $X = t$, $Y = V = t'$ and U equal to the vector Y of the pair $\{Y, t'\}$ under consideration. Then using $L_{t'}t = 0$, Lemma 3 gives

$$0 = -G(\underline{R}_{Wt}t, \underline{R}_{Yt'}t') - 2G(\underline{R}_{Yt'}t', W) - \frac{1}{2} G(\underline{R}_{tW}t', \underline{R}_{tt'}Y) \quad ;$$

but by (5.9), (5.10) and the Bianchi identity, $\underline{R}_{tt'}Y = 0$. Thus for $W \in [0]_t$

$$-2G(\underline{R}_{Yt'}t', W) = 0$$

and for $W \in [1]_t$

$$-3G(\underline{R}_{Yt'}t',W) = 0 ,$$

giving

$$\underline{R}_{Yt'}t' = 0 . \qquad (5.11)$$

Equations (5.9), (5.10) and (5.11) applied to equation (5.6) give

$$L_{\frac{t+t'}{\sqrt{2}}}Y = \frac{1}{2}Y ,$$

contradicting the fact that the only possible eigenvalues of $L_{\frac{t+t'}{\sqrt{2}}}$ are 0 and 1. Thus (M,G) has constant curvature 0 or 1.

We now show that if (M,G) has constant curvature 1, dim M = 2. Suppose dim M $\geq$ 3 and let {t,U,X} be an orthonormal triple. Now in Lemma 3 set Y = t, V = U, W = X and use the form of the curvature tensor $\underline{R}_{XY}Z = G(Y,Z)X - G(X,Z)Y$, then Lemma 3 gives the contradiction $-2 = -1/2$.

Finally it remains to prove the converse, that if (M,G) is flat or of dimension 2 and constant curvature 1, then (T_1M,g) is locally symmetric. In the flat case this is virtually evident for a number of reasons. By way of a remark we note that from (2.4), (TM,g) is flat and hence, since ξ is horizontal, (T_1M,g) satisfies $R_{XY}\xi = \bar{R}_{XY}\xi = 0$. We then note the following result, ([2], [1] p. 121, see also [3]).

__Theorem.__ Let M^{2n+1} be a contact metric manifold and suppose that $R_{XY}\xi = 0$ for all vector fields X and Y. Then M^{2n+1} is locally the product of a flat (n+1)-dimensional manifold and an n-dimensional manifold of positive constant curvature 4.

For (M,G) of dimension 2 and constant curvature 1, a direct computation shows that (T_1M,g') has constant curvature 1/4 and (T_1M,g) has constant curvature 1.

References

[1] Blair, D.E., Contact Manifolds in Riemannian Geometry, Lecture Notes in Mathematics, Vol. 509, Springer, Berlin, 1976.

[2] Blair, D.E., Two remarks on contact metric structures, Tôhoku Math. J. 29, 319-324 (1977).

[3] Blair, D.E. and Patnaik, J.N., Contact manifolds with characteristic vector field annihilated by the curvature, Bull. Inst. Math. Acad. Sinica 9, 533-545 (1981).

[4] Dombrowski, P., On the geometry of the tangent bundle, J. Reine Angew. Math. 210, 73-88 (1962).

[5] Kowalski, O., Curvature of the induced Riemannian metric on the tangent bundle of a Riemannian manifold, J. Reine Angew. Math. 250, 124-129 (1971).

[6] Okumura, M., Some remarks on space with a certain contact structure, Tôhoku Math. J. 14, 135-145 (1962).

[7] Tachibana, S. and Okumura, M., On the almost complex structure of tangent bundles of Riemannian spaces, Tôhoku Math. J. 14, 156-161 (1962).

[8] Takahashi, T., Sasakian φ-symmetric spaces, Tôhoku Math. J. 29, 91-113 (1977).

[9] Tanno, S., Locally symmetric K-contact Riemannian manifolds, Proc. Japan Acad. 43, 581-583 (1968).

[10] Tanno, S., Geodesic flows on C_L-manifolds and Einstein metrics on $S^3 \times S^2$, Minimal Submanifolds and Geodesics, Kaigai, Tokyo 283-292 (1978).

[11] Tashiro, Y., On contact structures of tangent sphere bundles, Tôhoku Math. J. 21, 117-143 (1969).

GEOMETRY AND TOPOLOGY (pp. 31–57)
edited by G. M. Rassias & G. M. Stratopoulos

Essential tori and Klein bottles in non-orientable 3-manifolds of genus two

Iván Cardona*
Department of Mathematics ,
University of Puerto Rico , Río Piedras , Puerto Rico , 00931

and

Wolfgang Heil
Department of Mathematics ,
Florida State University , Tallahassee , Florida , 32306

Abstract. Let M be a closed non-orientable 3-manifold with a Heegaard splitting of genus two. We show that , if every essential Klein bottle (or torus) is separating in M and if M has a non-trivial canonical system of 2-sided tori and Klein bottles , then there is a canonical system such that its intersection with one of the handlebodies in the Heegaard splitting consists of at most two essential disks. We use these results to give a complete list of all the non-simple non-orientable 3-manifolds with a Heegaard splitting of genus two which are not Seifert fibered.

§1. INTRODUCTION

M. Ochiai [11] showed that $P^2 \times S^1$ is the only closed 3-manifold with a Heegaard splitting of genus two that contains a 2-sided projective plane , this was done by extending a theorem of W. Haken to include projective planes (See [4] , Corollary II.7). In [7] T. Kobayashi showed that if a 3-manifold M with a Heegaard splitting of genus two contains a non-separating essential torus then there is a non-separating essential torus (in M) which intersects one of the handlebodies of the Heegaard splitting in a single disk. Then in [8] he extended these results and gave a complete list of the irreducible orientable 3-manifolds with Heegaard splittings of genus two that have a non-trivial canonical system of tori. We generalize this to the non-orientable case and show that the clasification is similar. First one can show :

* This paper contains some of the results of Iván Cardona's Ph. D. Thesis at Florida State University (1987).

Theorem 1. *Let* M *be a closed , connected* 3-*manifold with a Heegaard splitting* $(V_1,V_2 ; F)$ *of genus two . If* M *contains a* 2-*sided non-separating incompressible Klein bottle* K , *then there is a* 2-*sided non-separating incompressible Klein bottle (or torus)* $\hat{K}$ *such that* $\hat{K} \cap V_1$ *is a single essential disk in* V_1 .

and

Theorem 2. *Let* M *be a closed , connected* 3-*manifold with a Heegaard splitting* $(V_1,V_2 ; F)$ *of genus two such that each essential Klein bottle (or torus) in* M *is separating . If* M *contains a non-trivial canonical system of tori and Klein bottles , then there is a canonical system* $W=W_1 \cup W_2 \cup \ldots \cup W_k$ *such that* $W \cap V_1$ *consists of at most two essential disks in* V_1.

Now , since any non-orientable compact P^2- irreducible 3-manifold M is Haken , Thurston's Geometrization conjecture (see [15]) is true for M. That is , there is a canonical system of disjoint incompressible tori and Klein bottles which cuts M into pieces which admit a geometric structure. In particular , for a non-orientable closed 3-manifold with a Heegaard splitting of genus two either

(i) $M \approx S^2 \times_{\varphi} S^1 \# L_n$, where L_n is a lens space , S^3 or $S^2 \times S^1$,

(ii) $M \approx P^2 \times S^1$,

(iii) M is a Seifert fibered space ,

(iv) M admits a complete hyperbolic structure or

(v) M is decomposed by 2-sided incompressible tori and Klein bottles into pieces which admit a geometric structure.

In this paper , we study the manifolds with a canonical system of tori and Klein bottles (case (v) above). We use the above theorems to give a list of all non-simple non-orientable 3-manifolds with a Heegaard splitting of genus two which are not Seifert fibered spaces. K. Morimoto [10] has independently obtained similar results.

§ 2. PRELIMINARIES

We will work in the piecewise linear (PL) category throughout this paper. *Heegaard splitting* , *2-sided* surface , *incompressible* surface , *irreducible* manifold and P^2-*irreducible* manifold are defined in [3]. *Seifert fibered* manifold is defined in [13]. Two surfaces F_1 and F_2 are *parallel* in M if $F_1 \approx F_2 \approx F$ and there is an embedding $h: F \times [-1,1] \to M$ with $h(F \times \{-1\}) = F_1$, $h(F \times \{1\}) = F_2$ and $h(F \times [-1,1]) \cap \partial M = h(\partial F \times [-1,1])$. A 2-sided incompressible surface in M is *essential* if it is not parallel to a surface on ∂M. A 3-manifold is *simple* if every incompressible torus or Klein bottle is boundary parallel. A *Haken* manifold is a 3-manifold which is prime and contains a 2-sided incompressible surface ($\neq S^2$).

Let W be a 2-sided incompressible surface in the 3-manifold M such that :

(a) Each component of W is a torus or a Klein bottle.

(b) Each component of the closure of M cut along W is a simple space or doubly covered by a Seifert fibered space.

(c) W is minimal with respect to inclusion among all surfaces satisfying (a) and (b).

Then W is called a *canonical system of tori and Klein bottles* in M. Notice that if M is simple or a Seifert fibered space , then a canonical system for M is empty.

By [5] or [6] any closed orientable Haken manifold M has a canonical system of tori. If M is a non-orientable P^2-irreducible manifold , then M is a Haken manifold. The orientable double cover $\tilde{M}$ of M is Haken and will have a canonical system of tori. This canonical system can be isotoped to be invariant under the covering involution (see [14]). Hence , M has a canonical system of tori and Klein bottles (This system can be chosen to be a subset of the image , under the covering map , of the canonical system of the orientable double cover of M).

Let K be a Klein bottle (or a torus) and let $K'' = cl(K - (D_1 \cup D_2 \cup \ldots \cup D_n))$, where $\{D_1, D_2, \ldots, D_n\}$ is a collection of disjoint disks in K. If α is an essential arc in K'' , then

(i) We say that α is of *type* 1 if and only if α joins distinct components of $\partial K''$.

(ii) We say that α is of *type* 2 if and only if α joins one component of $\partial K''$ and there is an arc β in $\partial K''$ with $\partial\beta=\partial\alpha$ such that $\alpha\cup\beta$ bounds a disk in K .

(iii) We say that α is of *type* 3 if and only if α joins one component of $\partial K''$ and there is an arc β in $\partial K''$ with $\partial\beta=\partial\alpha$ such that $\alpha\cup\beta$ is orientation preserving and non-separating in K.

(iv) We say that α is of *type* 4 if and only if α joins one component of $\partial K''$ and there is an arc β in $\partial K''$ with $\partial\beta=\partial\alpha$ such that $\alpha\cup\beta$ is orientation preserving and separating in K.

(v) We say that α is of *type* 5 if and only if α joins one component of $\partial K''$ and there is an arc β in $\partial K''$ with $\partial\beta=\partial\alpha$ such that $\alpha\cup\beta$ is orientation reversing in K.

Observe that if K is a torus α can only be of types 1 , 2 or 3 . By Lemma 1 in [9] any essential arc in K″ must be of type 1,2,3,4 or 5. A *hierarchy* for a compact surface S is a sequence (S_0 , α_0) , (S_1 , α_1) , ... , (S_n , α_n) where α_i (i=0,1, ... ,n) is an essential arc in S_i , $S_0=S$, S_{i+1} is equal to S_i cut along α_i and each component of S_{n+1} is a disk. Let M be a manifold with a Heegaard splitting $(V_1,V_2 ; F)$ and S be any surface in M. Let Δ be a disk in M such that $\Delta\cap S=\alpha$ is an arc in $\partial\Delta$, $\Delta\cap F=\beta$ is an arc in $\partial\Delta$, $\partial\alpha=\partial\beta$ and $\alpha\cup\beta=\partial\Delta$. *An isotopy of type A* at α is performed by sliding α across Δ and past β. Such an isotopy shoves a part of S that is on one side of F in M through F and into the other side of F in M. The result of this isotopy , to that part of S which was on the same side of F as α before the isotopy , is equivalent to the result of cutting along α. An isotopy of type A has an inverse which is also of type A. See [4] (proof of II.9).

§ 3. ESSENTIAL ANNULI AND MOBIUS BANDS IN THE GENUS TWO HANDLEBODY

The following five lemmas show how essential annuli and Möbius bands are embedded in a genus two handlebody. The proofs use techniques from lemmas 3.2-3.5 of [8].

Lemma 3. *If* A *is an essential annulus in the non-orientable handlebody of genus two* V, *then either*

(i) A *cuts* V *into a non-orientable handlebody of genus two* V′ *and there is a complete system of meridian disks* $\{D_1,D_2\}$ *of* V′ *such that* $D_1 \cap A$ *is an essential arc of* A ,

(ii) A *cuts* V *into an orientable handlebody of genus two* V′ *and there is a complete system of meridian disks* $\{D_1,D_2\}$ *of* V′ *such that* $D_1 \cap A$ *is an essential arc of* A ,

(iii) A *cuts* V *into a solid Klein bottle* V_1 *and an orientable handlebody of genus two* V_2 *and there is a complete system of meridian disks* $\{D_1,D_2\}$ *of* V_2 *such that* $D_1 \cap A=\emptyset$ *and* $D_2 \cap A$ *is an essential arc of* A ,

(iv) A *cuts* V *into a solid torus* V_1 *and a non-orientable handlebody of genus two* V_2 *and there is a complete system of meridian disks* $\{D_1,D_2\}$ *of* V_2 *such that* $D_1 \cap A=\emptyset$ *and* $D_2 \cap A$ *is an essential arc of* A *or*

(v) A *cuts* V *into a solid Klein bottle* V_1 *and a non-orientable handlebody of genus two* V_2 *and there is a complete system of meridian disks* $\{D_1,D_2\}$ *of* V_2 *such that* $D_1 \cap A=\emptyset$ *and* $D_2 \cap A$ *is an essential arc of* A. *See Fig.* 1.

Lemma 4. *Let* $\{A_1,A_2,A_3\}$ *be a system of pairwise disjoint non-parallel essential annuli in the non-orientable handlebody of genus two* V *, then either*

(i) $A_1 \cup A_2 \cup A_3$ *cuts* V *into a solid Klein bottle* V_1 *, a solid torus* V_2 *and an orientable handlebody of genus two* V_3 *such that* $A_i \subset \partial V_1$ *(for* i=1,2 *or* 3, *say* 3), $A_1 \cup A_2 \cup A_3 \subset \partial V_2$, $A_1 \cup A_2 \subset \partial V_3$

Figure 1

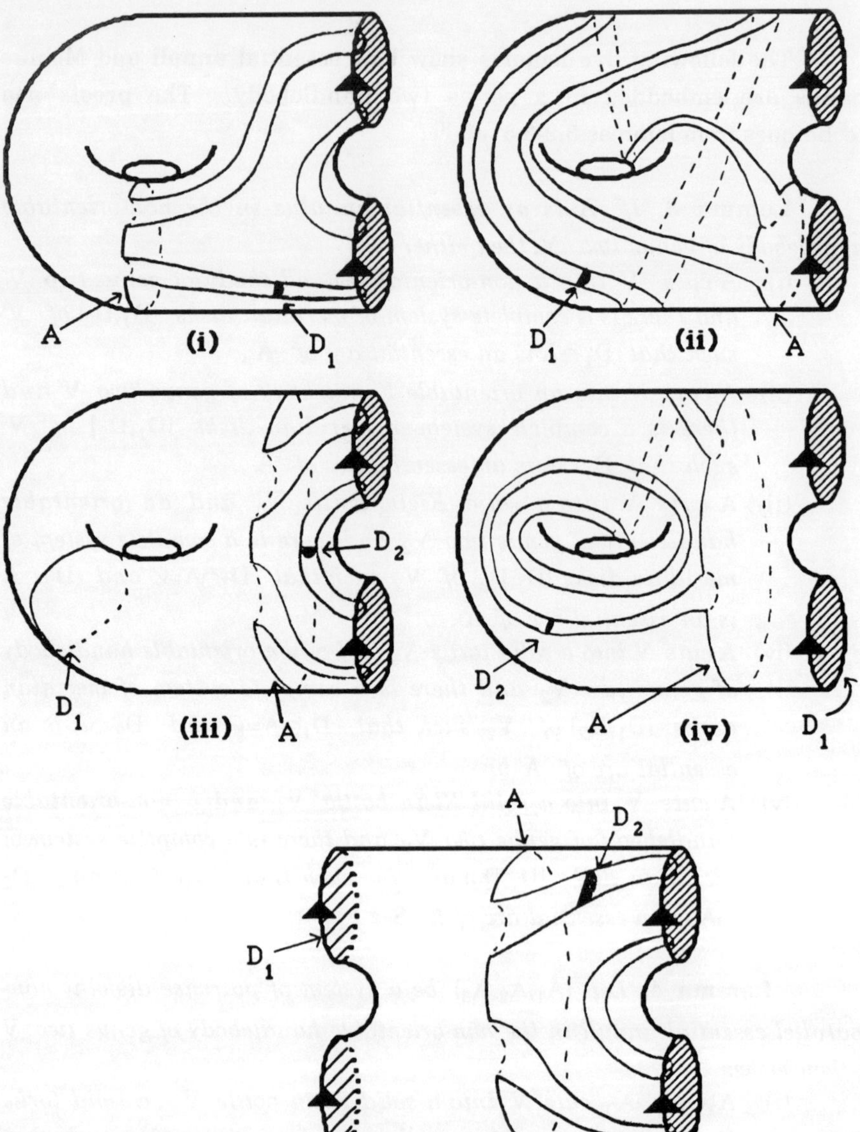

and there is a meridian disk D *of* V_2 *such that* $D \cap A_i$ (i=1,2,3) *is an essential arc of* A_i *or*

(ii) $A_1 \cup A_2 \cup A_3$ *cuts* V *into two solid tori* V_1 , V_2 *and an orientable handlebody of genus two* V_3 *such that* $A_i \subset \partial V_1$ *(for* i=1,2 *or* 3, *say* 3), $A_1 \cup A_2 \cup A_3 \subset \partial V_2$, $A_1 \cup A_2 \subset \partial V_3$ *and there is a meridian disk* D *of* V_2 *such that* $D \cap A_i$ (i=1,2,3) *is an essential arc of* A_i.

Also, in any case, there is a complete system of meridians disks $\{D_1, D_2\}$ *of* V_3 *such that* $D_i \cap A_j = \varnothing$ $(i \neq j)$ *and* $D_i \cap A_i$ *is an essential arc of* A_i (i=1,2). *See Fig.* 2.

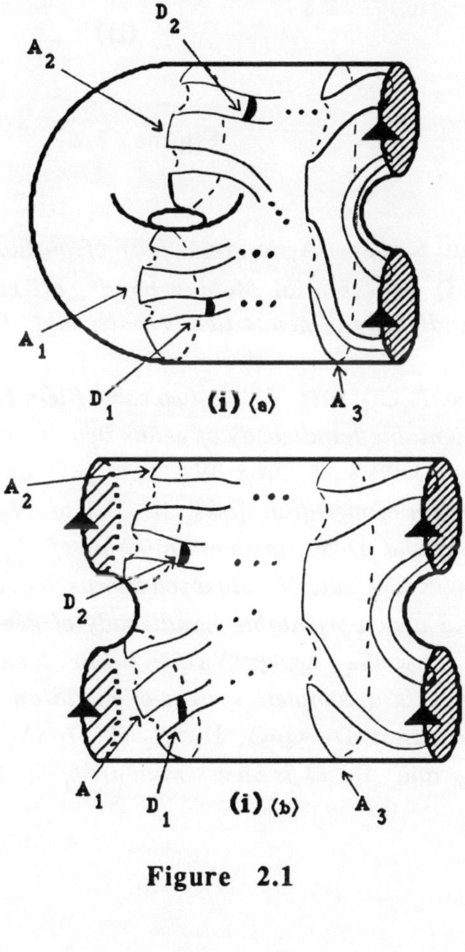

Figure 2.1

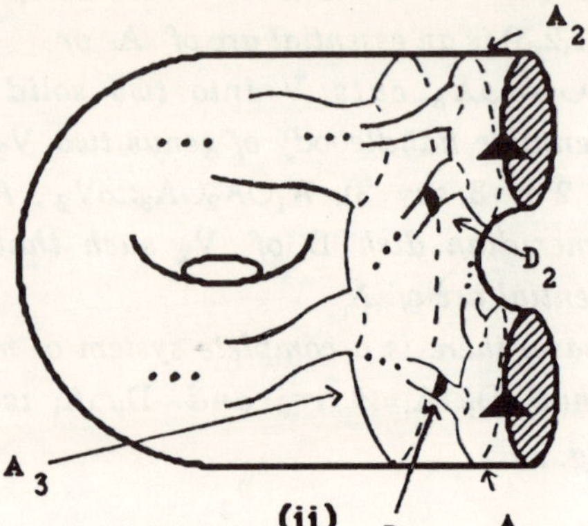

(ii)

Figure 2.2

Lemma 5. *Let* $\{A_1,A_2\}$ *be a pair of mutually disjoint essential annuli and* Q *an essential Möbius band , all embedded in the non-orientable handlebody of genus two* V *such that* $Q \cap A_i=\emptyset$ (i=1,2), *then either*

(i) $A_1 \cup A_2 \cup Q$ *cuts* V *into two solid Klein bottles* V_1 , V_2 *and an orientable handlebody of genus two* V_3 *such that* $A_1 \cup Q \subset \partial V_1$, $A_2 \cup Q \subset \partial V_2$, $A_1 \cup A_2 \subset \partial V_3$, $Q \cap V_3=\emptyset$ *and there is a complete system of meridian disks* $\{D_1,D_2\}$ *of* V_3 *such that* $D_i \cap A_j=\emptyset$ $(i \neq j)$ *and* $D_i \cap A_i$ *is an essential arc of* A_i (i=1,2) *or*

(ii) $A_1 \cup A_2 \cup Q$ *cuts* V *into a solid torus* V_1 , *a solid Klein bottle* V_2 *and a non-orientable handlebody of genus two* V_3 *such that* $A_1 \cup A_2 \subset \partial V_1$, $A_i \cup Q \subset \partial V_2$ *(for* i=1 *or* 2, *say* 1) , $A_2 \cup Q \subset \partial V_3$ *and there is a complete system of meridian disk* $\{D_1,D_2\}$ *of* V_3 *such that* $D_1 \cap Q=\emptyset$, $D_2 \cap A_2=\emptyset$, $D_1 \cap A_2$ *is an essential arc of* A_2 *and* $D_2 \cap Q$ *is an essential arc of* Q. *See Fig.* 3.

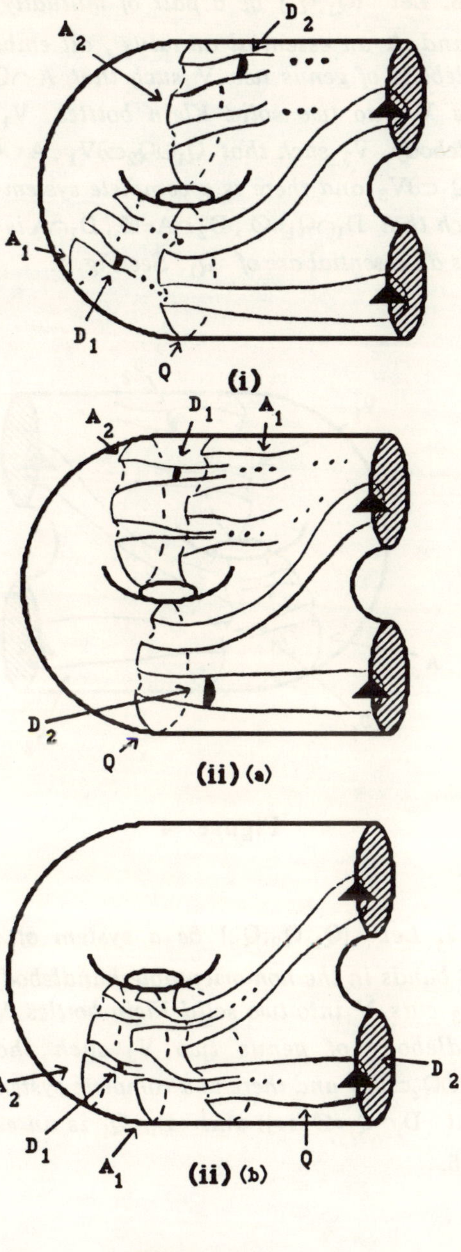

Figure 3

Lemma 6. *Let* $\{Q_1,Q_2\}$ *be a pair of mutually disjoint essential Möbius bands and* A *an essential annulus , all embedded in the non-orientable handlebody of genus two* V *such that* $A\cap Q_i=\emptyset$ (i=1,2) , *then* $A\cup Q_1\cup Q_2$ *cuts* V *into two solid Klein bottles* V_1 , V_2 *and a non-orientable handlebody* V_3 *such that* $Q_1\cup Q_2\subset\partial V_1$, $A\cup Q_i\subset\partial V_2$ (*for* i=1 *or* 2, *say* 2) , $A\cup Q_1\subset\partial V_3$ *and there is a complete system of meridian disks* $\{D_1,D_2\}$ *of* V_3 *such that* $D_1\cap Q_1=\emptyset$, $D_2\cap A=\emptyset$, $D_1\cap A$ *is an essential arc of* A *and* $D_2\cap Q_1$ *is an essential arc of* Q_1. *See Fig.* 4.

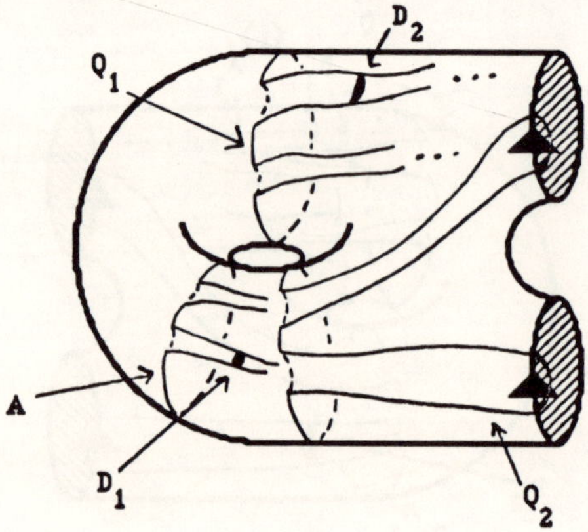

Figure 4

Lemma 7. *Let* $\{Q_1,Q_2,Q_3\}$ *be a system of mutually disjoint essential Möbius bands in the non-orientable handlebody of genus two* V , *then* $Q_1\cup Q_2\cup Q_3$ *cuts* V *into two solid Klein bottles* V_1 , V_2 *and a non-orientable handlebody of genus two* V_3 *such that* $Q_1\cup Q_3\subset\partial V_1$, $Q_2\cup Q_3\subset\partial V_2$, $Q_1\cup Q_2\subset\partial V_3$ *and there is a complete system of disks* $\{D_1,D_2\}$ *of* V_3 *such that* $D_i\cap Q_j=\emptyset$ $(i\neq j)$ *and* $D_i\cap Q_i$ *is an essential arc of* Q_i (i=1,2). *See Fig.* 5.

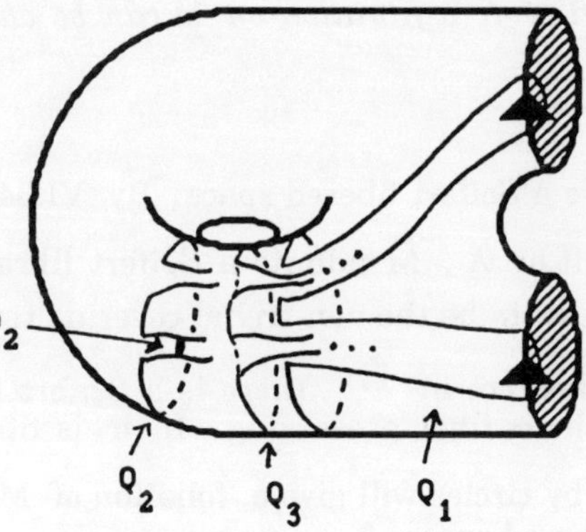

Figure 5

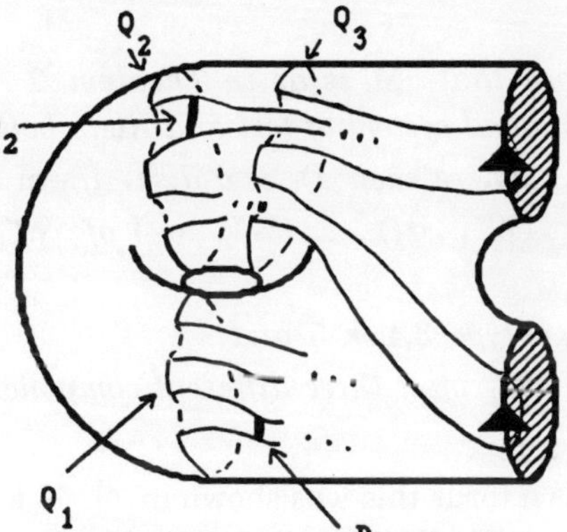

§ 4. PROOF OF THEOREM 2

First we need the following lemmas.

Lemma 8. *Let* M *be a compact manifold such that its orientable double cover* $\tilde{M}$ *is simple , irreducible with* $\partial\tilde{M}$ *the union of incompresible tori. If* M *contains an essential annulus or Möbius band*

A , *then* M ($\tilde{M}$ *resp.*) *is a generalized Seifert fibered space. Also , for each component* A_i *of the lift of* A *a fibration on* $\tilde{M}$ *can be chosen such that* ∂A_i *is a regular fiber.*

Proof. By 6.1 in [8] $\tilde{M}$ is a Seifert fibered space. By VI.34 in [4], given a component A_i of the lift of A , $\tilde{M}$ admits a Seifert fibration such that ∂A_i is a regular fiber. Let h be the non-trivial covering transformation of $\tilde{M}$, then by the main theorem in [14], there is a Seifert fibration on $\tilde{M}$ with respect to which h is fiber preserving. If h is fiber preserving , then the foliation of $\tilde{M}$ by circles will give a foliation of M by circles and (by [2]) M is a Seifert fibered space. ∎

Lemma 9. *Suppose that* M *is as in Theorem* 2 . *Suppose that* $W=W_1\cup \ldots \cup W_k$ *is a canonical system of tori and Klein bottles for* M *and that* $W \cap V_1=D_1\cup\ldots\cup D_m$, *where each* D_i *is a disk. If* $m \geq 3$, *then there is a hierarchy* $(W''_0, \alpha_0), (W''_1, \alpha_1), \ldots, (W''_n, \alpha_n)$ *of* $W''_0=W \cap V_2$ *such that*

(i) α_i (i=0, 1,2) *is of type* 3,4 *or* 5 *and*

(ii) $\partial\alpha_0$, $\partial\alpha_1$ *and* $\partial\alpha_2$ *meet three different components of* $\partial W''_0$.

Sketch of proof. If W is a torus this was shown in [8]. If k=1 and W is a Klein bottle K we let $K'=K\cap V_1$ and $K''=K\cap V_2$. Since K is incompressible, we may assume K′ and K″ are incompressible in V_1 and V_2 respectively. Furthermore , by moving K by a sequence of isotopies of type A , we may assume that each component of K′ is an essential disk. Let $c(F\cap K)=m$ be the number of components of $F \cap K$. As in [4] ,p. 24, there is a hierarchy $(K''_0, \alpha_0), (K''_1, \alpha_1), \ldots, (K''_n, \alpha_n)$ for $K''= K''_0$ which gives rise to a sequence of isotopies of K in M , where the first isotopy is of type A at α_0 , the second of type A at α_1 , … , and the (n+1)st of type A at α_n. Let $\bar{\alpha} = \alpha_0\cup\ldots\cup\alpha_n$ and let $K\cap V_1=D_1\cup\ldots\cup D_m$.

We can assume that α_0 (α_1 resp.) is of type 3,4 or 5 and that it meets ∂D_1 (∂D_2 resp.) (See lemmas 3.1-3.3 in [7]). Now by using arguments of Lemma 1 in [12] and *lemma* 2 one can show that α_2 must be of type 3 , 4 or 5 and does not meet ∂D_1 or ∂D_2. The proof for $k \geq 2$ is similar. The arguments are rather tedious and appear in detail in [1] . ∎

Given any canonical system of tori and Klein bottles W' we may assume that each of $W' \cap V_i$ (i=1,2) is incompressible and that each component of $W' \cap V_1$ is a disk. From now on W will be such that , among all canonical systems which meet V_1 in disks , the number of components of $W \cap V_1$ is minimal.

Sketch of proof of Theorem 2. We may assume each of $W \cap V_i$ (i=1,2) is incompressible and $W \cap V_1 = D_1 \cup \ldots \cup D_m$. Suppose $m \geq 3$. By *lemma* 9 we may assume that there is a hierarchy (W''_0, α_0), (W''_1, α_1) , ... , (W''_n, α_n) of $W''_0 = W \cap V_2$, where each of α_0 , α_1 and α_2 is of type 3,4 or 5 , that α_0 meets ∂D_1 , that α_1 meets ∂D_2 and that α_2 meets ∂D_3 . Moreover , if W'' has two or more components we may assume that α_0 and α_1 lie in different components of W''. Let $W^{(1)}$ be the image of W after an isotopy of type A at α_0 , $W^{(2)}$ be the image of $W^{(1)}$ after an isotopy of type A at α_1 and $W^{(3)}$ be the image of $W^{(2)}$ after an isotopy of type A at α_2.

Then we have the following possibilities:

(i) $W^{(3)} \cap V_1 = A_1 \cup A_2 \cup A_3 \cup D_4 \cup \ldots \cup D_m$

(ii) $W^{(3)} \cap V_1 = \begin{cases} A_1 \cup A_2 \cup Q_3 \cup D_4 \cup \ldots \cup D_m \\ A_1 \cup Q_2 \cup A_3 \cup D_4 \cup \ldots \cup D_m \\ Q_1 \cup A_2 \cup A_3 \cup D_4 \cup \ldots \cup D_m \end{cases}$

(iii) $W^{(3)} \cap V_1 = \begin{cases} A_1 \cup Q_2 \cup Q_3 \cup D_4 \cup \ldots \cup D_m \\ Q_1 \cup A_2 \cup Q_3 \cup D_4 \cup \ldots \cup D_m \\ Q_1 \cup Q_2 \cup A_3 \cup D_4 \cup \ldots \cup D_m \end{cases}$

(iv) $W^{(3)} \cap V_1 = Q_1 \cup Q_2 \cup Q_3 \cup D_4 \ldots \cup D_m$

Each of the three posibilities in case (ii) arise when α_2 is of type 5 , α_1 is of type 5 and α_0 is of type 5 , respectively. Similarly , each of the three posibilities in case (iii) arise when α_0 is not of type 5 , α_1 is not of type 5 and α_2 is not of type 5 , respectively. In all cases A_i (i=1,2,3) is an essential annulus and Q_i (i=1,2,3) an essential Möbius band (in V_1).

First , suppose that W has only one component (a torus or a Klein bottle). Then , W cuts M into two components M_1 and M_2. Notice that , since M contains a non-trivial canonical system of tori and Klein bottles , its double cover $\tilde{M}$ is not a Seifert fiber space.

Suppose that $W^{(3)} \cap V_1$ is as in case (i) above. Suppose that two of $\{A_1,A_2,A_3\}$, say A_1 and A_2 , are parallel in V_1.

If A_1 (hence A_2) satisfies conclusion (i) of *lemma* 3 , then there is an annulus A' on ∂V_1 such that $A' \cap (A_1 \cup A_2) = \partial A'$ and there is a Möbius band Q on ∂V_1 such that $Q \cap (A_1 \cup A_2) = Q \cap A_1 = \partial Q = C$, where C is a component of ∂A_1 and $A' \cap Q = \partial A' \cap \partial Q = C$. We may assume $A' \subset M_1$ and $Q \subset M_2$. By the minimality of W , A' is essential in M_1. If Q is parallel to a Möbius band on ∂W , then W is a Klein bottle and since W is incompressible α_0 is of type 4. In any case , since $m \geq 3$, we see that there is a Klein bottle $\hat{W}$ isotopic to W with $c(\hat{W} \cap V_1) < c(W \cap V_1)$ a contradiction. thus , Q is essential in M_2. Hence, by *lemma* 8 , the lift (in $\tilde{M}$) of M_1 (M_2 resp.) , $\tilde{M}_1$ ($\tilde{M}_2$ resp.) admits a Seifert fibration such that given any component C_i of the lift of C , C_i is a regular fiber. Thus , we can paste $\tilde{M}_1$(or a component of $\tilde{M}_1$) to $\tilde{M}_2$ (or a component of $\tilde{M}_2$) along one of the components of $\tilde{W}$ (the lift of W) such that the resulting space is a Seifert fibered space. This contradicts the fact that $\tilde{W}$ is a canonical system of tori for $\tilde{M}$.

If A_1 (hence A_2) satisfies conclusion (ii) of *lemma* 3 , then , by changing subscripts if necessary , there are two annuli A' , A'' on ∂V_1 such that $A' \cap (A_1 \cup A_2) = \partial A'$, $A'' \cap (A_2 \cup A_3) = \partial A''$ and $A' \cap A'' = \partial A' \cap \partial A'' = C$,where C is a component of ∂A_2. We may assume $A' \subset M_1$ and $A'' \subset M_2$. By the minimality of W , A' (A'' resp.) is essential in M_1(M_2 resp.). By *lemma* 8 , the lift (in $\tilde{M}$) of M_1 (M_2 resp.) , $\tilde{M}_1$ ($\tilde{M}_2$ resp.) admits a Seifert fibration such that given any component C_i of the lift of C , C_i is a regular fiber. Again , we can paste $\tilde{M}_1$(or a component of $\tilde{M}_1$) to $\tilde{M}_2$ (or a component of $\tilde{M}_2$) along one of the components of $\tilde{W}$ (the lift of W) such that the resulting space is a Seifert fibered space. This contradicts the fact that $\tilde{W}$ is a canonical system of tori for $\tilde{M}$.

If A_1 (hence A_2) satisfies (iii)-(v) of *lemma* 3 , we use similar arguments to arrive at a contradiction.

Now , if A_1 , A_2 and A_3 are mutually non-parallel , then $\{A_1,A_2,A_3\}$ satisfies one of the conclusions of *lemma* 4 (changing subscripts). If $\{A_1,A_2,A_3\}$ satisfies conclusion (i) of *lemma* 4 , then there is a Möbius band Q on ∂V_1 such that $Q\cap(A_1\cup A_2\cup A_3)=Q\cap A_3=C$, where C is a component of ∂A_3. Also , there is an annulus A' on ∂V_1 such that $A'\cap(A_1\cup A_2\cup A_3)=A'\cap A_i=\partial A'$ (for i=1 or 2 , say 2) and $A'\cap Q=\partial A'\cap\partial Q=C$. We may assume $A'\subset M_1$ and $Q\subset M_2$. By the minimality of W , A' is essential in M_1 and Q is essential in M_2. As before , we see this is impossible. If $\{A_1,A_2,A_3\}$ satisfies conclusion (ii) of *lemma* 4 , then there are two annuli A' , A'' on ∂V_1 such that $A'\cap(A_1\cup A_2\cup A_3)=A'\cap A_3=\partial A'\cap\partial A_3=C$, where C is a component of ∂A_3 , $A''\cap (A_1\cup A_2\cup A_3)=A''\cap(A_1\cup A_i)= \partial A''$ (for i=1 or 2 , say 2) and $A'\cap A''=C$. We may assume that $A'\subset M_1$, $A''\subset M_2$ and as above A' is essential in M_1 , A'' is essential in M_2 and we get a contradiction.

Cases (ii) , (iii) and (iv) are handled similarly by using *lemmas* 5 , 6 and 7. ∎

Corollary 10. W *has at most two components.*

§ 5. THE NON-SIMPLE , NON-ORIENTABLE 3-MANIFOLDS WITH A HEEGAARD SPLITTING OF GENUS TWO

Let V be a handlebody (orientable or non-orientable) and let α be a properly embedded arc in V. We say that α is *trivially embedded* , if there is a disk $D\subset V$ such that $D\cap\partial V=\beta$, where β is an arc and

$cl(\partial D - \beta) = \alpha$ (See Fig. 6). It can be shown that if α' is another trivially embedded arc in V, then there is an ambient isotopy h_t ($t \in [0,1]$) of V such that $h_1(\alpha) = \alpha'$. $S^2 \times_\varphi S^1$ is the space obtained from $S^2 \times I$ by identifying $(x,0) \in S^2 \times \{0\}$ with $(\varphi(x),1) \in S^2 \times \{1\}$, where $\varphi: S^2 \to S^2$ is the antipodal map. Let k be a knot in $S^2 \times_\varphi S^1$ and let $\{V_1, V_2\}$ be a Heegaard splitting of genus one for $S^2 \times_\varphi S^1$. If

(i) $k \cap V_i = \alpha_i$ (i=1,2) is a trivially embedded arc in V_i (i=1,2), we say that k is a *one bridge knot* in $S^2 \times_\varphi S^1$. See Fig. 6.

(ii) $k \cap V_i = k$ (i=1,2) and k is not trivial in V_i and orientation preserving on ∂V_i, we say that k is a *Klein bottle knot* in $S^2 \times_\varphi S^1$. See Fig. 6.

Let $L = k_1 \cup k_2$ be a link in $S^2 \times_\varphi S^1$ such that $k_1 \cap V_i = k_1$ (i=1,2) is orientation preserving on ∂V_i and $k_2 \cap V_i = \alpha_i$ (i=1,2) is a trivially embedded arc in V_i (i=1,2). Let D_i (i=1,2) be the disk (in V_i), associated with α_i, given by the definition of a trivially embedded arc. If $D_i \cap k_1 = \emptyset$ (i=1,2), we say that L is a *link of type* I in $S^2 \times_\varphi S^1$. Notice that in a link of type I in $S^2 \times_\varphi S^1$ one component is a one-bridge knot and the other component is a Klein bottle knot. In a link of type I the meridian of the one bridge knot component is called the *meridian* of $M(L)$ (the exterior of L in $S^2 \times_\varphi S^1$) and the curve on $\partial M(L)$ which runs parallel to the Klein bottle knot component will be called the **core** of $M(L)$. Let k be a knot in a lens space L_n, we say (following [8]) that k is a *one-bridge knot* in L_n if there is a pair $\{V_1, V_2\}$ of solid tori in L_n such that $L_n = V_1 \cup V_2$, $V_1 \cap V_2 = \partial V_1 = \partial V_2$ and $k \cap V_i = \alpha_i$ (i=1,2) is a trivially embedded arc in V_i (i=1,2).

We will use the following notations.

C(1) : the class of the exteriors of links of type I in $S^2 \times_\varphi S^1$ which have incompressible boundary and where each component of the link is orientation preserving in $S^2 \times_\varphi S^1$.

C(2) : the class of the exteriors of links of type I in $S^2 \times_\varphi S^1$ which have incompressible boundary and where each component of the link is orientation reversing in $S^2 \times_\varphi S^1$.

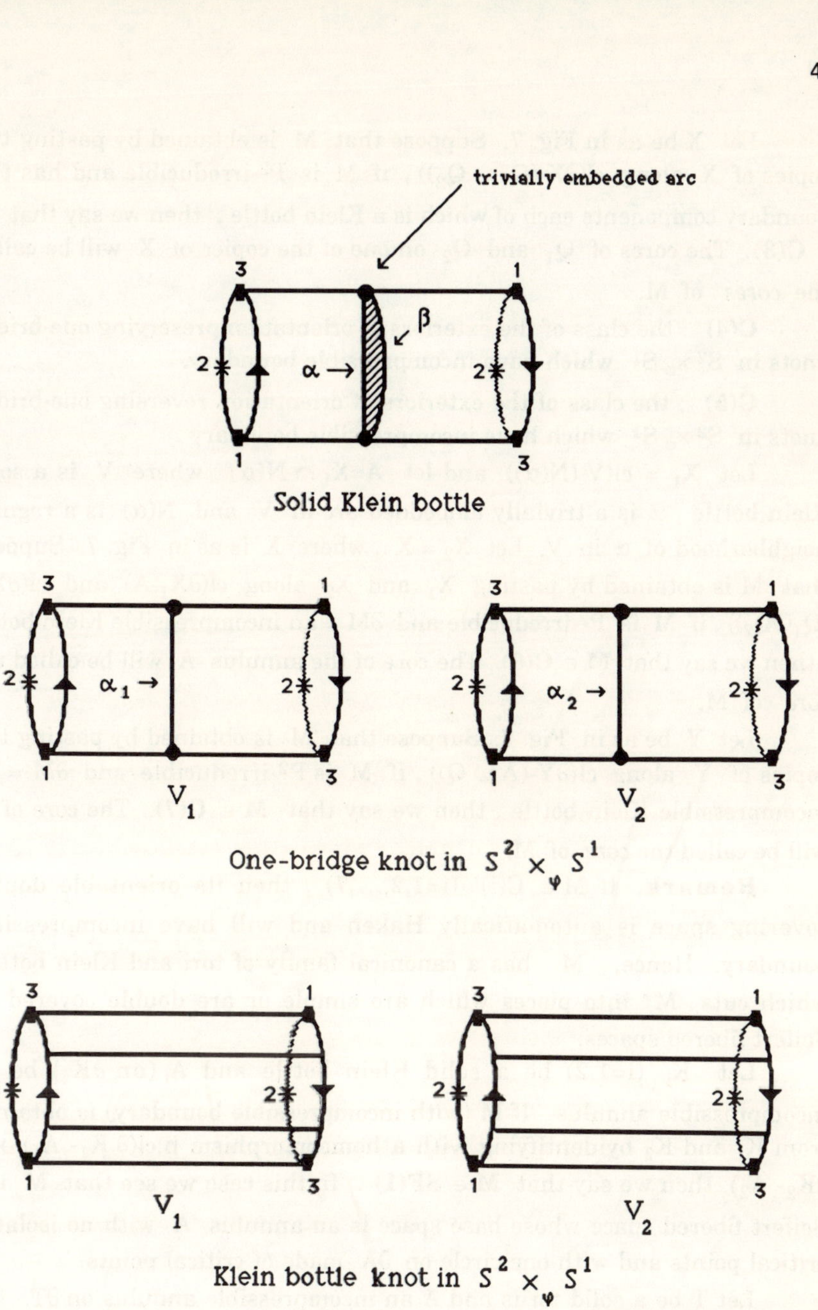
trivially embedded arc
3
1
2
β
α →
Solid Klein bottle
α1 →
α2 →
V1
V2
One-bridge knot in S2 ×φ S1
Klein bottle knot in S2 ×φ S1

Figure 6

Let X be as in Fig. 7. Suppose that M is obtained by pasting two copies of X along $cl(\partial X-(Q_1 \cup Q_2))$, if M is P^2-irreducible and has two boundary components each of which is a Klein bottle , then we say that **M** $\in$ **C(3)**. The cores of Q_1 and Q_2 on one of the copies of X will be called the *cores* of M.

C(4) : the class of the exteriors of orientation preserving one-bridge knots in $S^2 \times_\varphi S^1$ which have incompressible boundary.

C(5) : the class of the exteriors of orientation reversing one-bridge knots in $S^2 \times_\varphi S^1$ which have incompressible boundary .

Let $X_1 = cl(V-(N(\alpha))$ and let $A=X_1 \cap N(\alpha)$, where V is a solid Klein bottle , α is a trivially embedded arc in V and $N(\alpha)$ is a regular neighborhood of α in V. Let $X_2 = X$, where X is as in Fig. 7. Suppose that M is obtained by pasting X_1 and X_2 along $cl(\partial X_1\text{-}A)$ and $cl(\partial X_2\text{-}(Q_1 \cup Q_2))$, if M is P^2-irreducible and ∂M = an incompressible Klein bottle , then we say that **M** $\in$ **C(6)**. The core of the annulus A will be called the *core* of M.

Let Y be as in Fig. 7. Suppose that M is obtained by pasting two copies of Y along $cl(\partial Y-(A \cup Q))$, if M is P^2-irreducible and ∂M = an incompressible Klein bottle , then we say that **M** $\in$ **C(7)**. The core of A will be called the *core* of M.

Remark. If M $\in$ C(i) (i=1,2,...,7) , then its orientable double covering space is automatically Haken and will have incompressible boundary. Hence, M has a canonical family of tori and Klein bottles which cuts M into pieces which are simple or are double covered by Seifert fibered spaces.

Let K_i (i=1,2) be a solid Klein bottle and A_i (on ∂K_i) be an incompressible annulus. If M (with incompressible boundary) is obtained from K_1 and K_2 byidentifying with a homeomorphism $h: cl(\partial K_1 - A_1) \to cl(\partial K_2 - A_2)$ then we say that **M** $\in$ **SF(1)** . In this case we see that M is a Seifert fibered space whose base space is an annulus A with no isolated critical points and with one circle on ∂A made of critical points.

Let T be a solid torus and A an incompressible annulus on ∂T. Let K be a solid Klein bottle and Q_1 , Q_2 be a system of pairwise disjoint , incompressible Möbius bands on ∂K . If M (with incompressible boundary) is obtained from T and K by identifying with a homeomorphism h:cl(∂T- A)$\rightarrow$cl(∂K- ($Q_1 \cup Q_2$)) then we say that **M $\in$ SF(2)** . In this case we see that M is a Seifert fibered space whose base space is a disk D with one isolated critical point and with one interval on ∂D made of critical points.

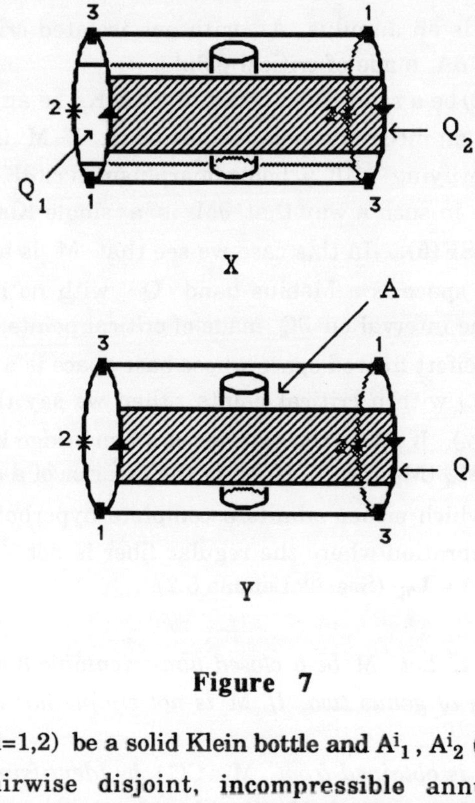

Figure 7

Let K_i (i=1,2) be a solid Klein bottle and $A^i{}_1$, $A^i{}_2$ (both on ∂K_i) be a system of pairwise disjoint, incompressible annuli. If M (with incompressible boundary) is obtained from K_1 and K_2 by identifying with a homeomorphism h:cl(∂K_1- ($A^1{}_1 \cup A^1{}_2$))$\rightarrow$cl(∂K_2- ($A^2{}_1 \cup A^2{}_2$)) in such a way that ∂M is a single torus , then we say that **M $\in$ SF(3)** . In this case we see that M is a Seifert fibered space whose base space is a Möbius band PQ (with one hole) with no isolated critical points and with one circle on ∂PQ made of critical points.

Let K_1 be a solid Klein bottle and A_1 , A_2 be a system of pairwise disjoint, incompressible annuli on ∂K_1. Let T be a solid torus and A'_1 be an incompressible annulus on ∂T. Let K_2 be a solid Klein bottle and A'_2 be an incompressible annulus on ∂K_2. If M is obtained from K_1, K_2 and T by identifying with a homeomorphism $h: cl(\partial K_1 - (A_1 \cup A_2)) \to cl((\partial T \cup \partial K_2) - (A'_1 \cup A'_2))$ in such a way that ∂M is a single torus, then we say that **M ∈ SF(4)**. In this case we see that M is a Seifert fibered space whose base space is an annulus A with one isolated critical point and with one circle on ∂A made of critical points.

Let K_i (i=1,2) be a solid Klein bottle, A_i (on ∂K_i) be an incompressible annulus and Q_i be an incompressible Möbius band. If M is obtained from K_1 and K_2 by identifying with a homeomorphism $h: cl(\partial K_1 - (A_1 \cup Q_1)) \to cl(\partial K_2 - (A_2 \cup Q_2))$ in such a way that ∂M is a single Klein bottle, then we say that **M ∈ SF(5)**. In this case we see that M is a Seifert fibered space whose base space is a Möbius band Q with no isolated critical points and with one interval on ∂Q made of critical points.

If M is a Seifert fibered space whose base space is a disk (annulus, Möbius band resp.) with n critical points, then we say that **M ∈ D(n)** (**A(n)**, **Mö(n)** resp.). If M is the exterior of a two bridge knot (link resp.) we say that **M ∈ $\mathbf{M_K}$** (**$\mathbf{M_L}$** resp.). If M is the exterior of a one-bridge knot in a lens space (which either admits a complete hyperbolic structure or admits a Seifert fibration where the regular fiber is not a meridian loop) then we say that **M ∈ $\mathbf{L_K}$** (See [8] ,Lemma 5.2).

Theorem 11. *Let* M *be a closed non-orientable* 3-*manifold with a Heegaard splitting of genus two. If* M *is not simple nor Seifert fibered, then either*

(i) M *is obtained from* $M_1 \in C(1)$ *by identifying the meridian and the core of* M_1 (*the identification done so as to yield a non-orientable space*),

(ii) M *is obtained from* $M_1 \in A(n)$ (n=0,1,2) *and* $M_2 \in M_L$ *by identifying their boundaries such that the regular fiber of* M_1 *is identified with the meridian loop of* M_2 (*the identification done so as to yield a non-orientable space*),

(iii) M *is obtained from* $M_1 \in C(2)$ *by identifying the meridian and the core of* M_1,

(iv) M *is obtained from* $M_1 \in C(3)$ *by identifying the cores of* M_1 ,

(v) M *is obtained from* $M_1 \in SF(1)$ *and* $M_2 \in L_K$ *by identifying their boundaries such that the regular fiber of* M_1 *is identified with the meridian loop of* M_2 ,

(vi) M *is obtained from* $M_1 \in D(2)$ *and* $M_2 \in C(4)$ *by identifying their boundaries such that the regular fiber of* M_1 *is identified with the meridian of* M_2 ,

(vii) M *is obtained from* $M_1 \in SF(1)$ *and* $M_2 \in C(n)$ (n=4,5) *by identifying their boundaries such that the regular fiber of* M_1 *is identified with the meridian of* M_2 ,

(viii) M *is obtained from* $M_1 \in SF(n)$ (n=1,2) *and* $M_2 \in C(6)$ *by identifying their boundaries such that the regular fiber of* M_1 *is identified with the core of* M_2 ,

(ix) M *is obtained from* $M_1 \in SF(n)$ (n=3,4) *and* $M_2 \in M_K$ *by identifying their boundaries such that the regular fiber of* M_1 *is identified with the meridian loop of* M_2 ,

(x) M *is obtained from* $M_1 \in Mö(n)$ (n=0,1,2) *and* $M_2 \in M_K$ *by identifying their boundaries such that the regular fiber of* M_1 *is identified with the meridian of* M_2 ,

(xi) M *is obtained from* $M_1 \in SF(5)$ *and* $M_2 \in C(7)$ *by identifying their boundaries such that the regular fiber of* M_1 *is identified with the core of* M_2 ,

(xii) M *is obtained from* $M_1 \in D(2)$, $M_2 \in SF(1)$ *and* $M_3 \in M_L$ *by identifying their boundaries such that the regular fibers of* M_1 *and* M_2 *are identified with the meridians of* M_2 *or*

(xiii) M *is obtained from* $M_1 \in SF(1)$, $M_2 \in SF(1)$ *and* $M_3 \in M_L$ *by identifying their boundaries such that the regular fibers of* M_1 *and* M_2 *are identified with the meridians of* M_2 .

Proof. M is P^2-irreducible , otherwise $M \approx S^2 \times_\varphi S^1 \# L_n$ where L_n is a lens space ($\neq S^3$) (this follows by Corollary II.10 in [4]) , $M \approx S^2 \times_\varphi S^1$ or (if

M contains a two-sided projective plane) $M \approx P^2 \times S^1$ (by [11]) and none of these admits an essential Klein bottle or torus.

Let $(V_1, V_2; F)$ be a Heegaard splitting of genus two for M. We have the following possibilities.

Case 1. Suppose that M contains a 2-sided non-separating incompressible torus. By Theorem 1 in [7] we may assume that M contains a 2-sided non-separating incompressible torus T such that $T \cap V_1 = D$, where D is an essential disk in V_1. By the argument in [7], Lemma 4.2, it follows that D is a meridian disk of V_1.

Case 1.1. Suppose D cuts V_1 into a solid Klein bottle. Since $T \cap V_2$ is incompressible in V_2, by looking at the intersections of $T \cap V_2$ with any complete system of meridian disks of V_2, we have a hierarchy (T''_0, α_0), (T''_1, α_1) for $T''_0 = T \cap V_2$. Let $T^{(1)}$ be the image of T after we push α_0 into V_1 with an isotopy of type A at α_0.

Let $A_i = T^{(1)} \cap V_i$ (i=1,2), then A_i is an essential annulus in V_i and since D cuts V_1 into a solid Klein bottle A_1(hence A_2) satisfies conclusion (i) of *lemma* 3. Thus, A_i cuts V_i into a non-orientable handlebody of genus two V'_i. Let A^j_i (j=1,2) be the two copies of A_i on $\partial V'_i$ (i=1,2). The attaching homeomorphism of the Heegaard splitting $(V_1, V_2; F)$ induces a homeomorphism h between $cl(\partial V'_1 - (A^1_1 \cup A^2_1))$ and $cl(\partial V'_2 - (A^1_2 \cup A^2_2))$. Each V'_i (i=1,2) is homeomorphic to a solid Klein bottle minus a regular neighborhood of a trivially embedded arc. The homeomorphism h identifies the boundaries of these solid Klein bottles as in the definition a link of type I in $S^2 \times_\varphi S^1$. Thus, if we attach V'_1 to V'_2 using h we get $M_1 \in C(1)$. We get M from M_1 by identifying the components of ∂M_1 such that the meridian of M_1 is identified with the core of M_1. Hence M is as in **(i)** above.

Case 1.2. Suppose D cuts V_1 into a solid torus. We have a hierarchy (T''_0, α_0), (T''_1, α_1) for $T''_0 = T \cap V_2$. Let $T^{(1)}$ be the image of T after an isotopy of type A at α_0 and $A_i = T^{(1)} \cap V_i$ (i=1,2). Then A_i is an essential annulus in V_i and since D cuts V_1 into a solid torus A_1(hence A_2) satisfies conclusion (ii) of *lemma* 3. Thus, A_i cuts V_i into an orientable handlebody of genus two V'_i. Let A^j_i (j=1,2) be the two copies of

A_i on $\partial V'_i$ (i=1,2). The attaching homeomorphism of the Heegaard splitting induces a homeomorphism h between $cl(\partial V'_1 - (A^1_1 \cup A^2_1))$ and $cl(\partial V'_2-(A^1_2 \cup A^2_2))$.

In this case M can be obtained by a construction similar to that in 7], Theorem 2. As in 7] we see that $M = M_1 \cup M_2$, where $M_1 \in A(n)$ (n=0,1,2) and $M_2 \in M_L$. Hence M is as in **(ii)** above.

Case 2. Suppose M contains a 2-sided non-separating incompressible Klein bottle. Then, by the theorem in **§1**, we may assume that M contains a 2-sided non-separating incompressible Klein bottle K such that $K \cap V_1 = D$, where D is an essential disk. Again it can be shown that D is a meridian disk. Let $(K''_0, \alpha_0), (K''_1, \alpha_1), \ldots, (K''_n, \alpha_n)$ be a hierarchy for $K''_0 = K \cap V_2$.

We show that D cuts V_1 into a solid Klein bottle. If α_0 is of type 3, then K''_1 is an annulus, so α_1 must be of type 5. If α_0 is of type 4, then K''_1 consists of two Möbius bands, so α_1 must be of type 5. Hence, either α_0 or α_1 is of type 5. Suppose α_0 is of type 3 or 4, then α_1 is of type 5 and since K is non-separating the annulus $A=K^{(1)} \cap V_1$ ($K^{(1)}$ is the image of K after an isotopy of type A at α_0) satisfies conclusion (i) or (ii) of *lemma* 3. A can't satisfy conclusion (ii), otherwise by doing an isotopy of type A at α_1 we get an orientation reversing simple closed curve on $\partial(V_1\text{-}N(A))$ this is impossible since $V_1\text{-}N(A)$ is an orientable handlebody of genus two. Therefore, A satisfies conclusion (i) of *lemma* 3 and, by this, it follows that D cuts V_1 into a Klein bottle. If α_0 is of type 5, then we can show that D cuts V_1 into a solid Klein bottle.

Next, we show that α_0 can't be of type 4. If α_0 is of type 4, then $K^{(1)} \cap V_1$ is an essential annulus A which satisfies conclusion (i) of *lemma* 3 and $K^{(1)} \cap V_2$ consists of two essential Möbius bands Q_1, Q_2. Thus, $cl(\partial V_1\text{-}A)$ consists of two components, each of which has one boundary component, on the other hand the components of $cl(\partial V_2\text{-}(Q_1 \cup Q_2))$ have two boundary components each, impossible. So α_0 is not of type 4.

Case 2.1. Suppose α_0 is of type 3. Let $A_i = K^{(1)} \cap V_i$ (i=1,2). Then A_i is an essential annulus in V_i and, by the above argument, A_i satisfies (in V_i) conclusion (i) of *lemma* 3. Thus, A_i cuts V_i into a non-orientable handlebody of genus two V'_i. Let A^j_i (j=1,2) be the copies of A_i on $\partial V'_i$. The attaching homeomorphism of the Heegaard splitting induces a homeomorphism h between $cl(\partial V'_1-(A^1{}_1\cup A^2{}_1))$ and $cl(\partial V'_2-(A^1{}_2\cup A^2{}_2))$. Each V'_i (i=1,2) is homeomorphic to a solid Klein bottle minus the regular neighborhood of a trivially embedded arc. The homeomorphism h identifies the boundaries of these Klein bottles as in the definition of a link of type I in $S^2\times_\varphi S^1$. Thus , if we attach V'_1 and V'_2 using h we get $M_1\in C(2)$.

We get M from M_1 by identifying the components of ∂M_1 such that the meridian of M_1 is identified with the core of M_1. Hence M is as in **(iii)**.

Case 2.2. Suppose α_0 is of type 5. Let $Q_i=K^{(1)}\cap V_i$ (i=1,2). Then Q_i is an essential Möbius band in V_i. Thus , Q_i cuts V_i into a non-orientable handlebody of genus two V'_i. Let $Q^j{}_i$ (j=1,2) be the copies of Q_i on $\partial V'_i$. Again, the attaching homeomorphism of the Heegaard splitting induces a homeomorphism h between $cl(\partial V'_1-(Q^1{}_1\cup Q^2{}_1))$ and $cl(\partial V'_2-(Q^1{}_2\cup Q^2{}_2))$. If we attach V'_1 and V'_2 using h we get $M_1\in C(3)$ (See Fig. 7).

We get M from M_1 by identifying the components of ∂M_1 such that the cores of M_1 are identified. Hence M is as in **(iv)**.

From now on we assume that every essential Klein bottle (or torus) is separating in M. Suppose that M has a canonical system of tori and Klein bottles W such that $W\cap V_1$ consists of disks and is minimal. By *Theorem* 2 , we may assume $W\cap V_1$ consists of at most two essential disks.

Case 3. Suppose that W=T is a single torus , that $T\cap V_1$ is an essential disk (in V_1). M is decomposed into two components M_1, M_2 by the canonical system. There is a hierarchy (T''_0, α_0) , (T''_1, α_1) , ..., (T''_n, α_n) for $T''_0 = T\cap V_2$. Let $T^{(1)}$ be the image of T after an isotopy of type A at α_0. Since T is a torus, α_0 is of type 3 and $A_i=T^{(1)}\cap V_i$ is an essential annulus in V_i. Since T is separating in M , A_i (i=1,2) satisfies one of the conclusions (iii)-(v) of *lemma* 3.

Case 3.1. Suppose $A_1\subset V_1$ satisfies conclusion (iii) of *lemma* 3. Then $A_2\subset V_2$ must also satisfy conclusion (iii) . So, for each i, A_i cuts V_i into a solid Klein bottle $V^1{}_i$ and an orientable handlebody of genus two $V^2{}_i$.

Attaching $V^1{}_1$ to $V^1{}_2$ along $cl(\partial V^1{}_i\text{-}A_i)$ we get $M_1 \in SF(1)$ (changing the subscripts of the M_i' s , if necessary). Attaching $V^2{}_1$ to $V^2{}_2$ along $cl(\partial V^2{}_i\text{-}A_i)$ we get $M_2 \in L_K$ (by [8]). M is obtained from M_1 , M_2 by identifying their boundaries such that the regular fiber of M_1 is identified with the meridian of M_2. Hence M is as in **(v)**.

Case 3.2. Suppose $A_1 \subset V_1$ satisfies conclusion (iv) of *lemma* 3. Then $A_2 \subset V_2$ must also satisfy conclusion (iv) . So, for each i, A_i cuts V_i into a solid torus $V^1{}_i$ and a non-orientable handlebody of genus two $V^2{}_i$. Attaching $V^1{}_1$ to $V^1{}_2$ along $cl(\partial V^1{}_i\text{-}A_i)$ we get $M_1 \in D(2)$. Notice that each $V^2{}_i$ is homeomorphic to a solid Klein bottle minus the regular neighborhood of a trivially embedded arc removed. Thus , attaching $V^2{}_1$ to $V^2{}_2$ along $cl(\partial V^2{}_i\text{-}A_i)$ we get $M_2 \in C(4)$. M is obtained from M_1 , M_2 by identifying their boundaries such that the regular fiber of M_1 is identified with the meridian of M_2. Hence M is as in **(vi)**.

Case 3.3. Suppose $A_1 \subset V_1$ satisfies conclusion (v) of *lemma* 3. Then $A_2 \subset V_2$ also satisfies conclusion (v) . So A_i cuts V_i into a solid Klein bottle $V^1{}_i$ and a non-orientable handlebody of genus two $V^2{}_i$. Attaching $V^1{}_1$ to $V^1{}_2$ we get $M_1 \in SF(1)$ and attaching $V^2{}_1$ to $V^2{}_2$ we get $M_2 \in C(4)$. M is obtained from M_1 , M_2 by identifying their boundaries such that the regular fiber of M_1 is identified with the meridian of M_2. Hence M is as in **(vii)**.

Case 4. Suppose that W=K is a single Klein bottle and that $K \cap V_1$ is an essential disk (in V_1). In this case we obtain conclusions **(vii)** and **(viii)**.

Case 5. Suppose that W=T is a single torus and that $T \cap V_1 = D_1 \cup D_2$, where D_1 , D_2 are disjoint essential disks in V_1. In this case we obtain conclusions **(ix)** and **(x)**.

Case 6. Suppose that W=K is a single Klein bottle and that $K \cap V_1$ consists of two disks D_1 , D_2. In this case we obtain conclusions **(viii)** and **(xi)**.

Case 7. Suppose that $W = T_1 \cup T_2$, where T_i (i=1,2) is a torus and that $(T_1 \cup T_2) \cap V_1$ consists of two disks. In this case we obtain conclusions **(xii)** and **(xiii)**.

It is interesting to notice that the canonical system can't be so that it has two components and at least one of them is a Klein bottle. This can be done by examining the components of M cut along W as we have done so far. In each case we arrive at a contradiction (see [1] for more details). ∎

Similar results were obtained independently by Kanji Morimoto at Kobe University ([10]). He also shows that for M as in Theorem 11 , if M contains a 2-sided incompressible Klein bottle , then $H_1(M)$ is either Z , Z + Z_2 or Z + Z_{4p} , where p is an integer.

REFERENCES

1] Cardona , I. , "Canonical systems of tori and Klein bottles in non-orientable 3-manifolds of genus two" , Ph. D. Thesis Florida State University (1987).

2] Epstein , D. B. A. , "Periodic flows on three-manifolds" , Ann. of Math 95 (1972).

3] Hempel , J. , "3-manifolds" , Ann. of Math. Studies 86 , Princeton University Press , Princeton N.J. (1976).

4] Jaco , W. , "Lectures on three manifold topology" , Providence R. I. , Amer. Math. Soc. (1980).

5] Jaco , W. and Shalen , P. , "Seifert fibered spaces in 3-manifolds" , Memoirs of the Amer. Math. Soc. (1978).

6] Johannson , K. , "Homotopy equivalences of 3-manifolds with boundaries" , Lecture Notes in Math. Springer Verlag 761 (1979).

7] Kobayashi , T. , "Non-separating incompressible tori in 3-manifolds" , J. Math. Soc. Japan 36 (1984) , p. 11-22.

8] Kobayashi , T. , "Structures of the Haken manifolds with Heegard splittings of genus two" , Osaka J. Math. 21 (1984) , p. 437-455.

9] Lickorish , W. B. R. , "Homeomorphisms of non-orientable two manifolds" , Proc. Camb. Phil. Soc. 59 (1963) , p. 307-317.

10] Morimoto , K. , "Structures of non-orientable closed 3-manifolds of genus two" , preprint.

11] Ochiai , M. , "2-sided embeddings of projective planes into 3-manifolds" , Trans. Amer. Math. Soc. 274 (1982) , p. 641-650.

12] Ochiai , M. , "On Haken's theorem and its extension" , Osaka J. Math. 20 (1983), p. 461-468.

13] Scott , P. , "The geometries on 3-manifolds" , Bull. Lond. Math. Soc. 15 (1983) , p. 401.

14] Tollefson , J. , "Involutions of Seifert Fiber spaces" , Pacific J. Math. 74 (1978), p. 519-529.

15] Thurston , W. P. , "Three dimensional manifolds , Kleinian groups and hyperbolic geometry " , Bull. Amer. Math. Soc. 6 (1983) , p.357-381.

GEOMETRY AND TOPOLOGY (pp. 58–73)
edited by G. M. Rassias & G. M. Stratopoulos

ABSTRACT

by

T. A. Chapman

The main result of this paper asserts that if N is a high-dimensional compact PL manifold, then any PL concordance on $N\times[0,1]$ is PL isotopic to a product of a PL concordance on N with the identity on [0,1]. The proof turns out to be an easy application of recent results from the controlled category concerning controlled h-cobordisms and controlled concordances.

Destabilization Of Piecewise Linear Concordances

by
T.A. Chapman

Department of Mathematics
Lexington, Kentucky 40506

Introduction

Let N be a compact Piecewise Linear (PL) manifold, let $I=[0,1]$, and identify N with $\{0\}\times N$ in $I\times N$. A *PL concordance* on N is a PL homeomorphism $f: I\times N \rightarrow I\times N$ for which $f|N=id$. The purpose of this paper is to give a proof of the following result.

Destabilization Theorem. *There exists an integer n_0 so that if $dim\ N \geq n_0$, then any PL concordance f on $N\times I$ is PL isotopic rel $N\times I$ to a product $f'\times id_I$, where f' is a PL concordance on N.*

More generally let $C(N)$ be the group of all PL cocordances on N, and define the suspension map Σ: $C(N) \rightarrow C(N\times I)$ by $\Sigma(f)=f\times id_I$. Then the Destabilization Theorem essentially says that Σ is 0-connected if $dim\ N \geq n_0$. This is a special case of the PL Stability Conjecture, which asserts that for every $k\geq 0$ there exists an integer n_k for which Σ is k-connected if $dim\ N \geq n_k$.

This conjecture was originally established by Hatcher [4], but his proof was found to have several flaws. More recently the differentiable version of this conjecture has been established by Igusa by using his work on the elimination of higher singularities [7]. When this is combined with the result of Burghelea and Lashof on the equivalence of PL and differentiable stability for smoothable manifolds [1], it follows that the PL Stability Conjecture is true for N smoothable.

The strategy involved in Hatcher's approach was to set up and manipulate parametrized families of PL handlebody structures on N, with parameter in a polyheron B (i.e., fibered PL handlebody structures over a base B). This was no easy task, for the order in which these handles were attached could be different for different values of the parameter. Also there were isolated values of the parameter for which these handlebody structures were allowed to collapse to N.

Our proof of the Destabilization Theorem uses Hatcher's ideas to the extent that we set up and manipulate 1- parameter families of PL handlebody structures on N. However our handlebody structures differ from those used by Hatcher in the following essential way: *The order in which the handles are attached is constant for different values of the parameter.* Without permitting handle rearrangements we cannot collapse these handlebody structures for isolated values of the parameter. Our strategy is to avoid these collapses by working in the controlled category of [2]. Then near the end of the proof we turn the controlled information back into actual collapses by using a result from [2] on controlled h-cobordisms and a result from [3] on controlled concordances.

Even though the proof that we give here seems to only establish the case $k=0$ of the PL Stability Conjecture, it is hoped that some modification of it will succeed in establishing the general case.

Here is a brief description of the remaining sections which appear in this paper. Section 2 is devoted to the introduction and discussion of some basic concepts. This includes the important notion of a fibered handle decomposition of a fibered h-cobordism. In Section 3 we establish the Fibered Handle Trading Theorem. This turns out to be a routine 1- parameter version of the usual handle-trading process. Section 4 uses the Fibered Handle Trading Theorem to establish the Factorization Lemma, which asserts that a high-dimensional fibered h-cobordism can be factored into the product of a low-dimensional one with a cell. Finally in Section 5 we use the Factorization Lemma to prove the Destabilization Theorem. It is here that the results from [2] and [3] on controlled h-cobordisms and controlled concordances are used.

Finally we make a comment about the integer n_0. In the course of the argument we shall see that $n_0=20$ will suffice. This number results from the need to have sufficient maneuvering room in order that certain unfibered constructions can be readily converted into fibered constructions. It is reasonable to expect that a careful analysis of the argument will show that n_0 can be chosen less than 20. Whatever the case may be, its eventual size is not important for its role in the Destabilization Theorem.

2. Definitions And Notation

Throughout this paper B will always be [0,1]. By a *PL manifold bundle* we will mean a PL locally trivial bundle $E \rightarrow B$ with fiber a compact PL manifold. For any $b \in B$ we denote the b-level of E by $(E)_b$. The boundary of this bundle is defined to be $\delta E = \cup\{\partial(E)_b \mid b \in B\}$, and it is obvious that $\delta E \rightarrow B$ is also a PL manifold bundle. Let N be a compact PL manifold such that $B \times N$ is a subpolyhedron of E for which each $\{b\} \times N$ has codimension 0 in $\partial(E)_b$. It follows from regular neighborhood theory that each level $\{b\} \times N$ is PL collared in $(E)_b$, thus Hudson's isotopy extension theorem [5, p.147] implies that $B \times N$ is fiber preserving (f.p.) PL collared in E. When this is true we say that $(E, B \times N) \rightarrow B$ is a fibered cobordism. It is defined to be a fibered h-cobordism if each level $((E)_b, \{b\} \times N)$ is an ordinary h-cobordism, i.e., there are deformation retractions of $(E)_b$ to $\{b\} \times N$ and $\overline{\partial(E)_b - \{b\} \times N}$. We also define $\delta^- E = B \times N$ and $\delta^+ E = \overline{\delta E - (B \times N)}$, and note that $\delta^+ E \rightarrow B$ is a PL manifold bundle. It follows from the general theory of fibrations that there exist $f.p.$ deformation retractions of E to $\delta^- E$ and to $\delta^+ E$.

Next we give the definition of a fibered handle attached to a PL manifold bundle. Let $E \rightarrow B$ be a compact PL manifold bundle of dimension $n+2$ and let $\phi \colon B \times \partial B^r \times B^{n+1-r} \rightarrow \delta^+ E$ be a $f.p.$ PL embedding, where $B^r = [-1,1]^r$. Using ϕ as an attaching map we form $E \cup H^r$, where H is a copy of $B \times B^r \times B^{n+1-r}$. We know that each level $(E \cup H)_b$ is a manifold, thus by Hudson's isotopy extension theorem $E \cup H \rightarrow B$ must be a PL manifold bundle. H is called a *fibered r-handle,* and we say that $E \cup H$ has been formed by attaching a fibered r-handle to δE. If we additionally assume that $(E, B\times N) \to B$ is a fibered cobordism, then we will also require H to be attachd to $\delta E - (B\times N)$.

If $(E, B\times N) \to B$ is a fibered cobordism, then a *fibered handle decomposition* of E on $B\times N$ is a presentation $E = C \cup S_1 \cup \ldots \cup S_p$, where C is a *f.p. PL* collar on $B\times N$ and each S_i is a union of disjoint fibered handles of the same index (index S_i) which are attached to the cobordism $(C\cup S_1 \cup \cdots \cup S_{i-1}, B\times N) \to B$. We also assume that it is *monotonic,* i.e., index $S_i \leq$ index S_{i+1}. For additional notation let $E^k = C \cup (\cup\{S_i \mid \text{index } S_i \leq k\})$ (the k-skeleton of E) and similarly let $E_0 = C$, $E_i = C \cup S_1 \cup \cdots \cup S_i$. In analogy with the unfibered case it is easy to carry out fibered handle modifications by *f.p.* homeomorphisms of attaching maps. By this we mean that if $h : E_i \to E_i$ is a *f.p. PL* homeomorphism which is the identity on $B\times N$, then we obtain a new fibered handle decomposition of E:

$$E = h(C) \cup h(S_1) \cup \cdots \cup h(S_i) \cup S_{i+1} \cup \cdots \cup S_p.$$

We now show how to construct fibered handle decompositons. Starting with a fibered cobordism $(E, B\times N) \to B$ we choose $b \in B$ and consider the cobordism $(W,N) = ((E)_b, \{b\}\times N)$. Using the fact that $E \to B$ is a PL trivial bundle we can find a *f.p. PL* homeomorphism $g : B\times W \to E$ which is the identity on $B\times N$. If $W = C' \cup S'_1 \cup \cdots \cup S'_p$ is an ordinary handle decomposition of W on N, then we obtain the following fibered handle decompostion of E on $B\times N$:

$$E = g(B\times C')\cup g(B\times S'_1)\cup \cdots \cup g(B\times S'_p).$$

In the sequel we will be dealing with fibered handle decompositions which are cover controlled in the following sense. Let $(E, B\times N) \to B$ be a fibered cobordism and let α be an open cover of E. A fibered handle decomposition $E = C\cup S_1 \cdots \cup S_p$ is *α-small* provided that each level of each fibered handle lies in some element of α, and the *f.p.* collar C is of the form $B\times N\times [0,1]$ where each $\{x\}\times[0,1]$ lies in some element of α. It is clear from the above paragraph that given any α there exists a fibered handle decomposition which is α-small. Situations will arise in which we will be forced to increase the size of α, but the increase must depend only on *dim E*. In order to describe this more precisely we define $St^0(\alpha)=\alpha$, and proceeding inductively we define $St^{m+1}(\alpha)$ to be the open cover consisting of all sets of the form $U\cup(\cup\{V\in\alpha \mid U\cap V\neq\emptyset\})$, where $U\in St^m(\alpha)$. Then we say that a fibered handle decomposition of E is *α-controlled* if there is an integer m (depending only on *dim E*) for which it is $St^m(\alpha)$-small. Also if $f: B\times X \to E$ is any *f.p.* map, then we say that it is *α-controlled* if there is an integer m as above for which each level of f lies in some member of $St^m(\alpha)$.

3. The Fibered Handle Trading Theorem

For notation let $(E, B\times N^n) \to B$ be a fibered h-cobordism an let $E = C\cup S_1\cup \cdots \cup S_p$ be a fibered handle decomposition such that S_1 is a union of k fibered r-handles. We are also given an open cover α of E such that the given decomposition is α-small and such that there exists a *f.p.* deformation retraction of E to $B\times N$ which is *f.p.* α-homotopic to the identity.

Fibered Handle Trading Theorem. *If* $r \leq n-8$, *then there is an* α-*controlled f.p. trade of the fibered r-handles in* S_1 *for fibered* $(r+2)$-*handles. By this we mean that there is a new* α-*controlled fibered handle decomposition of* E *which has* k *fewer fibered r-handles and* k *additional fibered* $(r+2)$-*handles.*

Remarks 1. We will only sketch the proof of this result. Complete details will not be given because of its similarity with ordinary (unfibered) handle trading.

2. Observe that the theorem is obvious if we omit the cover control. To see this just use the *PL* triviality of the bundle $(E, B \times N) \rightarrow B$ along with ordinary handle trading in the fiber. In this case the integer 8 can be reduced to the more well-known 3.

Proof. The specific treatment of ordinary handle trading that we have in mind for adaptation to the fibered and controlled categories is the largely geometric one given by Wall [10]. In fact it was shown by Quinn that this geometric approach easily adapts to the controlled category [9]. For our proof it will suffice to show how to trade a single fibered r-handle in S_1 for a fibered $(r+2)$-handle. To deal with the general case one similarly treats all of the fibered r-handles in S_1 simultaneously. Also our required cover control follows more - or - less automatically from the constructions.

For notation let $H^r = B \times B^r \times B^{n+1-r}$ be the fibered r-handle in S_1 that we wish to trade. Decompose ∂B^{r+1} into r-balls, $\partial B^{r+1} = B^r \cup B_1^r$, which intersect in a $(r-1)$-sphere. Also let $E^{r+2} = E_i$, for some $i > 1$. The key to trading the handle H is the existence of an α-controlled *f.p. PL* embedding $g : B \times B^{r+1} \rightarrow \delta^+ E_i - (B \times N)$ such that $g(B \times B_1^r) \subset \delta^+ C$ and $g|B \times B^r$ is an inclusion $j : B \times B^r \times \{*\} \hookrightarrow H$, for some $* \in \partial B^{n+1-r}$. Of course this also includes the usual fibered handle modifications of the given fibered handle decomposition by $f.p.$ PL homeomorphisms of attaching maps. Once g has been constructed it can be used in the standard way to complete the trade. By this we mean that we can attach to $\delta^+ E_i$ a cancelling pair (K^{r+2}, J^{r+1}) of fibered handles so that (J,H) is also a cancelling pair. When this latter pair is cancelled we obtain a new α-controlled fibered handle decomposition in which H has been traded for K. So all that remains is to establish the existence of the embedding g.

It is easy to construct an α-controlled $f.p.$ map $f : B \times B^{r+1} \to \delta^+ E_1 \cup (E_i - E_1)$ such that $f(B \times B_1^r) \subset \delta^+ C \cap \delta^+ E_1$ and $f | B \times B^r = j$. One obtains f by generalizing to the fibered category the proof of the Corollary in [10, p. 602]. The cases $r \leq 2$ can be quickly taken care of as follows. First approximate f with an α-controlled $f.p.$ PL embedding $g' : B \times B^{r+1} \to \delta^+ E_1 \cup (E_i - E_1)$ such that $g'(B \times B_1^r) \subset \delta^+ C \cap \delta^+ E_1$ and $g' | B \times B^r = j$. Then by a combiation of general position moves and fibered handle modifications we can push the image of g' into $\delta^+ E_i - (B \times N)$ to obtain the desired embedding g.

In order to deal with the cases $3 \leq r \leq n-8$ we have to replace f with an α-controlled $f.p.$ PL embedding $g' : B \times B^{r+1} \to \delta^+ E_i - (B \times N)$ such that $g'(B \times \partial B^{r+1}) \subset \delta^+ E_1$ and $g' | B \times \partial B^{r+1}$ is $f.p.$ homotopic to $f | B \times \partial B^{r+1}$ in $\delta^+ E_1$ via an α-controlled homotopy. One obtains g' by generalizing to the fibered category the proof of Theorem 3 in [10]. The technique of proof consists mainly of standard PL singular set manipulations along with fibered handle modifications. Similarly we can generalize to the fibered category the proof of Stallings' lemma which is also given in [10]. This gives us a $f.p.$ PL isotopy $g'|B\times\partial B^{r+1}\simeq u$ in δ^+E_1 such that $u(B\times B_1^r)\subset\delta^+C\cap\delta^+E_1$ and $u|B\times B^r=j$. We can extend this isotopy to a $f.p.$ PL isotopy of δ^+E_1 to itself. This enables us to simultaneously carry out a fibered handle modification and a $f.p.$ PL isotopy of g' to our desired embedding g. □

4. The Factorization Lemma

We say that a PL manifold bundle $E\to B$ has a *factorization* if there exists a PL manifold bundle $E'\to B$ and a $f.p.$ PL homeomorphism $h: E'\times I\to E$. If α is an open cover of E, then this factorization is *α-small* if each $h(\{x\}\times I)$ lies in some elements of α. It is *α-controlled* if it is $St^m(\alpha)$-small, where m depends only on $dim\ E$. In the statement below we are given a fibered h-cobordism $(E,\ B\times N^n)\to B$ and an open cover α of E for which there are $f.p.$ deformation retractions of E to δ^-E and δ^+E that are $f.p.$ α-homotopic to their respective identities. We also assume that $B\times N$ has a given α-small factorization which we write as $B\times N'\times I=B\times N$.

Factorization Lemma. *If $n\geq 20$, then there exists a fibered h-cobordism $(E',\ B\times N')\to B$ and an α-controlled $f.p.$ PL homeomorphism $h: E'\times I\to E$ which is the identity on $B\times N$.*

Proof. We will not pay close attention to the controlled version since all of the steps that we will go through are α-controlled. Our strategy for the proof is to set up a fibered handle decomposition which can be inductively factored. Using the Fibered Handle Trading Theorem there is a fibered handle decomposition of E on $B \times N$ whose handles have indices ≥ 4. By turning this decomposition upside down we obtain the dual decomposition. Applying the Fibred Handle Trading Theorem to this dual decompostion we can trade "down" the handles having indices ≥ 9. When this is done we obtain a fibered handle decomposition of the form $E = C \cup S_1 \cup \cdots \cup S_p$, where $4 \leq$ index $S_i \leq 8$.

In our proof we will actually establish something stronger. We will show that (after a fibered handle modification) our desired factorization $h: E' \times I \to E$ can be constructed so that E' has a fibered handle decomposition $E' = C' \cup S'_1 \cup \cdots \cup S'_p$ which satisfies $h(S'_i \times I) = S_i$. So not only is E being factored, but all of its handles are also being factored. Thus our strategy is to inductively factor the E_i. Starting with the factorization of E_0 it is obvious that we can write $C = [0,1] \times B \times N$, where $B \times N \equiv \{0\} \times B \times N$. Defining $C' = [0,1] \times B \times N'$ we have $C = C' \times I$ and $E = (C' \times I) \cup S_1 \cup \cdots \cup S_p$.

For our inductive hypothesis we assume that E_i has been factored in the above manner. By choosing the factoring homeomorphism to be the identity we can write $E_i = E'_i \times I$ and $E_{i+1} = (E'_i \times I) \cup S_{i+1}$. We will show that after a fibered handle modification we can similarly factor E_{i+1}. It suffices to assume that S_{i+1} is a single handle H with attaching map $\phi: B \times \partial B^r \times B^{n+1-r} \to \delta^+ E_i - (B \times N)$, where $4 \leq r \leq 8$. Factoring B^{n+1-r} into $B^{n-r} \times I$ and identifying $\partial B^r \times I$ with $\partial B^r \times \{0\} \times I$ in $\partial B^r \times B^{n+1-r}$ we let $\phi_0 = \phi | B \times \partial B^r \times I$. Our first step is to find a *f.p. PL* isotopy in $\delta^+ E_i - (B \times N)$ from ϕ_0 to a product $\psi_0 \times id_I$, where $\psi_0: B \times \partial B^r \to \delta^+ E'_i - (B \times N')$ is a *f.p. PL* embedding. By turning the given handle decomposition of E'_i upside down (i.e., forming the dual) we may view E'_i as formed by attaching fibered handles with indices $\geq n-7$ to $\delta^+E'_i$. Thus $(E'_i, \delta^+E'_i)$ is *f.p.* highly connected. Since $\delta^+E_i = (\delta^+E'_i \times I) \cup (E'_i \times \partial I)$ we can use the high connectivity of $(E'_i, \delta^+E'_i)$ to obtain *f.p.* homotopy of ϕ_0 into $\delta^+E'_i \times I$. Since $2(r+1)+2 \leq n$ we can use unfibered *PL* unknotting of polyhedra [8, p.111] to replace this *f.p.* homotopy with a *f.p. PL* isotopy of ϕ_0 into $[\delta^+E'_i - (B \times N')] \times I$. Now it is easy to follow this isotopy of ϕ_0 with a further isotopy to our desired product $\psi_0 \times id$. The isotopy $\phi_0 \simeq \psi_0 \times id$ takes place in $\delta^+E_i - (B \times N)$ and can be extended to a *f.p.* ambient isotopy of δ^+E_i. Thus after a fibered handle modification we may assume that $\phi_0 = \psi_0 \times id$.

We now want to find a *f.p. PL* isotopy in $\delta^+E_i - (B \times N)$ from ϕ to an embedding into $[\delta^+E'_i - (B \times N')] \times I$ which is of the form $\psi \times id$, where $\psi\colon B \times \partial B^r \times B^{n-r} \to \delta^+E'_i - (B \times N')$ is a *f.p. PL* embedding extending ψ_0. Choose a fibered regular neighborhood $R \subset \delta^+E'_i - (B \times N')$ of $\psi_0(B \times \partial B^r)$ by taking an ordinary regular neighborhood of any level of $\psi_0(B \times \partial B^r)$ and then using the triviality of the bundle $\delta^+E'_i \to B$ to extend it over all of B. This gives us a compact *PL* manifold M^{n-1} containing ∂B^r in its interior which collapses to ∂B^r and a *f.p. PL* homeomorphism $\theta\colon B \times M \to R$ which agrees with ψ_0 on $B \times \partial B^r$. For any $b \in B$ the levels $(R)_b \times I$ and $\phi(\{b\} \times \partial B^r \times B^{n+1-r})$ are regular neighborhoods of $\phi(\{b\} \times \partial B^r \times I)$ in the relative sense of [6]. Thus there exists a *PL* homeomorphism of $(\delta^+E_i)_b$ onto itself taking $\phi(\{b\} \times \partial B^r \times B^{n+1-r})$ onto $(R)_b \times I$ and fixing $\phi(\{b\} \times \partial B^r \times I)$. Using Hudson's isotopy extension theorem this homeomorphism can be extended to a $f.p.$ PL homeomorphism $g: \delta^+E_i \rightarrow \delta^+E_i$ such that $g\,\phi(B\times\partial B^r\times B^{n+1-r})=R\times I$ and $g|\phi(B\times\partial B^r\times I)=id$. By a repetition of this argument we can adjust g so that it additionally satisfies $g\,\phi(B\times\partial B^r\times B^{n-r}\times\{-1\})=R\times\{-1\}$. Note that by construction g is $f.p.$ PL isotopic to the identity rel $\phi(B\times\partial B^r\times I)$.

Since $g\,\phi(B\times\partial B^r\times B^{n-r}\times\{-1\})=R\times\{-1\}$ there exists a $f.p.$ PL homeomorphism $\psi: B\times\partial B^r\times B^{n-r}\rightarrow R$ such that $g\,\phi=\psi\times id$ on $B\times\partial B^r\times B^{n-r}\times\{-1\}$. Using the well-known fact that concordances on 2-connected manifolds are trivial it follows that for any $b\in B$ the b-levels $(g\,\phi)_b$ and $(\psi\times id)_b$ are PL isotopic rel $\{b\}\times\partial B^r\times I$. The trivial bundle structure allows us to extend this to a $f.p.$ PL isotopy $g\phi\simeq\psi\times id$ rel $B\times\partial B^r\times I$. Finally by piecing together the isotopics $\phi\simeq g\phi\simeq\psi\times id$ we obtain our desired isotopy $\phi\simeq\psi\times id$. Since this isotopy extends to an ambient isotopy we can carry out a fibered handle modification to obtain $\phi=\psi\times id$.

For the last part of the proof we use ψ as an attaching map to attach a fibered r-handle H' to $\delta^+E'_i$ to form $E'_{i+1}\cup S'_{i+1}$, where $S'_{i+1}=H'$. Then $H'\times I$ is a fibered r-handle attached to δ^+E_i with attaching map $\psi\times id=\phi$. Thus there exists a $f.p.$ PL homeomorphism of $H'\times I$ onto H which extends via the identity to a $f.p.$ PL homeomorphism of $E'_{i+1}\times I$ onto E_{i+1}. □

5. Proof Of The Destabilization Theorem

The notation is the same as given in Section 1. The strategy is to apply the Factorization Lemma to the fibered h-cobordism $(B\times I\times N\times I,\ B\times N\times I)\rightarrow B$, and this entails the choice of an appropriate open cover α of $B\times I\times N\times I$. Our basic requirement is that a typical element V of α must satisfy

$$V\cap([0,\frac{1}{3})\times I\times N\times I)=[0,\frac{1}{3})\times I\times U\times I$$

and

$$V\cap((\frac{2}{3},1]\times I\times N\times I)=(\frac{2}{3},1]\times f(I\times U'\times I),$$

where U and U' have small diameters. With this restriction choose α so that there are *f.p.* deformation retractions of $B\times I\times N\times I$ to $B\times N\times I$ which are *f.p.* α-homotopic to their respective identities. It follows from the Factorization Lemma that there exists a fibered h-cobordism $(E', B\times N)\rightarrow B$ and a *f.p.* PL homeomorphism $h: E'\times I\rightarrow B\times I\times N\times I$ which is $St^m(\alpha)$-small and the identity on $B\times N\times I$. This means that the arcs $h_0(\{x\}\times I)$ and $f^{-1}h_1(\{x\}\times I)$ have small diameters when projected to N.

Now consider the composition

$$r_0: (E')_0 \xrightarrow{i} (E')_0\times I \xrightarrow{h_0} I\times N\times I \xrightarrow{p} N,$$

where $i(x)=(x,0)$ and $p=proj$. It is easy to see that r_0 is a retraction and there is a homotopy $r_0\simeq id$ rel N whose composition with r_0 is an ϵ-homotopy in N, where ϵ is small. In the language of [2] this means that r_0 is an ϵ-homotopy equivalence. Thus by the Controlled h-Cobordism Theorem of [2] there exists a PL homeomorphism $g_0: (E')_0\rightarrow I\times N$ such that $g_0|N=id$ and qg_0 is close to r_0, where $q=proj: I\times N\rightarrow N$. This easily implies that ph_o is close to $p(g_0\times id)$, and therefore $h_0(g_0\times id)^{-1}$ is a concordance on $N\times I$ for which $ph_0(g_0\times id)^{-1}$ is close to p. Using the Controlled Concordance Theorem of [3] there is a PL isotopy $h_0(g_0 \times id)^{-1} \simeq id$ rel $N \times I$, thus h_0 is PL isotopic to $g_0 \times id$ rel $N \times I$.

Since $(E', B \times N) \to B$ is a trivial cobordism with fiber $(E')_0 \cong I \times N$, we may write $E' = B \times I \times N$. Then h becomes a $f.p.$ PL homeomorphism of $B \times I \times N \times I$ to itself and g_0 becomes a PL homeomophism of $I \times N$ to itself such that h_0 is PL isotopic to $g_0 \times id$ rel $N \times I$. Next consider the composition

$$r_1 : I \times N \xrightarrow{i} I \times N \times I \xrightarrow{h_1} I \times N \times I \xrightarrow{f^{-1}} I \times N \times I \xrightarrow{p} N.$$

By repeating the above argument there exists a PL homeomorphism $g_1 : I \times N \to I \times N$ such that $g_1 | N = id$ and $f^{-1} h_1$ is PL isotopic to $g_1 \times id$ rel $N \times I$. Stringing together all of the above isotopies we obtain the following PL isotopy of f to a product:

$$f \simeq h_1(g_1 \times id)^{-1} \simeq h_0(g_1 \times id)^{-1} \simeq (g_0 \times id)(g_1 \times id)^{-1} = g_0 g_1^{-1} \times id \qquad \square$$

References

1. D. Burghelea and R. Lashoff, *Stability of Concordances and the Suspension Homomorphism,* Annals of Math. (2) **105** (1977), 449-472.

2. T.A. Chapman, *Controlled Simple Homotopy Theory and Applications,* Lecture Notes in Math. 1009, Springer-Verlag (1983), Berlin

3. ________________, *Controlled Concordances,* Teubner-Text zur Math. **70** (1984), 17-35.

4. A.E. Hatcher, *Higher Simple Homotopy Theory,* Annals of Math (2) **102** (1975), 101-137.

5. J.F.P. Hudson, *Piecewise Linear Topology,* W.A. Benjamin, New York, 1969.

6. J.F.P. Hudson and E.C. Zeeman, *On Regular Neighborhoods,* Proc. London Math. Soc. (3) **14** (1964), 719-745.

7. K. Igusa, *Stability of Pseudoisotopies,* Aarhus University Seminar Notes 1 (1982), 27-37.

8. T.B. Rushing, *Topological Embeddings,* Academic Press, New York, 1973.

9. Frank Quinn, *Ends of Maps, I,* Annals of Math. 110 (1979), 275-331.

10. C.T.C. Wall, *Geometric Connectivity I,* J. London Math. Soc. (2) **3** (1971), 597-604.

Department of Mathematics
University of Kentucky
Lexington, Kentucky 40506

GEOMETRY AND TOPOLOGY (pp. 74–85)
edited by G. M. Rassias & G. M. Stratopoulos

On a Conjecture on CR-submanifolds

Bang-yen Chen
Department of Mathematics
Michigan State University
East Lansing, Michigan 48824, U.S.A.
and
Bao-Qiang Wu
Department of Mathematics
Xuzhou Teacher's College
Jiangsu, People's Republic of China

ABSTRACT. It was conjectured that every mixed foliate CR-submanifold N of a complex hyperbolic space H^m is either a complex submanifold or a totally real submanifold [1,4]. In this article, we prove that the conjecture is true if the holomorphic distribution $\mathcal{D}$ of N is of (complex) odd-dimension.

1. INTRODUCTION.

A submanifold N of a Kaehler manifold $\tilde{M}$ is called a CR-submanifold if (1) the maximal complex subspace $\mathcal{D}_x$ of the tangent space $T_x\tilde{M}$ contained in T_xN, $x \in N$, defines a differentiable distribution $\mathcal{D}$, called the holomorphic distribution, on N and (2) the orthogonal complementary distribution $\mathcal{D}^\perp$ of $\mathcal{D}$ is a totally real distribution, i.e., $J\mathcal{D}_x^\perp \subset T_x^\perp N$, $x \in N$, where J denotes the complex structure of $\tilde{M}$ and $T_x^\perp N$ the normal space of N in $\tilde{M}$ at x.

A CR-submanifold N is called mixed totally geodesic if $\sigma(\mathcal{D},\mathcal{D}^\perp) = \{0\}$, where σ denotes the second fundamental form of N in $\tilde{M}$. It is known that the totally real distribution $\mathcal{D}^\perp$ is always integrable (see, [3,4]). A CR-submanifold N is called *mixed foliate* if it is mixed totally geodesic and if its holomorphic distribution $\mathcal{D}$ is integrable.

It was proved in [2] that a mixed foliate CR-submanifold in $\mathbb{C}P^m$ is either a complex submanifold or a totally real submanifold. In [4], it was shown that mixed foliate CR-submanifolds in $\mathbb{C}^m$ are exactly CR-products (in the sense of [4]). CR-products in $\mathbb{C}^m$ are completely classified in [4].

In [5], the first-named author proved that every mixed foliate CR-submanifold of dimension ≤ 5 in a complex hyperbolic space H^m is either a complex submanifold or a totally real submanifold. In [1,5], we have the following.

<u>Conjecture</u>. Let N be a mixed foliate CR-submanifold of a complex hyperbolic space H^m. Then N is either a complex submanifold or a totally real submanifold.

The main purpose of this article is to prove the following theorem which settles the conjecture when $\dim_{\mathbb{C}} \mathcal{D} = h$ is odd.

Theorem 1. *Let* N *be a mixed foliate CR-submanifold of* H^m. *If* $h = \dim_{\mathbb{C}} \mathcal{D}$ *is odd, then* N *is a complex submanifold.*

2. PRELIMINARIES.

For simplicity, we assume that H^m is the (complex) m-dimensional complex hyperbolic space with constant holomorphic sectional curvature -4. Let N be a mixed foliate CR-submanifold. Then by definition the holomorphic distribution $\mathcal{D}$ of N is integrable and the second fundamental form σ of N in H^m satisfies $\sigma(\mathcal{D},\mathcal{D}^{\perp}) = \{0\}$. We denote by $\langle\ ,\ \rangle$ the metric tensor of H^m as well as the induced metric on N. Let D and A denote the normal connection and the Weingarten map of N in H^m, respectively. Let N^T be a leaf of $\mathcal{D}$. Then N^T is a complex submanifold of H^m. Denote by σ^T, D^T, A^T and ∇^T the second fundamental form, the normal connection, the Weingarten map and the Levi-Civita connection of N^T (in H^m), respectively and by σ', D', A' the corresponding quantities for N^T in N. Then we have

$$\sigma^T(X,Y) = \sigma'(X,Y) + \sigma(X,Y) \tag{2.1}$$

for X, Y tangent to N^T. Since $\sigma(\mathcal{D},\mathcal{D}^\perp) = \{0\}$, we also have

$$A^T_{JZ} = A_{JZ} \tag{2.2}$$

for Z in $\mathcal{D}^\perp$. Since N^T is a complex submanifold of H^m, the complex structure J satisfies

$$\sigma^T(JX,Y) = J\sigma^T(X,Y) = \sigma^T(X,JY) \quad , \tag{2.3}$$

$$A^T_{JZ} = JA^T_Z \; , \quad JA^T_{JZ} = -A^T_{JZ}J \; , \quad JA^T_Z = -A^T_Z J \; , \tag{2.4}$$

for X, Y tangent to N^T and Z normal to N^T.

For any vector X tangent to N we put

$$JX = PX + FX \tag{2.5}$$

where PX and FX are the tangential and the normal components of JX, respectively. For a vector ξ normal to N, we put

$$J\xi = t\xi + f\xi \tag{2.6}$$

where $t\xi$ and $f\xi$ are the tangential and the normal components of $J\xi$, respectively.

Since H^m is of constant holomorphic sectional curvature -4, the curvature tensor $\tilde{R}$ of H^m is given by

$$\tilde{R}(X,Y)Z = \langle X,Z\rangle Y - \langle Y,Z\rangle X + \langle JX,Z\rangle JY \tag{2.7}$$
$$- \langle JY,Z\rangle JX - 2\langle X,JY\rangle JZ$$

for X,Y,Z tangent to H^m.

We need the following results of [5].

Lemma 1. *Let* N *be a mixed foliate CR-submanifold of* H^m. *Then*

(a) $D^T_X JZ = D_X JZ = F\nabla_X Z$,

(b) $D^T_X Z = D'_X Z = -tD_X JZ$,

(c) $\operatorname{Im} \sigma^T = \mathcal{D}^\perp \oplus J\mathcal{D}^\perp$,

(d) A^T_Z , $A^T_{JZ} \in O(2h)$, *and*

(e) $A^T_Z A^T_W + A^T_W A^T_Z = 0$,

for X *tangent to* N^T *and orthonormal vectors* Z, W *in* $\mathcal{D}^\perp$.

Lemma 2. *Let* N *be a mixed foliate proper CR-submanifold of* H^m. *Then*

(a) *each leaf* N^T *of* $\mathcal{D}$ *lies in a complex (h+p)-dimensional totally geodesic complex submanifold* H^{h+p} *of* H^m,

(b) *each leaf* N^T *is an Einstein-Kaehler submanifold of* H^{h+p} *with Ricci tensor given by*

$$S^T(X,Y) = -2(h+p+1)\langle X,Y\rangle \quad , \tag{2.8}$$

(c) $h + 1 \geq p \geq 2$; $h \geq 2$,

where $p = \dim_{\mathbb{R}} \mathcal{D}^\perp$.

3. PROOF OF THEOREM 1.

Let N be a mixed foliate CR-submanifold of H^m. Assume that N is not a complex submanifold of H^m. Then $p = \dim_{\mathbb{R}} \mathcal{D}^\perp$ is positive. From Lemma 2, either N is totally real $(h = 0)$ or $p \geq 2$. If N is totally real, $h = 0$. Thus h is not odd. Thus, we may assume that $p \geq 2$. From Lemma 1, we have

$$A^T_Z A^T_W + A^T_W A^T_Z = 0 \tag{3.1}$$

for orthonormal vectors Z, W in $D^\perp$. Let $Z_1,\dots,Z_p$ be an orthonormal basis of $\mathcal{D}^\perp$. We put

$$A^T_\alpha = A^T_{Z_\alpha} , \quad A^T_{\alpha*} = A^T_{JZ_\alpha} , \quad \alpha = 1,\cdots,p \quad . \tag{3.2}$$

Then, from property (d) of Lemma 1, each $A^T_{\alpha*}$ has eigenvalues 1 and -1 with the same multiplicity h. Let $X_1,\cdots,X_h$ be orthonormal eigenvectors of $A^T_{\alpha*}$ with eigenvalue 1. Then (2.4) implies that $JX_1,\cdots,JX_h$ are eigenvectors of $A^T_{\alpha*}$ with eigenvalue -1. With respect to basis $\{X_1, \cdots, X_h, JX_1, \cdots, JX_h\}$, we have

$$A^T_{\alpha *} = \begin{pmatrix} I_h & 0 \\ 0 & -I_h \end{pmatrix}, \quad J = \begin{pmatrix} 0 & -I_h \\ I_h & 0 \end{pmatrix}, \tag{3.3}$$

where I_h denotes the $h \times h$ identity matrix. Thus, by (2.4) we have

$$A^T_{\alpha} = \begin{pmatrix} 0 & -I_h \\ -I_h & 0 \end{pmatrix}. \tag{3.4}$$

In particular, if we choose $\alpha = 1$, we obtain

$$A^T_1 = \begin{pmatrix} 0 & -I_h \\ -I_h & 0 \end{pmatrix}, \quad A^T_{1*} = \begin{pmatrix} I_h & 0 \\ 0 & -I_h \end{pmatrix}. \tag{3.5}$$

For (2.4) and (3.1) we have

$$A^T_{\alpha} A^T_{\beta *} - A^T_{\beta *} A^T_{\alpha} = 0, \quad \alpha \neq \beta, \quad \alpha, \beta = 1, \cdots, p. \tag{3.6}$$

By using (3.1), (3.5) and (3.6) we find

$$A^T_2 = \begin{pmatrix} A & 0 \\ 0 & -A \end{pmatrix}, \quad A^T_{2*} = \begin{pmatrix} 0 & A \\ A & 0 \end{pmatrix}. \tag{3.7}$$

Moreover, because $A^T_2 \in O(2h)$ (Lemma 1), we also have

$$A \in O(h), \quad {}^tA = A. \tag{3.8}$$

Now, we prove the following.

Lemma 3. *Let* N *be a mixed foliate proper CR-submanifold of* H^m. *Then* $p \geq 3$.

Proof. Under the hypothesis, if $p < 3$, then $p = 2$. From above, we have a basis $X_1, \cdots, X_h$, $JX_1, \cdots, JX_h$, Z_1, Z_2, JZ_1, JZ_2 such that A^T_1, A^T_2, A^T_{1*} and A^T_{2*} take the forms of (3.5), (3.7) and (3.8).

Now, we put

$$V = \mathrm{Span}\{X_1, \cdots, X_h\} \quad . \tag{3.9}$$

Then $TN^T = V \oplus JV$. Since $A \in O(h)$ and ${}^tA = A$, there is an orthonormal basis of V, also denoted by $X_1,\cdots,X_h$ such that, with respect to it, A takes the following form:

$$A = \begin{pmatrix} I_r & 0 \\ 0 & -I_{h-r} \end{pmatrix} . \tag{3.10}$$

<u>Case</u> 1: $r = h$. In this case, we have

$$-A_1^T = A_{2*}^T = \begin{pmatrix} 0 & I_h \\ I_h & 0 \end{pmatrix} , \quad A_{1*}^T = A_2^T = \begin{pmatrix} I_h & 0 \\ 0 & -I_h \end{pmatrix} . \tag{3.11}$$

We put

$$W = \frac{1}{\sqrt{2}} (Z_1 + JZ_2), \quad JW = \frac{1}{\sqrt{2}} (-Z_2 + JZ_1) \quad . \tag{3.12}$$

Then we have

$$A_W^T = A_{JW}^T = 0 \quad . \tag{3.13}$$

Therefore, from Lemma 1, we find

$$\mathrm{Im}\, \sigma^T = \mathrm{Span}\{\bar{W}, J\bar{W}\} \ , \quad \bar{W} = \frac{1}{\sqrt{2}} (Z_1 - JZ_2) \quad . \tag{3.14}$$

From property (a) of Lemma 1, we have

$$D_X^T JZ_1 = \lambda(X) JZ_2 \ , \quad D_X^T JZ_2 = -\lambda(X) JZ_1 \tag{3.15}$$

for some function λ. Thus, by applying property (b) of Lemma 1, we have

$$D_X^T Z_1 = \lambda(X) Z_2 \ , \quad D_X^T Z_2 = -\lambda(X) Z_1 \quad . \tag{3.16}$$

Combining (3.15) and (3.16) we obtain

$$D^T_X W = \frac{1}{\sqrt{2}}(D^T_X Z_1 + D^T_X JZ_2) = -\lambda(X)JW \quad , \tag{3.17}$$

$$D^T_X JW = \frac{1}{\sqrt{2}}(D^T_X JZ_1 - D^T_X Z_2) = \lambda(X)W \quad . \tag{3.18}$$

Consequently, we see that the normal subbundle $\nu = \mathrm{Span}\{W,JW\}$ is a parallel subbundle of the normal bundle of N^T in $H^{h+p} = H^{h+2}$. Moreover, since $\mathrm{Im}\,\sigma^T$ is orthogonal to ν, $\mathrm{Im}\,\sigma^T$ is also a parallel subbundle of the normal bundle of N^T in H^{h+2}. Thus, by applying Theorem 2.3 of [6], we know that N^T is a complex submanifold of a complex totally geodesic submanifold H^{h+1} of H^{h+2}. Since N^T is Einsteinian according to Lemma 2, a result of [7] implies that N^T is totally geodesic. Thus, the Ricci tensor S^T of N^T is given by $S(X,Y) = -2(h + 1)\langle X,Y\rangle$ which contradicts to Lemma 2.

<u>Case</u> 2: $r = 0$. This is also impossible by an argument similar to that of Case 1.

<u>Case</u> 3: $r > 0$ and $h - r > 0$. In this case, we decompose V and JV into orthogonal decompositions:

$$V = V' \oplus V'' \ , \quad JV = JV' \oplus JV'' \quad , \tag{3.19}$$

where V' and V'' are the eigenspaces of

$$A = \begin{pmatrix} I_r & 0 \\ 0 & -I_{h-r} \end{pmatrix}$$

with eigenvalues 1 and -1, respectively. Then according to (3.5), (3.7), (3.10) and Lemma 1 we have

$$\begin{aligned}
\sigma^T(X',T) &= -\langle JX',T\rangle Z_1 + \langle X',T\rangle JZ_1 + \langle X',T\rangle Z_2 + \langle JX',T\rangle JZ_2 \ , \\
\sigma^T(X'',T) &= -\langle JX'',T\rangle Z_1 + \langle X'',T\rangle JZ_1 - \langle X'',T\rangle Z_2 - \langle JX'',T\rangle JZ_2 \ , \\
\sigma^T(Y',T) &= \langle JY',T\rangle Z_1 - \langle Y',T\rangle JZ_1 - \langle Y',T\rangle Z_2 - \langle JY',T\rangle JZ_2 \ , \\
\sigma^T(Y'',T) &= \langle JY'',T\rangle Z_1 - \langle Y'',T\rangle JZ_1 + \langle Y'',T\rangle Z_2 + \langle JY'',T\rangle JZ_2
\end{aligned} \tag{3.20}$$

for T tangent to N^T, $X' \in V'$, $X'' \in V''$, $Y' \in JV'$ and $Y'' \in JV''$. From Lemma 1, we also have

$$D^T_X Z_1 = \lambda(X) Z_2 \ , \quad D^T_X Z_2 = -\lambda(X) Z_1 \ , \quad D^T_X JZ_1 = \lambda(X) JZ_2$$
$$D^T_X JZ_2 = -\lambda(X) JZ_1 \ , \tag{3.21}$$

for X tangent to N^T.

Since N^T is a complex submanifold of H^{h+p}, (2.7) gives

$$(\bar{\nabla}_X \sigma^T)(Y,Z) = (\bar{\nabla}_Y \sigma^T)(X,Z) \quad , \tag{3.22}$$

where $(\bar{\nabla}_X \sigma^T)(Y,Z) = D^T_X \sigma^T(Y,Z) - \sigma^T(\nabla^T_X Y,Z) - \sigma^T(Y,\nabla^T_X Z)$. Let $X \in V'$, $Y \in V''$ and $Z \in JV'$. Then (3.20) and (3.21) imply

$$(\bar{\nabla}_X \sigma^T)(Y,Z) = -\sigma^T(\nabla^T_X Y,Z) - \sigma^T(Y,\nabla^T_X Z) \tag{3.23}$$
$$= \langle J\nabla^T_X Y,Z\rangle Z_1 + \langle \nabla^T_X Y,Z\rangle JZ_1 + \langle \nabla^T_X Y,Z\rangle Z_2 + \langle \nabla^T_X Y,JZ\rangle JZ_2$$
$$+ \langle JY,\nabla^T_X Z\rangle Z_1 - \langle Y,\nabla^T_X Z\rangle JZ_1 + \langle Y,\nabla^T_X Z\rangle Z_2 + \langle JY,\nabla^T_X Z\rangle JZ_2 \ ,$$

$$(\bar{\nabla}_Y \sigma^T)(X,Z) = -Y\langle JX,Z\rangle Z_1 - \lambda(Y)\langle JX,Z\rangle Z_2 \tag{3.24}$$
$$+ Y\langle JX,Z\rangle JZ_2 - \lambda(Y)\langle JX,Z\rangle JZ_1$$
$$- \langle JZ,\nabla^T_Y X\rangle Z_1 + \langle Z,\nabla^T_Y X\rangle JZ_1 + \langle Z,\nabla^T_Y X\rangle Z_2 + \langle JZ,\nabla^T_Y X\rangle JZ_2$$
$$+ \langle JX,\nabla^T_Y Z\rangle Z_1 - \langle X,\nabla^T_Y Z\rangle JZ_1 - \langle X,\nabla^T_Y Z\rangle Z_2 - \langle JX,\nabla^T_Y Z\rangle JZ_2 \ .$$

By comparing the Z_2-components of both sides of (3.22) and using (3.23) and (3.24), we find

$$0 = -\lambda(Y)\langle JX,Z\rangle + \langle Z,\nabla^T_Y X\rangle - \langle X,\nabla^T_Y Z\rangle \ . \tag{3.25}$$

Since $\langle X,Z\rangle = 0$, (3.25) gives

$$2\langle \nabla^T_Y X,Z\rangle - \lambda(Y)\langle JX,Z\rangle = 0 \ . \tag{3.26}$$

Similarly, by comparing the JZ_1-components of both sides of (3.22) we obtain

$$2\langle\nabla^T_Y X, Z\rangle - \lambda(Y)\langle JX, Z\rangle = 2\langle\nabla^T_X Y, Z\rangle \quad . \tag{3.27}$$

From (3.26) and (3.27) we get

$$\langle\nabla^T_X Y, Z\rangle = 0 \quad , \tag{3.28}$$

for any X in V', Y in V'' and Z in JV'. From (3.28) we also have

$$\langle\nabla^T_X Z, Y\rangle = 0 \quad . \tag{3.29}$$

Therefore, we obtain

$$\nabla^T_{V'} V'' \perp JV' \quad \text{and} \quad \nabla^T_{V'} JV' \perp V'' \quad . \tag{3.30}$$

Since $\nabla^T J = 0$, (3.30) gives

$$\nabla^T_{V'} JV'' \perp V' \quad \text{and} \quad \nabla^T_{V'} V' \perp JV'' \quad . \tag{3.31}$$

Similarly, by considering X in V', Y in JV'' and Z in JV', (3.22) gives

$$\nabla^T_{V'} V' \perp V'' \ , \quad \nabla_{V'} JV' \perp JV'' \ , \tag{3.32}$$

Hence we also have

$$\nabla^T_{V'} V'' \perp V' \ , \quad \nabla^T_{V'} JV'' \perp JV' \quad . \tag{3.33}$$

We put

$$U' = V' \oplus JV' \quad \text{and} \quad U'' = V'' \oplus JV'' \quad . \tag{3.34}$$

Then, by (3.30), (3.31), (3.32) and (3.33), we get

$$\nabla^T_{V'} U' \perp U'' \ , \quad \nabla^T_{V'} U'' \perp U' \quad . \tag{3.35}$$

Similarly, we may also prove that

$$\nabla^T_{JV'} U' \perp U'' \ , \quad \nabla^T_{JV} U'' \perp U' \quad . \tag{3.36}$$

Consequently, both U′ and U″ are integrable and parallel complex distributions on N^T. Thus, N^T is locally the Riemannian product of two Kaehler manifolds. On the other hand, it is well-known that H^m admits no Kaehler submanifold which is the Riemannian product of two Kaehler manifolds. Thus we obtain a contradiction. Consequently, Lemma 3 is proved.

Lemma 4. *Let* N *be a mixed foliate proper CR-submanifold of* H^m. *If* $p = \dim_{\mathbb{R}} \mathcal{D}^{\perp} \geq 3$, *then* $h = \dim_{\mathbb{C}} \mathcal{D}$ *is even.*

Proof. Under the hypothesis, if $p \geq 3$, then as before, there exists an orthonormal basis $X_1,\cdots,X_h$, $JX_1,\cdots,JX_h$, $Z_1,\cdots,Z_p$, $JZ_1,\cdots,JZ_p$ such that

$$A_1^T = \begin{pmatrix} 0 & -I_h \\ -I_h & 0 \end{pmatrix}, \quad A_{1*}^T = \begin{pmatrix} I_h & 0 \\ 0 & -I_h \end{pmatrix}, \tag{3.37}$$

$$A_2^T = \begin{pmatrix} A & 0 \\ 0 & -A \end{pmatrix}, \quad A_{2*}^T = \begin{pmatrix} 0 & A \\ A & 0 \end{pmatrix}, \tag{3.38}$$

where

$$A = \begin{pmatrix} I_r & 0 \\ 0 & -I_{h-r} \end{pmatrix}.$$

Since $A_3^T A_1^T + A_1^T A_3^T = 0$, we also have

$$A_3^T = \begin{pmatrix} B & 0 \\ 0 & -B \end{pmatrix}, \quad A_{3*}^T = \begin{pmatrix} 0 & B \\ B & 0 \end{pmatrix}, \tag{3.39}$$

where $B \in O(h)$ and ${}^tB = B$. Since Lemma 1 implies

$$A_2^T A_3^T + A_3^T A_2^T = 0, \quad A_2^T A_{3*}^T - A_{3*}^T A_2^T = 0, \tag{3.40}$$

B takes the following form:

$$B = \begin{pmatrix} 0 & E \\ {}^tE & 0 \end{pmatrix} \tag{3.41}$$

for some $(r\times(h-r))$-matrix E. Since $B \in O(h)$, we obtain

$$E\,{}^tE = I_r \quad \text{and} \quad {}^tEE = I_{h-r} \quad . \tag{3.42}$$

This is impossible unless E is a square matrix. Thus, we have $r = 0$, $h = 2r$ or $h = r$. If $r = 0$, then

$$A_2^T = \begin{pmatrix} -I_h & 0 \\ 0 & I_h \end{pmatrix} . \tag{3.43}$$

Thus, from (3.38) and (3.39) we find $B = 0$. This implies that $A_3^T = 0$ which contradicts to Lemma 1. Similarly, we may prove that $h \neq r$. Therefore, we obtain $h = 2r$. This completes the proof of Lemma 4.

According to Lemma 4, if h is odd, then $p \leq 2$. But this is impossible according to Lemma 3 unless N is either a complex submanifold or a totally real submanifold. The later case cannot occur since h is odd. (Q.E.D.)

4. A FURTHER RESULT.

For mixed foliate CR-submanifolds with $p \neq 3$, we have the following.

Theorem 2. *Let N be a mixed foliate CR-submanifold of H^m. If $p \neq 3$ and $h \not\equiv 0 \pmod 4$, then the conjecture is true.*

Proof. If N is a mixed foliate proper CR-submanifold of H^m, then, by Lemma 3, $p \geq 3$, and by Lemma 4, h is even; say $h = 2r$. Moreover, there is an orthonormal basis $X_1, \cdots, X_h$, $JX_1, \cdots, JX_h$, $Z_1, \cdots, JZ_p$ such that A_1^T, A_{1*}^T, A_2^T and A_{2*}^T are given by (3.36) and (3.37). Let $X_1, \cdots, X_h$ be chosen in such a way that

$$X_{r+1} = A_3^T X_1, \cdots, X_{2r} = A_3^T X_r \quad .$$

Then we also have

$$A_3^T = \begin{pmatrix} B & 0 \\ 0 & -B \end{pmatrix} , \quad A_{3*}^T = \begin{pmatrix} 0 & B \\ B & 0 \end{pmatrix} , \tag{4.1}$$

where

$$B = \begin{pmatrix} 0 & I_r \\ I_r & 0 \end{pmatrix} . \tag{4.2}$$

If $p \neq 3$, then $p \geq 4$. Thus there exist A_4^T and A_{4*}^T such that

$$A_4^T = \begin{pmatrix} F & 0 \\ 0 & -F \end{pmatrix} , \quad A_{4*}^T = \begin{pmatrix} 0 & F \\ F & 0 \end{pmatrix} \tag{4.3}$$

where

$$F = \begin{pmatrix} 0 & G \\ {}^tG & 0 \end{pmatrix} , \quad G \in O(r) . \tag{4.4}$$

Since $A_3^T A_4^T + A_4^T A_3^T = 0$, we obtain

$$G = -{}^tG , \tag{4.5}$$

i.e., G is skew-symmetric. Because $G \in O(r)$, (4.5) implies that r is even. Thus, $h = 0 \pmod 4$. (Q.E.D.)

REFERENCES

[1] Bejancu, A., Geometry of CR-submanifolds, Reidel Publishing Co., Dordrecht, 1986.

[2] Bejancu, A., Kon, M., and Yano, K., CR-submanifolds of a complex space form, J. Differential Geometry 16, 137-145, (1981).

[3] Blair, D.E., and Chen, B.Y., On CR-submanifolds of Hermitian manifolds, Israel J. Math., 34, 353-363, (1979).

[4] Chen, B.Y., CR-submanifolds of a Kaehler manifold, I, J. Differential Geometry, 16, 305-322, (1981).

[5] Chen, B.Y., CR-submanifolds of a Kaehler manifold, II, J. Differential Geometry, 16, 493-509, (1981).

[6] Chen, B.Y., and Ogiue, K., Some extrinsic results for Kaehler submanifolds, Tamkang J. Math., 4, 207-213, (1973).

[7] Chern, S.S., On Einstein hypersurfaces in a Kaehlerian manifold of constant holomorphic sectional curvature, J. Differential Geometry 1, 21-31, (1967).

[8] Ogiue, K., Differential geometry of Kaehler submanifolds, Advances in Math., 13, 73-114, (1974).

GEOMETRY AND TOPOLOGY (pp. 86–99)
edited by G. M. Rassias & G. M. Stratopoulos

CLOSED SUBSPACES OF THE SEQUENCE SPACE

Paul Cherenack

Department of Mathematics, University of Cape Town

Cape Town, South Africa

ABSTRACT:

Closed linear subspaces (those defined by polynomial equations) of the sequence space are provided with an echelon form and are then classified and shown to be closed under sum (the most significant result of the paper). Products of closed linear subspaces of the sequence space are proven to be cofinal in products of subspaces of the sequence space. A directed family of closed linear subspaces of the sequence space having a countable cofinal subfamily is seen to cover a closed linear subspace H only when one of its elements contains H. An echelon form for arbitrary subspaces of the sequence space is introduced.

0. INTRODUCTION

This paper considers some basic results about subspaces of the sequence space

$$k^{\mathbb{N}} = \{(a_i) \mid a_i \in k, i \in \mathbb{N}\}$$

where k is a field. These results derive from the author's investigation into the algebraic geometry of $k^{\mathbb{N}}$. See Cherenack[1),2),3)].

In section one we show that a closed linear subspace H of $k^{\mathbb{N}}$ (one defined by polynomial equations) corresponds under a linear isomorphism α where α and α^{-1} have polynomial coordinates to either a finite dimensional subspace of $k^{\mathbb{N}}$ or to $k^{\mathbb{N}}$ itself. The proof here and later proofs are expedited by introducing an echelon form for such H. The major result of the paper however proven in section two asserts that if H and K are closed linear subspaces of $K^{\mathbb{N}}$, then $H + K$ is also. This fact enables the author to eliminate the assumption[2] that the closed linear subspaces of a vector subspace of $k^{\mathbb{N}}$ are directed by inclusion. Here a directed family of closed linear subspaces of $k^{\mathbb{N}}$ (under inclusion) is a set D of closed linear subspaces of $k^{\mathbb{N}}$ such that $A,B \in D$ implies there is a $C \in D$ with $A \cup B \subset C$. In section three we demonstrate that a closed linear subspace in a countable product $\mathop{\mathsf{X}}_{i\in\mathbb{N}} H_i'$ of subspaces of $k^{\mathbb{N}}$ is a subspace of a countable product $\mathop{\mathsf{X}}_{i\in\mathbb{N}} H_i$ of closed linear subspaces H_i of $k^{\mathbb{N}}$ such that $\mathop{\mathsf{X}}_{i\in\mathbb{N}} H_i \subset \mathop{\mathsf{X}}_{i\in\mathbb{N}} H_i'$. Suppose that D is a directed family of closed linear subspaces of $k^{\mathbb{N}}$ and that D has a countable subset S such that for $H \in D$ there is a $H' \in S$ with $H \subset H'$. Then S is called cofinal in D and in this situation we show for a closed linear subspace H of $k^{\mathbb{N}}$ that when $H = \cup\{H \cap K \mid K \in D\}$ then $H \subset L$ for some $L \in D$. Thus for a directed family of closed linear subspaces of $k^{\mathbb{N}}$ containing a countable cofinal subfamily to cover a closed linear subspace of $k^{\mathbb{N}}$ one of its members must already cover that subspace. Finally in section five we define a notion of echelon form for subspaces of $k^{\mathbb{N}}$ which could prove useful in lifting results from closed linear subspaces of $k^{\mathbb{N}}$ to arbitrary subspaces.

For related treatments of the linear algebra of the sequence space the reader is referred to Jarvinen[4] (Chapter 2).

§1. ECHELON FORM FOR CLOSED LINEAR SPACES

By a linear P–function we mean a degree one linear polynomial in the variables $\{x_i\}_{i\in\mathbb{N}}$ of $k^{\mathbb{N}}$. A closed linear subspace of $k^{\mathbb{N}}$ will be a subset of $k^{\mathbb{N}}$ defined by some collection of linear P–functions. A linear P–isomorphism $f: H \to K$ between two closed linear subsets of $k^{\mathbb{N}}$ is a linear isomorphism such that the coordinates of f and f^{-1} are linear P–functions. Let H denote a closed linear subspace of $k^{\mathbb{N}}$ defined by the complete set L_H of linear P–functions vanishing on H. A set V of variables is linearly independent on H if there is no element $\ell = \alpha_1 x_{p_1} + \alpha_2 x_{p_2} + \dots + \alpha_m x_{pm} \neq 0$ in L_H with $\{x_{p_1}, \dots, x_{pm}\} \subset V$. Let $\{V_i\}_{i\in I}$ be a collection of sets such that

(a) V_i is a set of variables linearly independent on H_i for $i \in I$.

(b) For $i,j \in I$ either $V_i \subset V_j$ or $V_j \subset V_i$.

Then one can readily show :

1.1 Lemma $V = \bigcup_{i\in I} V_i$ is a set of variables linearly independent on H.

By Zorn's lemma there is a maximal set W of variables linearly independent on H. Write the variable set for $k^{\mathbb{N}}$ in the following way :

$$\{x_i\}_{i\in\mathbb{N}} = W \cup W'$$

where

(i) $W \cap W' = \varphi$

(ii) $W = \{x_i\}_{i\in C}$

(iii) $W' = \{x_i\}_{i\in C'}$.

Then one has a unique equation

$$\ell_i = x_i - \sum_{j\in C} \alpha_j^i x_j \quad (i \in C') \quad \text{in} \quad L_H .$$

1.2 Definition The set $E_H = \{\ell_i \mid i \in C'\}$ is called an echelon form for H (or more properly for a set of equations defining H). W will be called the set of free variables and W' the bound variables relative to E_H .

From the echelon form above one sees :

1.3 Lemma The projection of H onto the closed linear subspace of $k^{\mathbb{N}}$ defined by the equations $x_i = 0$ for $i \in C'$ is a linear P–isomorphism which extends to a linear P–isomorphism of $k^{\mathbb{N}}$.

Remark: Let H be the closed linear subset of $k^{\mathbb{N}}$ defined by the linear equations $x_{i+1} = x_i$ for $i \in \mathbb{N}$. These linear equations cannot be reduced to echelon form by standard techniques. An echelon form for H consists of the equations $x_i = x_1$ for $i > 1$.

From Lemma 1.3 one obtains :

1.4 Proposition Every closed linear subspace of $k^{\mathbb{N}}$ is linearly P–isomorphic to k^t where $t = 0$ or $t \in \mathbb{N}$ or $t = \mathbb{N}$.

One also has the following converse :

1.5 Proposition Every finite dimensional vector subspace of $k^{\mathbb{N}}$ is defined by a denumerable number of polynomial linear equations.

Proof: Suppose $\dim_k V = 1$ and V has a basis element $(a_1, a_2, \ldots, a_n, \ldots)$. Then, it is easy to see that

$$a_i x_j - a_j x_i = 0$$

for $i, j = 1, 2, \ldots$ are linear equations for V denumerable in number.

Suppose that $V = W \oplus U$ where $\dim_k W = 1$ and $\dim_k U = n$. We may suppose that U is defined by a denumerable number of linear polynomial equations.

Suppose that W has a basis element $(a_1, a_2, \ldots, a_n, \ldots)$. Since permutation of coordinates is linear P–isomorphism of $k^{\mathbb{N}}$ we may assume that $a_1 \neq 0$. The linear P–isomorphism sending $(c_1, c_2, \ldots, c_n, \ldots)$ to $(c_1, a_1c_2 - a_2c_1, \ldots, a_1c_n - a_nc_1, \ldots)$ sends $(a_1, a_2, \ldots, a_n, \ldots)$ to $(1, 0, \ldots, 0, \ldots)$ and hence we may assume that $(a_1, a_2, \ldots, a_n, \ldots) = (1, 0, \ldots, 0, \ldots)$. But then V is defined by a denumerable number of polynomial equations found by eliminating from the equations defining U those involving x_1. □

One can construct subspaces of $k^{\mathbb{N}}$ without difficulty which are isomorphic to $k^{\mathbb{N}}$ as vector spaces but not closed linear by embedding $k^{\mathbb{N}} \times V$ into $k^{\mathbb{N}}$ where V is a non–closed linear subspaces of $k^{\mathbb{N}}$.

§2. CLOSURE OF CLOSED LINEAR SUBSPACES UNDER SUM

Let H, K be closed linear subspaces of $k^{\mathbb{N}}$. It is apparent that $H \cap K$ is a closed linear subspace of $k^{\mathbb{N}}$. More significantly one has :

2.1 Theorem $H+K$ is a closed linear subspace of $k^{\mathbb{N}}$.

Proof: Let $E_H = \{\ell_i^H\}_{i\in C}$ (resp. $E_K = \{\ell_i^K\}_{i\in D}$) be the echelon form of H (resp. K). With a change of coordinate guaranteed by Lemma 1.3 we write $\ell_i^K = x_i$ $(i \in D')$ amd we write $\ell_i^H = x_i - \sum_{j\in C} \alpha_j^i x_j$ $(i \in C')$. Set $J^1 = E_H \cup E_K$.

Case 1 : If x_i for $i \in C'$ is free for E_K we remove ℓ_i^H from J^1. Similarly if x_i for $i \in D'$ is free for E_H we remove $\ell_i^K = x_i$ from J^1. Call the resulting set J^2.

Case 2 : Let $F^2 = \{x_m \mid m \in C' \cap D'\}$ and $F^1 = \{x_m \in F^2 \mid \alpha_i^j \neq 0 \text{ for some } j\}$. Let $J^3 = J^2 - G$ where $G = F^1 \cup \{\ell_m^H \mid x_m \in F^1\}$. J^3 may contain too few equations and not define $H+K$.

Suppose that F^3 consists of those x_i $(i \in \mathbb{N})$ such that x_i is free for E_H. Set $F = F^3 \cup F^1$. Let for a subset T of $\{x_i\}_{i\in\mathbb{N}}$ the map $P_T : k^{\mathbb{N}} \to k^{\mathbb{N}}$ be the projection onto the closed linear subspace of $k^{\mathbb{N}}$ defined by $x_i = 0$ for $x_i \notin T$. Let L be the closed linear subspace of $k^{\mathbb{N}}$ defined by $x_i = 0$ for $x_i \notin F$.

2.2 Lemma Suppose that $P_F(H+K)$ is a closed linear subspace of L defined by a set of linear P–functions A_1. Then $H+K$ is a closed linear subspace of $k^{\mathbb{N}}$ defined by $A = A_1 \cup J^3$.

Proof: Let $Z(A) = \{(c_i) \in k^{\mathbb{N}} \mid f(c_i) = 0 \text{ for all } f \in A\}$. Note that $J^3 = \{x_m \mid x_m = \ell_m^H = \ell_m^K\}$. Clearly $H + K \subset Z(A)$. Consider $F_1 = \{x_m \mid x_m$ is free for E_K but not for $E_H\}$. Then if $(c_i) \in Z(A)$ one has $(c_i) = P_F(c_i) + P_{F_1}(c_i) + P_{J^3}(c_i)$ since F, F_1 and J^3 are pairwise disjoint with union $\mathbb{N}$. Since $J^3 \subset A$, $P_{J^3}(c_i) = 0$. Since $A_1 \subset A$, $P_F(c_i) = P_F(h_i) + P_F(k_i')$ where $h_i \in H$, $k_i' \in K$. Since K is defined by equations of the form $x_m = 0$ one has $P_F(k_i') \in K$. Thus $(c_i) = P_F(h_i) + k_i + P_{F_1}(c_i)$ with $h_i \in H$, $k_i \in K$. As $h_i = 0$ for $x_i \in J^3$,

$$\begin{aligned}(c_i) &= P_{F\cup J^3}(h_i) + k_i + P_{F_1}(c_i)\\ &= P_{F\cup J^3\cup F_1}(h_i) + k_i + P_{F_1}(c_i) - P_{F_1}(h_i)\\ &= h_i + k_i + P_{F_1}(c_i - h_i)\end{aligned}$$

since $F \cup J^3 \cup F_1 = \mathbb{N}$. Since x_m is free for $x_m \in F_1$ and where bound defined by an equation of the form $x_m = 0$, one has $k_i + P_{F_1}(c_i - h_i) \in K$. □

Thus we need only show that $P_F(H + K)$ is a closed subset of L. Suppose that $F^3 = F^4 \cup F^5$ where $F^4 \cap F^5 = \varphi$ and F^5 consists of those variables x_m such that x_m is free for E_H but not E_K. We relabel the variables and write $F^4 = \{y_i\}_{i \in \varphi \subset \mathbb{N}}$, $F^5 = \{z_i\}_{i \in \mu \subset \mathbb{N}}$ and $F^1 = \{w_i\}_{i \in \epsilon \subset \mathbb{N}}$ where φ, μ and ϵ are either of the form $\{1, \dots, n\}$ for some $n \in \mathbb{N}$ or equal $\mathbb{N}$. L can be identified with $k^\varphi \times k^\mu \times k^\epsilon$ and then $P_F(H + K)$ consists of all $((a_i + c_i)_{i\in\varphi}, (d_i)_{i\in\mu}, (b_i)_{i\in\epsilon})$ such that $b_i = \sum_{j\in\varphi} a_j^i a_j + \sum_{j\in\mu} d_j^i d_j$ with $i \in \epsilon$. Since the c_i are arbitrary it is clear that to complete the proof one must show that the projection of H onto the closed linear subspace $\{0\} \times k^\mu \times k^\epsilon$ of $k^\varphi \times k^\mu \times k^\epsilon$ is a closed linear subspace.

2.3 Lemma $P_{F5 \cup F1}(H)$ is a closed linear subspace of $\{0\} \times k^{\mu} \times k^{\epsilon}$.

Proof: One reduces the infinite matrix $((a_j^i)\ (d_j^i)\ |\ (b_i))$ (using block matrix notation) to the following echelon form by row operations in the usual way

$$((\alpha_m^j)\ (\gamma_m^j)\ |\ G_m(b_i)) \tag{1}$$

where if $N(m) = \min \{j \mid \alpha_m^j \neq 0\}$ then for $m < m'$ one has $N(m) < N(m')$. Let $N^*(m) = \min \{j \mid \gamma_m^j = 0\}$. Note that the row operations above can be made and will be made without making (α_m^j) or (γ_m^j) have an infinite number of non–zero entries for fixed $m \in \mathbb{N}$.

Let $\Gamma_1 = \{(\sum_{j \in \mu} \gamma_m^j z_j) - G_m(w_i) \mid N(m) = \infty \text{ and } N^*(m) \neq \infty\}$ and $\Gamma_2 = \{G_m(w_i) \mid N(m) = \infty \text{ and } N^*(m) = \infty\}$. We claim that $\Gamma = \Gamma_1 \cup \Gamma_2$ is a set of linear P–functions defining $P_{F5 \cup F1}(H)$ as a closed linear subset of $0 \times k^{\mu} \times k^{\epsilon}$. Clearly $Z(\Gamma) \supset P_{F1 \cup F5}(H)$. Let $z = ((d_j)_{j \in \mu}, (b_j)_{j \in \varphi}) \in Z(\Gamma)$. To see that $z \in P_{F1 \cup F5}(H)$, it suffices to show that the linear system of equations corresponding to the rows of the matrix (1) when $N(m) \neq \infty$ has a solution after substituting z. For this purpose by moving terms involving (γ_m^j) to the right hand sides of the equations, one can assume such terms are zero. A "back to front" substitution is now needed. Let $M(m) = \max \{j \mid \alpha_m^j \neq 0\}$. By further row operations recursively defined one can assume also that for all $n < m$ if $M(n) > M(m+1)$ then $\alpha_n^{M(m+1)} = 0$ $(j,m \in \mathbb{N})$. Note that under the row operations used $M(n)$ may decrease but will not increase. Suppose that a solution $(\alpha_j)_{j \in T(m)}$ (where $T(m)$ consists of those $j \in \mathbb{N}$ such that $j \leq \max \{M(p) \mid p \leq m\}$ and $j \neq M(n)$ for $n > m$) to the system of equations (by assumption $M(p) \neq 0$)

$$f_p = \sum_{j \in C} \alpha_p^j \alpha_j = G_p(b_i) \quad p = 1, 2, \ldots, m \tag{2}$$

of equations has been found. Since $x_{M(m+1)}$ is a free variables in f_{m+1} not present in f_p for $p = 1, 2, \ldots, m$ one can extend $(\alpha_j)_{j\in T(m)}$ to a solution $(\alpha_j)_{j\in T(m+1)}$. □

The proof of theorem 2.1 is thus complete. □

If $W \subset k^{\mathbb{N}}$, we let $W^{\perp}$ denote the collection of linear polynomial forms vanishing on W. If A is a collection of linear hyperplanes, we let

$$A^{\perp} = \{P \in k^{\mathbb{N}} \mid f(P) = 0 \text{ for each } f \in A\}.$$

One can then show :

2.4 Corollary If V_1 and V_2 are closed linear subspaces of $k^{\mathbb{N}}$, then $(V_1^{\perp} \cap V_2^{\perp})^{\perp} = V_1 + V_2$.

Proof: Suppose that $V_1 + V_2 = W^{\perp}$. Then $V_1, V_2 \subset W^{\perp}$ and since $\perp$ reverses inequalities one then has

$$W \subset V_1^{\perp} \cap V_2^{\perp}$$

and

$$(V_1^{\perp} \cap V_2^{\perp})^{\perp} \subset W^{\perp}.$$

That

$$V_1 + V_2 \subset (V_1^{\perp} \cap V_2^{\perp})^{\perp}$$

follows in a similar way without assuming that $V_1 + V_2 = W^{\perp}$.

□

§3 CLOSED LINEAR SUBSPACES IN PRODUCTS

Let H_i' be a vector subspace of $k^{\mathbb{N}}$ $(i \in \mathbb{N})$ and H a closed linear subspace of $(k^{\mathbb{N}})^{\mathbb{N}} = \{(\alpha_i^j) \mid i,j \in \mathbb{N}\}$. The coordinates of $(k^{\mathbb{N}})^{\mathbb{N}}$ will be denoted x_i^j $(i,j \in \mathbb{N})$. A closed linear subspace of $(k^{\mathbb{N}})^{\mathbb{N}}$ is thus one defined by linear homogeneous polynomials each containing only a finite number of x_i^j $(i,j \in \mathbb{N})$.

3.1 Proposition If $H \subset \bigtimes_{i \in \mathbb{N}} H_i'$ then there are closed linear subspaces H_i of $k^{\mathbb{N}}$ $(i \in \mathbb{N})$ such that $H \subset \bigtimes_{i \in \mathbb{N}} H_i \subset \bigtimes_{i \in \mathbb{N}} H_i'$.

Proof: For each $i \in \mathbb{N}$ choose a maximal set of linearly independent variables $\{x_i^j\}_{j \in C(i)}$ on H. Again this is possible by Zorn's lemma. Extend $\{x_i^j\}_{j \in C(i)}$ to a maximal set of free variables $W^i = \{x_p^j\}_{j,p \in D(i)}$ of free variables for an echelon form E_H^i of H (Zorn's lemma). Let $E_H^i = \{\ell_{pH}^j \mid x_p^j \notin W^i\}$ and H_i be defined by $A = \{\ell_{iH}^p \mid \ell_{iH}^p \in E_H^i\}$. Note that ℓ_{iH}^p for $\ell_{iH}^p \in E_H^i$ contains only variables of the form x_i^q $(q \in \mathbb{N})$ arising from our choice of $\{x_i^j\}_{j \in C(i)}$. Since the $\ell_{iH}^p \in A$ in part define H one has $H \subset \bigtimes_{i \in \mathbb{N}} H_i$. Suppose that the projection

$$P_i : (k^{\mathbb{N}})^{\mathbb{N}} \to k^{\mathbb{N}}$$

is defined by $P_i(\alpha_m^j) = (\alpha_i^j)$ for fixed $i \in \mathbb{N}$.

Consider $(\alpha_m^j) \in \bigtimes_{m \in \mathbb{N}} H_m$. From our choice of $C(i)$ and E_H^i there is a point (c_m^j) such that $c_m^j = a_m^j$ if $m = i$ and $(c_m^j) \in H$. Hence $P_i(a_m^j) = P_i(c_m^j) \in P_i(H) \subset H_i'$. Thus $(a_m^j) \in \bigtimes_{i \in \mathbb{N}} H_i'$. □

As above letting $\underline{n} = \{1, 2, ..., n\}$ one can show even more simply:

3.2 Proposition If $H \subset \mathop{\times}\limits_{i \in \underline{n}} H_i'$, then there are closed linear subspaces H_i of $k^{\mathbb{N}}$ $(i \in \mathbb{N})$ such that $H \subset \mathop{\times}\limits_{i \in \underline{n}} H_i \subset \mathop{\times}\limits_{i \in \underline{n}} H_i'$.

§4 COVERINGS OF CLOSED LINEAR SUBSPACES

Let D be a directed family of closed linear subspaces of $k^{\mathbb{N}}$.

4.1 Proposition Suppose that D has a countable cofinal subcollection S. Let H be a closed linear subspace of $k^{\mathbb{N}}$ such that $H = \cup \{H \cap K \mid K \in D\}$. Then $H \subset L$ for some $L \in S$.

Proof: By change of coordinate one can assume that $H = k^t$ where $t \in \mathbb{N}$ or $t = \mathbb{N}$. Let $S = \{H_i\}_{i \in \mathbb{N}}$. For $H_i, H_j \in S$ there is a $L \in D$ such that $H_i, H_j \subset L$. Since S is cofinal in D there is a $H_m \in S$ such that $H_m \supset L$ and hence $H_m \supset H_i, H_j$. Thus there is a countable cofinal subset S^* of S and hence D which we write $S^* = \{K_i\}_{i \in \mathbb{N}}$ where $K_i \subset K_{i+1}$ for $i \in \mathbb{N}$. Let $S' = \{K_i \cap H\}_{i \in \mathbb{N}}$ Then since S^* is cofinal in D, one has $H = \cup S'$.

Case 1: $t \in \mathbb{N}$. Since $(K_i \cap H) \subset (K_{i+1} \cap H)$ for $i \in \mathbb{N}$ there is a m for which $K_m \cap H = H$.

Case 2: $t = \mathbb{N}$. Suppose that $K_i^* = K_i \cap H$ is infinite dimensional $(i \in \mathbb{N})$. Choose free variables $W_i = \{x_{ji}\}_{ji \in \mathbb{N}}$ for K_i^* so that $W_i \subset W_{i+1}$ with corresponding echelon forms $E_{K_i \cap H}$ $(i \in \mathbb{N})$. Construct a linear P–isomorphism $\Sigma : k^{\mathbb{N}} \to k^{\mathbb{N}}$ where

(i) $\Sigma(x_i) = (F_j(x_i))$

(ii) For $x_j \in W_{m+1} - W_m$ one has $x_j' = F_j(x_i) = x_j + G_j(x_i)$ in $E_{K_m \cap H}$ with $G_j(x_i)$ in the linear span of W_m.

(iii)For $x_j \notin W_{m+1} - W_m$ for some $m \in \mathbb{N}$ let $x_j' = F_j(x_i) = x_j$.

Σ has an inverse $\Sigma^{-1}(x_j') = (F_j'(x_j'))$ where $F_j'(x_i) = x_j' - G_j(x_i) = x_j' - G_j'(x_i')$ where it is possible to write x_i in terms of x_n' $(n \in \mathbb{N})$ since at level m one must make at most m substitutions. It follows after this linear P–isomorphism that $H = k^{\mathbb{N}} = \cup K_i^*$ where $K_i^* = Z(U_i)$ with $U_i = \{x_{ti}\}_{ti \in \mathbb{N}}$ and $U_{i+1} \subset U_i$ for $i \in \mathbb{N}$. Consider $\underline{\alpha} = (\alpha_i)_{i \in \mathbb{N}}$ where $\alpha_i \neq 0$ for all $i \in \mathbb{N}$. Then $\alpha \notin K_i^*$ for any i unless $H = K_i^*$ for some $i \in \mathbb{N}$. The equation $H = K_i \cap H$ for some $i \in \mathbb{N}$ implies that $H \subset K_i \in S$. The case where K_i^* is not infinite dimensional can be handled by a readily seen modification. □

§5 ECHELON FORM FOR SUBSPACES

Let S be a subspace of $k^{\mathbb{N}}$. S is the directed union of its closed linear subspaces by Proposition 1.5 and Theorem 2.1. Let T denote the collection of closed linear subspaces of S.

5.1 Proposition There is a minimal subset V of $\{x_i\}_{i\in\mathbb{N}}$ (for the inclusion relation) such that

(a) V contains a maximal set of linearly independent variables for each $H \in T$.

(b) V is order minimal in the sense : Suppose that V' is a set of variables such that V' contains a maximal set of linearly independent variables $\{x_{jm}\}_{m\in M}$ for $H \in T$ (where $M = \mathbb{N}$ or $M = \{i \in \mathbb{N} \mid 1 \le i \le n\}$ for some $n \in \mathbb{N}$). Then V contains a maximal set of linearly independent variables $\{x_{jm}\}_{m\in M}$ for H with $im \le jm$ for $m \in M$.

Proof: V is constructed as follows :

Let j be the first integer such that $\{x_j\}$ is linearly independent for $H \neq 0$ $(H \in T)$. Set $(V \phi H)(1) = \{x_j\}$. Suppose that $(V \phi H)(i)$ has been defined for $i \le n$ and $(V \phi H)(i)$ differs from $(V \phi H)(i-1)$ by the addition of one element for $2 \le i \le n$. Set $(V \phi H)(n+1) = (V \phi H)(n) \cup \{x_s\}$ where if $\theta = \{i \mid (V \phi H)(n) \cup \{x_i\}$ is a set of linearly independent variables for $H\}$ then $s = \min \theta$. Note that if s fails to exist one sets $(V \phi H)(n+1) = (V \phi H)(n)$. Next let $(V \phi H) = \cup (V \phi H)(n)$ $(n \in \mathbb{N})$. Finally one sets $V = \cup (V \phi H)$ $(H \in T)$. □

The echelon form for S is then $E_S = \{E_H\}_{H\in T}$ where E_H is the echelon form for H with free variables $V \phi H$.

References

1. Cherenack, P., "A Cartesian closed extension of a category of affine schemes", Cahiers de Top. et Geom. Differentielle XXIII–3. 291–315 (1982).

2. Cherenack, P., "Internal hom–sets in an extensiuon of affine schemes over a field", in Algebraic Geometry, Proc. Summer Meet. Copenhagen (K. Lonsted, Ed.), Lecture Notes in Mathematics, 732. 30–42, Springer–Verlag, 1978.

3. Cherenack, P., "The topology of ind–affine sets", to appear in volume Nonlinear Analysis, World Scientific, Singapore.

4. Jarvinen, R.D., "Finite and Infinite Dimensional Linear Spaces", Marcel Dekker, 1981.

This work was supported by a grant from the S.A. Council for Scientific and Industrial Research to the Topology Research Group of the University of Cape Town.

GEOMETRY AND TOPOLOGY (pp. 100–104)
edited by G. M. Rassias & G. M. Stratopoulos

C^∞-HOMOGENEOUS CURVES ON ORIENTABLE CLOSED SURFACES

DONČO DIMOVSKI
UNIVERSITY OF SKOPJE, SKOPJE, YUGOSLAVIA

DUŠAN REPOVŠ
UNIVERSITY OF LJUBLJANA, LJUBLJANA, YUGOSLAVIA

EVGENIJ V. ŠČEPIN
STEKLOV MATHEMATICAL INSTITUTE, MOSCOW, USSR

Let M^2 be an oriented smooth closed surface. A Jordan curve on M^2 is a continuous map $K:[0,1] \longrightarrow M^2$ (or $K: R \longrightarrow M^2$) such that for each $t \in (0,1)$ (or $t \in R$) there is an $\varepsilon > 0$ such that K is one-to-one on the open ε-neighborhood of t. The orientation on a Jordan curve K on M^2 will always be the induced orientation by the chosen one on the surface M^2.

A Jordan curve K on M^2 is said to be C^∞-homogeneous in M^2 if for every two points $x,y \in K$ there exists a diffeomorphism $h_{x,y}: M^2 \longrightarrow M^2$ such that (i) $h_{x,y}(x)=y$; (ii) $h_{x,y}(K) \subset K$; (iii) $h_{x,y}$ is orientation preserving for all x and y (or it is orientation reversing for all x and y); and (iv) $h_{x,y}$ preserves the orientation of K for all x and y (or it reverses the orientation of K for all x and y).

PROPOSITION 1. Let K be a C^∞-homogeneous Jordan curve in the plane R^2. Then for every point $x \in K$ there exist a closed neighborhood $U \subset R^2$ of x and smooth curves $C_1, C_2 \subset U$ such that (i) $K \cap U$ separates U; (ii) C_1 and C_2 lie on the opposite sides of $K \cap U$; and (iii) $C_1 \cap K = \{x\} = C_2 \cap K$ (i.e. x is wedged between C_1 and C_2).

Proof. Let $x \in K$. Since K is C^∞-homogeneous Jordan curve in R^2 the Jordan Curve Theorem implies that there is a closed neighborhood $U \subset R^2$ of x in R^2 such that $K \cap U$ separates U. Let

$p_i \in U-K$, $i=1,2$, be any two points on the opposite sides of $K \cap U$. Let $p_i^* \in K$ be a point on K nearest to p_i, $i=1,2$ (there can be more than one, in general). Let C_i^* be a circle in R^2 centered at p_i^* with radius $d(p_i, p_i^*)$, $i=1,2$. Modify the circles C_i^* so that eventually $C_i^* \cap K = \{p_i^*\}$, $i=1,2$, while keeping them smooth at all points. Use the C^∞-homogeneity of the curve K to produce two diffeomorphisms $h_i : R^2 \longrightarrow R^2$ such that $h_i(p_i^*)=x$. Define C_i to be the component of $h_i(C_i^*)$ which contains the point x, $i=1,2$. It's is now easy to verify that the curves C_1 and C_2 satisfy the required properties.

A <u>1-parameter group of diffeomorphisms</u> of an oriented closed smooth surface M^2 is a continuous map $G: R \times M^2 \longrightarrow M^2$ such that (i) $G(t,x)=g^t(x)$ where g^t is a diffeomorphism of M^2; (ii) $g^{t+s}=g^t g^s$; and (iii) $g^0 = id_{M^2}$. In other words, a 1-parameter group of diffeomorphisms is a homomorphism $G:(R,+) \longrightarrow (Diff(M^2), o)$ such that $G: R \times M^2 \longrightarrow M^2$ is continuous.

PROPOSITION 2. <u>Let $G: R\times M^2 \longrightarrow M^2$ be a 1-parameter group of diffeomorphisms of M^2 and let $x \in M^2$ be an arbitrary point of M^2. Then the orbit $O_x = \{g^t(x) \mid t \in R\}$ is a C^∞-homogeneous curve in M^2 unless x is a fixed point.</u>

<u>Proof</u>. Any two diffeomorphisms g^t and g^s are homotopic via the map $G : [t,s] \times M^2 \longrightarrow M^2$, $t < s$, which implies that either both of them preserve the orientation of M^2 or they both reverse it. The same holds for the orientation of the orbits. Pick $x \in M^2$ and let $y \in O_x$. Then $H_y = \{t \mid g^t(y)=y\}$ is a subgroup of $(R,+)$ so there are 3 possibilities: H_y is either discrete or dense in R or trivial (i.e. 0). If H_y is dense in R it follows that $H_y = R$ hence $O_x = O_y = \{x\} = \{y\}$. If $O_x \neq \{x\}$ and $H_x = \{0\}$ then O_x is the image of an embedding of R in M^2. If H_x is a discrete group, i.e. $H_x = rZ$ for some $r>0$, then O_x is the image of an embedding of $[0,r)$ in M^2, i.e. of a map $f:[0,r] \longrightarrow R$, $f(0)=f(r)$, and f one-to-one on the interval $[0,r)$.

For every two points $y,z \in O_x$ there exist $t,s \in R$ such that $y=g^t(x)$ and $z=g^s(x)$. Consequently, $g^{s-t}(y)=g^{s-t}(g^t(x))=g^s(x)=z$ hence g^{s-t} is a required diffeomorphism of M^2 since we also have that

$g^{s-t}(0_x)=0_x$. This proves that the orbit 0_x is indeed C^∞-homogeneous in M^2 as asserted.

PROPOSITION 3. Let K be a Jordan curve in the plane R^2 and suppose that at a point $x \in K$, K is wedged between two smooth curves, i.e. that there are two smooth curves $C_1, C_2 \subset R^2$ such that $C_1 \cap K = C_2 \cap K = \{x\}$. Then K has a tangent at x.

Proof. Consider the secants $L_n \subset R^2$ of K based at x. For every $n \in N$, pick $q_n \in L_n \cap K$. We may choose the sequence (q_n) so that it converges to x. There are two possible cases to consider.

Case 1. For all but finitely many n, $q_n \notin C_1 \cup C_2$. Then for some subsequence $(q_{s(n)})$, $L_{s(n)}$ have the same slope as the tangent to C_1 (and hence to C_2) at x and therefore so does the limit.

Case 2. For some subsequence $(q_{s(n)})$, $q_{s(n)} \in C_1$ (resp. C_2). Then $L_{s(n)}$ is also a secant for C_1 (resp. C_2) at x hence the slopes of $L_{s(n)}$ must converge - to the derivative of C_1 (resp. C_2) at x. This implies that K is differentiable at x.

EXAMPLE. The following example shows that in Proposition 3 one cannot, in general, also prove that the curve K has a continuous derivative at x (hence much less that K is smooth at x). Let $P_1 = \{(x,x^2) \mid x \in R\} \subset R^2$ and $P_2 = \{(x,-x^2) \mid x \in R\} \subset R^2$. For every $n \in N$, let $A_n, A_n^* \in P_1$ and $B_n, B_n^* \in P_2$ be given by:

$$A_n = \{((2n)^{-1}, (4n^2)^{-1})\}, \qquad A_n^* = \{(-(2n)^{-1}, (4n^2)^{-1})\},$$

$$B_n = \{((2n-1)^{-1}, -(2n-1)^{-2})\}, \quad B_n^* = \{(-(2n-1)^{-1}, -(2n-1)^{-2})\},$$

and let $K^* = \bigcup_{n \in N} (\overline{A_n B_n} \cup \overline{A_n B_{n+1}} \cup \overline{A_n^* B_n^*} \cup \overline{A_n^* B_{n+1}^*}) \cup \{(0,0)\}$. Let K be the curve in the plane, obtained from K* by smoothing its corners A_n, A_n^*, B_n, B_n^* (without changing K* near the x-axis). Then the point $T=(0,0) \in K$ is wedged between the smooth curves P_1 and P_2 but the first derivative of K at T isn't continuous (hence, in particular, K isn't smooth at T). To see this, let $f: [-1,1] \longrightarrow R$ be the map whose graph is K and define the map $F=(f_1,f_2): [-1,1] \longrightarrow R^2$ by $f_1(t)=t$ and $f_2(t)=f(t)$. Then $df_1/dt\,|(0)=1$ and $df_2/dt\,|(0)=0$. Let $(t_n) \subset R$ be the sequence of points on the x-axis, defined by $F(t_n) = \overline{A_n B_n} \cap$ (x-axis). Then $\lim t_n = 0$ whereas $\lim df_2/dt\,|(t_n) = -2 \neq 0 = df_2/dt\,|(0)$.

REMARK. The curve K constructed above isn't C^∞-homogeneous in R^2. For suppose this were the case and pick any point $T^* \neq T$ on K. We would then have a diffeomorphism $h: R^2 \longrightarrow R^2$ such that $h(T)=T^*$. Now, K is clearly smooth at T^* hence it should also be smooth at the image of T^*, $h^{-1}(T^*)=T$. Contradiction.

QUESTION. Under the additional assumption in Proposition 3, that the curve K is C^∞-homogeneous in R^2, can one prove that K is then necessarily smooth at x, i.e. does "C^∞-homogeneous" imply "smooth"? (Note that the converse is true, i.e. every smooth curve in R^2 is locally flat at every point hence one can build diffeomorphisms of R^2 which interchange arbitrary pairs of points on K.)

After this paper was written, W.J.R.Mitchell brought to our attention the work of L.D.Loveland[2)] where he used a similar idea of wedging the curve or a sphere between balls: using entirely different methods from ours he proved e.g. that a curve $K \subset S^3$ is tame in S^3 (i.e. there is an ambient homeomorphism of S^3 which takes K onto a polygonal arc) if for some $t > 0$ at each $x \in K$ there are 3-dimensional balls $B_1, B_2 \subset S^3$, each with radius t, such that $B_1 \cap B_2 = (B_1 \cup B_2) \cap K = \{x\}$.

Our work, on the other hand, was inspired by a remark in V.I. Arnol'd's textbook[1)] (cf.Problem 1 on p.24). Note that our argument yields a very simple geometric proof of a special case of Theorem (5.2.3) in D.Montgomery-L.Zippin's monograph[3)].

ACKNOWLEDGEMENTS. The authors wish to acknowledge comments from A.Gray. The second author wants to thank the Soviet Academy of Sciences for its support during his visit at the Steklov Mathematical Institute in Spring 1988 when this research was started. This project is supported in part by a grant from the Research Council of Slovenia.

REFERENCES

(1) Arnol'd, V.I., "Ordinary Differential Equations"(Russian), Nauka, Moscow 1971.

(2) Loveland, L.D.,"Double tangent ball embeddings of curves in E^3", Pacif. J. Math. 104 (1983), 391-399.

(3) Montgomery, D. and Zippin, L., "Topological Transformation Groups", Interscience Publ. Inc., New York 1955, Interscience Tracts in Pure and Appl. Math. No.1.

GEOMETRY AND TOPOLOGY (pp. 105–133)
edited by G. M. Rassias & G. M. Stratopoulos

A MORSE INDEX THEOREM FOR NULL GEODESICS WITH SPACELIKE ENDMANIFOLDS

Paul E. Ehrlich
Department of Mathematics
University of Florida
Gainesville, Florida 32611
U.S.A.

Seon - Bu Kim
Department of Mathematics
Chonnam National University
Kwangju City
500 Korea
South Korea

Abstract: We study the Morse theory of a null geodesic segment with spacelike endmanifolds in an arbitrary Lorentzian manifold.

SECTION 1: INTRODUCTION.

Let β: $[0,1] \to (M,g)$ be a null geodesic segment in an arbitrary space-time of dimension $n \geqslant 3$ and let K_1,K_2 be spacelike submanifolds perpendicular to β at $\beta(0)$ and $\beta(1)$ respectively. The primary purpose of this paper is to extend the conjugate point Morse index theorem for a null geodesic segment β: $[0,1] \to (M,g)$ given in Beem and Ehrlich [3], cf. [4, section 9.3] to the focal case of variations perpendicular to K_1 and K_2, assuming that $t = 1$ is not a K_1-focal point to $t = 0$ along β.

To gain a perspective on the focal index theory, we review certain aspects of the conjugate point Morse theory for a unit speed geodesic c: $[a,b] \to (N,g_0)$ in a Riemannian manifold (N,g_0). In the conjugate point case, one considers the vector space $V_0^{\perp}(c)$ of continuous, piecewise smooth vector fields Y along c perpendicular to c with boundary conditions $Y(a) = Y(b) = 0$ and the associated index form I: $V_0^{\perp}(c) \times V_0^{\perp}(c) \to \mathbf{R}$ arising from the second variation of arc length and given by

$$I(X,Y) = \int_{t=a}^{b} (g_0(X',Y') - g_o(R(X,c')c',Y))dt. \qquad (1.1)$$

The index form (1.1) is negative semi-definite up to and including the first conjugate point to $t = a$ along c, so that it is sensible to define the index Ind(c) of c by

$$\text{Ind}(c) = \sup\{\dim A;\ A \text{ is an } \mathbf{R} \text{ vector subspace of } V_0^{\perp}(c) \text{ on which } I|_{A\times A} \text{ is positive definite}\}.$$

M. Morse obtained the important result that Ind(c) is finite and is in fact given by summing the multiplicites of conjugate points to $t = a$ along c in (a,b), i.e.,

$$\text{Ind}(c) = \sum_{t\in(a,b)} \dim J_t(c)$$

where $J_t(c)$ denotes the vector space of smooth Jacobi fields J along c with $J(a) = J(t) = 0$. Important in Morse's proof of the index theorem is the result that if c is restricted to [a,t] for any $t \in [a,b]$, then the restricted index form

$$f(t) := \text{Ind}(c|_{[a,t]}) \qquad (1.2)$$

is nondecreasing as a function of t and is left continuous. (Also used is the fact that the augmented or extended index $f_0(t) = \text{Ind}_0(c|_{[a,t]})$ is nondecreasing and right continuous, and finally, $f_0(t) - f(t) = \dim J_t(c)$.) Geometrically, if a family $Y^{(t)}$ of piecewise smooth Jacobi fields along $c|_{[a,t]}$ with $Y^{(t)}(t) = 0$ is considered, then as $t \to t_0$, one has $I(Y^{(t)},Y^{(t)}) \to I(Y^{(t_0)},Y^{(t_0)})$.

In Beem and Ehrlich [4, chapter 9], it was shown that the proof method of Morse for the conjugate point Riemannian index theory may be extended to a timelike geodesic segment or to a null geodesic segment in an arbitrary space-time. The key observation here was that the appropriate analogue of (1.2) defines a nondecreasing function with the same continuity properties as in the Riemannian case, provided that a quotient bundle construction suggested by General Relativity theory was employed in the null case.

The index theory for the variational problem with geodesic $c: [a,b] \to (N,g_0)$ perpendicular at $p = c(a)$ to a submanifold K_1 and at $q = c(b)$ to a submanifold K_2 of the Riemannian manifold (N,g_0) has been found to be more complicated using Morse's approach for two reasons. First, to generalize Morse's conjugate point approach and study the index of $c|_{[a,t]}$, a "geometrically meaningful" submanifold $K(t)$ perpendicular at t to c with $K(t) \to K_2$ as $t \to b$ needs to be constructed. Then K_1, $K(t)$-Jacobi fields need to be studied, rather than the simpler conjugate point condition $Y^{(t)}(t) = 0$ as $t \to t_0$ for a fixed $t_0 \in (0,1]$. This was done by Ambrose [2] using the geodesic flow of the normal bundle to K_1. This construction leads, however, to technical difficulties at the K_1-focal points at which the submanifolds $K(t)$ may lose dimension and differentiability. For this purpose, Ambrose [2] introduced a boundary condition as a pair of operators $S_1, S_2: (c'(t))^{\perp} \to (c'(t))^{\perp}$ satisfying certain properties valid even if t is a K_1-focal point. In the case that t is not a K_1-focal point, S_1 corresponds to the projection map of $(c'(t))^{\perp}$ onto $T_{c(t)}(K(t))$ and S_2 to the second fundamental form of $K(t)$ as a submanifold of (N,g_0).

A second complication in the focal point index theory is that the index function $f(t)$ defined by analogy to (1.2) is no longer so nicely behaved. In 1961, Ambrose [2] studied this problem by breaking up the index function as a sum of the orders of what Ambrose calls the K_1,K_2-conjugate points and a convexity term arising from the difference in second fundamental forms of K_1 and the $K(t)$'s. Using this approach, Ambrose showed that the index function was nondecreasing, but failed to have the left and right continuity properties of the index and augmented index in the simpler conjugate point problem. In 1977, Bolton [6] obtained an index theorem using subdivision arguments like Morse and Ambrose, but working with an index function which fails to be nondecreasing. In 1984, Kalish [9] gave a new proof of the Morse index theorem for a geodesic $c: [a,b] \to (N,g_0)$ in a Riemannian manifold with perpendicular endmanifolds K_1,K_2 provided that $t = b$ is not a K_1-focal point to $t = a$ along c. Kalish's proof avoids the use of subdivision arguments and hence avoids the need to define a family $K(t)$ of submanifolds perpendicular to $c(t)$ as in [2], [6].

It is this third approach of Kalish [9] which we show in the present paper may be adapted to the case of a null geodesic segment $\beta: [0,1] \to (M,g)$ in an arbitrary Lorentzian manifold (M,g) with spacelike endmanifolds K_1,K_2 orthogonal at $\beta(0),\beta(1)$ respectively to β, again supposing that $t = 1$ is not a K_1-focal point to $t = 0$ along β. The proof entails in Section 4 adopting Kalish's arguments to the null quotient bundle setting and to K_1-vector classes, (cf. [10], [4, section 9.3]). Also we establish in Section 3 the finiteness of the quotient bundle index, in general, using properties of K-vector classes in the quotient bundle as developed in [10]. Finally, we note that similar results hold for arbitrary timelike geodesic segments with spacelike endmanifolds in general space-times.

SECTION 2: PRELIMINARIES

Let (M,g) be a connected Lorentzian manifold of dimension $n \geq 3$. Thus M is a smooth manifold with a countable basis and a smooth Lorentzian metric g of signature $(-, +, \ldots, +)$. Let D denote the Levi-Civita connection determined by g. A nonzero tangent vector $v \in TM$ is said to be timelike (resp. nonspacelike, null, spacelike) if $g(v,v) < 0$ (resp. ≤ 0, $= 0$, > 0). An immersed smooth submanifold

$f: K \to (M,g)$ is said to be spacelike if the pull back metric f^*g for K is positive definite. As usual, we will identify K and f(K) and thus assume K is contained in M.

Let $\beta: [0,b] \to (M,g)$ be a fixed null geodesic segment and let K be a spacelike submanifold of dimension k of (M,g) with $\beta'(0)$ perpendicular to K at $p: = \beta(0)$. From this perpendicularity, $k \leq n - 2$. We denote by $V^\perp(\beta)$ the **R**-vector space of continuous, piecewise smooth vector fields Y along β with $g(Y,\beta') = 0$ and $V^\perp(\beta,K): = \{Y \in V^\perp(\beta);\ Y(0) \in T_pK\}$, $V_0^\perp(\beta,K) = \{Y \in V^\perp(\beta,K);\ Y(b) = 0\}$. Recall that a Jacobi field J along β is a smooth vector field satisfying the differential equation $J'' + R(J,\beta')\beta' = 0$. We will adopt the sign convention on the second fundamental tensor $S_{\beta'(0)}: T_pK \to T_pK$ that $S_{\beta'(0)}x: = -(D_xN)^T$ where $T: T_pM \to T_pK$ denotes the projection map and N is any local vector field normal to K with $N(p) = \beta'(0)$. With this sign convention, we make the usual

Definition 2.1: A (smooth) Jacobi field $J \in V^\perp(\beta,K)$ is a K-Jacobi field if $J(0) \in T_pK$ and $J'(0) + S_{\beta'(0)}J(0) \in (T_pK)^\perp$. Also $t_0 \in (0,b]$ is said to be a K-focal point along β if there exists a nontrivial smooth K-Jacobi field $J \in V^\perp(\beta,K)$ with $J(t_0) = 0$.

Requiring $J \in V^\perp(\beta,K)$ in Definition 2.1 makes sense because it is easily checked that if J is a smooth Jacobi field along β satisfying $J(0) \in T_pK$ and $J(t_0) = 0$ for some $t_0 > 0$, then J is perpendicular to β'. Secondly, it may be checked that if dim M = 2 and J is a smooth Jacobi field along β with $J(0) \in T_pK$ and $J(t_0) = 0$ for some $t_0 > 0$, then $J = 0$. Thus we have assumed $n \geq 3$ above. In some contexts (cf. [4], [7], [8]), $t_0 \in (0,b]$ is defined to be a focal point if there exists a smooth K-Jacobi field $J \in V^\perp(\beta)$ with $J(0) \in T_pK$, $J'(0) + S_{\beta'(0)}J(0) = 0$ and $J(t_0) = 0$. (In the terminology of Warner [11], such a K-Jacobi field is called a strong-K-Jacobi field). However, with the stronger initial condition

$J'(0) + S_{\beta'(0)}J(0) = 0$, one cannot obtain a "basis" of n - 1 Jacobi fields for the space $V^{\perp}(\beta,K)$ in either the Riemannian, timelike or null case when K has codimension greater than 1. Thus we adopt the weaker requirement that $J'(0) + S_{\beta'(0)}J(0) \in (T_pK)^{\perp}$ in formulating the definition of K-focal point, cf. Ambrose [1], [2].

The usual submanifold index form $I_{(b,K)}$: $V^{\perp}(\beta,K) \times V^{\perp}(\beta,K) \to \mathbf{R}$ may be defined for $X,Y \in V^{\perp}(\beta,K)$ by

$$I_{(b,K)}(X,Y) := g(S_{\beta'(0)}X(0),Y(0)) \tag{2.1}$$

$$- \int_{t=0}^{b} (g(X',Y') - g(R(X,\beta')\beta',Y))\, dt.$$

As in the conjugate point case (cf. [4, section 9.3]), this particular index form (2.1) is not immediately useful in characterizing K-focal points. For instance, if f: $[0,b] \to \mathbf{R}$ is any smooth function with $f(0) = 0$, then $Y = f\beta' \in V^{\perp}(\beta,K)$ and $I_{(b,K)}(X,Y) = 0$ for all $X \in V^{\perp}(\beta,K)$. But such a vector field $Y = f\beta'$ is a smooth K-Jacobi field in $V_0^{\perp}(\beta,K)$ iff $f = 0$. For this reason as well as to prove the index theorem, we will work with the quotient index form $\bar{I}_{(b,K)}$ as in [7], [3], [4, chapters 9,11]. We then need to briefly review this construction, as detailed for the conjugate point case in [4, Section 9.3].

Let

$$[\beta'(t)] = \{\lambda\beta'(t);\ \lambda \in \mathbf{R}\} \text{ for any } t \in [0,b],$$

$$[\beta'] = \bigcup\{[\beta'(t)];\ 0 \leq t \leq b\},$$

$$(\beta'(t))^{\perp} = \{v \in T_{\beta(t)}M;\ g(v,\beta'(t)) = 0\},$$

$$N(\beta) = \bigcup\{(\beta'(t))^{\perp};\ 0 \leq t \leq b\},$$

$$G(\beta(t)) = \frac{N(\beta(t))}{[\beta'(t)]},$$

and finally the quotient bundle

$$G(\beta) = \bigcup \{G(\beta(t));\ 0 \leq t \leq b\}.$$

We have projection maps $\pi\colon N(\beta(t)) \to G(\beta(t))$ given by $\pi(v) = v + [\beta'(t)]$ and $\pi\colon\ N(\beta) \to G(\beta)$ given by $\pi(X)(t) = X(t) + [\beta'(t)]$ for all $t \in [0,b]$. As in [4, section 9.3], let $\chi(\beta)$ denote the piecewise smooth sections of $G(\beta)$ and $\chi_0(\beta) = \{V \in \chi(\beta);\ V(b) = [\beta'(b)]\}$. In our context, let $\chi(\beta,K)$ denote the piecewise smooth sections $V\colon [0,b] \to G(\beta)$ such that $V = \pi(X)$ for some $X \in V^{\perp}(\beta,K)$ and $\chi_0(\beta,K) = \{V \in \chi(\beta,K);\ V(b) = [\beta'(b)]\}$. We then have the restriction of the projection map $\pi\colon V^{\perp}(\beta,K) \to \chi(\beta,K)$ still given by $\pi(X) = X + [\beta']$. Note also that for $X,Y \in V^{\perp}(\beta,K)$, we have $\pi(X) = \pi(Y)$ iff $X = Y + f\beta'$ for a piecewise smooth function $f\colon [0,b] \to \mathbf{R}$ with $f(0) = 0$. Thus $\pi(X) = \pi(Y)$ for $X,Y \in V^{\perp}(\beta,K)$ implies $X(0) = Y(0)$.

One then obtains a positive definite quotient metric $\bar{g}$, covariant derivative operator $\bar{D}$ and curvature tensor $\bar{R}$ for $\chi(\beta,K)$ as in [7], [4, section 9.3]. Given $V,W \in \chi(\beta,K)$, set $\bar{g}(V,W) = \bar{g}(X,Y)$, $V'(t) = \bar{D}_{\beta'}V(t) = \pi(D_{\beta'}X(t))$ and $\bar{R}(V,\beta')\beta' = \pi(R(X,\beta')\beta')$ where $V = \pi(X)$ and $W = \pi(Y)$. Moreover, we extend the second fundamental tensor $S_{\beta'(0)}$ to the quotient bundle construction as

Definition 2.2: Given $V \in \chi(\beta,K)$, define $\bar{S}_{\beta'(0)}V(0) \in \pi(T_pK)$ by $\bar{S}_{\beta'(0)}V(0) := \pi(S_{\beta'(0)}X(0)) = S_{\beta'(0)}X(0) + [\beta'(0)]$ where $V = \pi(X)$, $X \in V^{\perp}(\beta,K)$. (Our definition of $\bar{S}_{\beta'(0)}$ is well defined since if $Y \in V^{\perp}(\beta,K)$ with $\pi(X) = \pi(Y) = V$, then $X(0) = Y(0)$ as noted above).

Recall that a Jacobi class V in $G(\beta)$ is a smooth vector class satisfying the differential equation $V'' + \bar{R}(V,\beta')\beta' = [\beta']$. With the above machinery in hand, we may now make the following

Definition 2.3: A (smooth) Jacobi class $V \in \chi(\beta,K)$ is a K-Jacobi class if V satisfies the boundary conditions

(i) $V(0) \in \pi(T_pK)$,
(ii) $\bar{S}_{\beta'(0)}V(0) + V'(0) \in \pi((T_pK)^{\perp})$.

A K-Jacobi class is said to be a strong K-Jacobi class if $V(0) \in \pi(T_pK)$ and $\bar{S}_{\beta'(0)}V(0) + V'(0) = [\beta'(0)]$.

It is then easy to check that V is a K-Jacobi class along β iff there exists a K-Jacobi field $J \in V^{\perp}(\beta,K)$ (as defined in Defn. 2.1) with $V = \pi(J)$. Thus these two definitions are consistent. Second, it may be shown that if $k = \dim K = n - 2$, then every K-Jacobi class along β is a strong K-Jacobi class (cf. [10, Proposition 2.12]). Finally, it is also consistent with Definition 2.1 to adopt the following

Definition 2.4. $t_o \in (0,b]$ is a K-focal point along β if there exists a nontrivial K-Jacobi class $V \in \chi(\beta,K)$ with $V(t_0) = [\beta'(t_o)] \in G(\beta(t_o))$.

The quotient index form $\bar{I}_{(b,K)}: \chi(\beta,K) \times \chi(\beta,K) \to \mathbf{R}$ may now be defined for $V,W \in \chi(\beta,K)$ by

$$\bar{I}_{(b,K)}(V,W) := \bar{g}(\bar{S}_{\beta'(0)}V(0),W(0)) - \int_{t=0}^{b} (\bar{g}(V',W') - \bar{g}(\bar{R}(V,\beta')\beta',W))\,dt \tag{2.2}$$

and it follows that if $V = \pi(X), W = \pi(Y)$ with $X,Y \in V^{\perp}(\beta,K)$, then $\bar{I}_{(b,K)}(V,W) = I_{(b,K)}(X,Y)$.

Working along the lines of [7], [4, p. 315], Kim [10, Theorem 3.3] has established the following basic result that the quotient bundle index form characterizes K-focal points.

Theorem 2.5: For $W \in \chi_0(\beta,K)$ the following are equivalent:

(a) W is a K-Jacobi class in $\chi_0(\beta,K)$,

(b) $\bar{I}_{(b,K)}(W,Z) = 0$ for any $Z \in \chi_0(\beta,K)$.

As in the conjugate point Morse theorem for null geodesic segments in [3], we want ultimately to relate the index of $\bar{I}_{(b,K)}$ to K-Jacobi fields and K-focal points along β even though the index theorem will be proven using K-Jacobi classes. Thus we need to establish the following analogue of Lemma 9.75 in [4, p. 323].

Lemma 2.6: Let $W \in \chi(\beta,K)$ be a K-Jacobi class with $W(t_0) = [\beta'(t_0)]$ for some $t_0 > 0$. Then there exists a unique K-Jacobi field $J \in V^{\perp}(\beta,K)$ with $W = \pi(J)$ and $J(t_0) = 0$.

Proof. First there exists a smooth K-Jacobi field $Z \in V^{\perp}(\beta,K)$ with $W = \pi(Z)$. Thus $Z(t_0) = \lambda\beta'(t_0)$ for some $\lambda \in \mathbf{R}$. Then $J(t) = Z(t) - \frac{\lambda}{t_0} t\beta'(t)$ is a Jacobi field with $J(t_0) = 0$ which is checked also to be a K-Jacobi field.

Now suppose $J_1, J_2 \in V^{\perp}(\beta,K)$ are two smooth K-Jacobi fields with $V = \pi(J_1) = \pi(J_2)$ and $J_1(t_0) = J_2(t_0) = 0$. Then $J_1 = J_2 + f\beta'$ where f is a smooth function with $f'' = 0$ and $f(0) = f(t_0) = 0$. Thus $f = 0$ and $J_1 = J_2$. □

Now let

$$J_{t_0}(\beta) = \{K - \text{Jacobi fields } J \in V^{\perp}(\beta,K);\ J(t_0) = 0\}$$

and

$$\bar{J}_{t_0}(\beta) = \{K - \text{Jacobi classes } V \in \chi(\beta,K);\ V(t_0) = [\beta'(t_0)]\}.$$

As a consequence of Lemma 2.6, we obtain

Corollary 2.7: The projection map $\pi: J_{t_0}(\beta) \to \bar{J}_{t_0}(\beta)$ is an isomorphism.

Finally the following generalization of a well known conjugate point result (cf. Warner [12, p. 603] is needed for the proof of the Morse Index Theorem in Section 4. Let

$$A_{\beta(t)} = \{V(t);\ V \in \chi(\beta,K) \text{ is a K-Jacobi class}\} \subset G(\beta(t))$$

and

$$B_{\beta(t)} = \{V'(t);\ V \in \chi(\beta,K) \text{ is a K-Jacobi class with } V(t) = [\beta'(t)]\} \subset G(\beta(t)).$$

Then for any $t \in [0,b]$,

$$A_{\beta(t)} \oplus B_{\beta(t)} = G(\beta(t)). \tag{2.3}$$

Especially for $t = 0$, this equation reduces to
$A_{\beta(t)} = \pi(T_pK)$, $B_{\beta(t)} = \pi((T_pK)^{\perp})$.

SECTION 3: MAXIMALITY OF JACOBI CLASSES AND THE FINITENESS OF THE QUOTIENT BUNDLE INDEX FORM.

We wish to obtain a Morse index theorem in the case of an abitrary null geodesic segment $\beta: [0,b] \to (M,g)$ which is perpendicular at $p = \beta(0)$ to a spacelike submanifold K_1 and perpendicular at $q = \beta(b)$ to a second spacelike submanifold K_2. In this setting, the relevant space of vector fields to consider will be denoted by

$$V^{\perp}(\beta,K_1,K_2) = \{Y \in V^{\perp}(\beta);\ Y(0) \in T_pK_1 \text{ and } Y(b) \in T_qK_2\}$$

and the index form $I_{(b,K_1,K_2)}: V^{\perp}(\beta,K_1,K_2) \times V^{\perp}(\beta,K_1,K_2) \to \mathbf{R}$ is given for $X,Y \in V^{\perp}(\beta,K_1,K_2)$ by

$$I_{(b,K_1,K_2)}(X,Y) := g(S^1(X(0)),Y(0)) - g(S^2(X(b)),Y(b)) \tag{3.1}$$

$$- \int_{t=0}^{b} (g(X',Y') - g(R(X,\beta')\beta',Y))\, dt$$

where $S^1: T_pK_1 \to T_pK_1$ and $S^2: T_qK_2 \to T_qK_2$ denote the second fundamental forms

$$S^1(v) = S_{\beta'(0)}v \text{ and } S^2(w) = S_{\beta'(b)}w$$

for $v \in T_pK_1$, $w \in T_qK_2$ associated to K_1,K_2 in the directions $\beta'(0),\beta'(b)$ respectively. In the special case that $K_2 = \{q\}$, one has $I_{(b,K_1,K_2)} = I_{(b,K_1)}$. We may further restrict the projection map $\pi: V^{\perp}(\beta,K) \to \chi(\beta,K_1)$ of the quotient bundle $G(\beta)$ defined in Section 2 and set

$$\chi(\beta,K_1,K_2) = \pi(V^{\perp}(\beta,K_1,K_2))$$

$$= \{V \in \chi(\beta,K_1);\ V(b) \in \pi(T_qK_2)\}$$

As discussed in Section 2, the index form $I_{(b,K_1,K_2)}$ will not characterize Jacobi fields J which are K_1-Jacobi fields at $\beta(0)$ and K_2-Jacobi fields at $\beta(b)$. Thus as in the conjugate point null index theorem ([3]), we work with a quotient index form

$$\overline{I}_{(b,K_1,K_2)}: \chi(\beta,K_1,K_2) \times \chi(\beta,K_1,K_2) \to \mathbb{R}$$

given for $V,W \in \chi(\beta,K_1,K_2)$ by

$$\overline{I}_{(b,K_1,K_2)}(V,W) = \overline{g}(\overline{S}^1(V(0)),W(0)) - \overline{g}(\overline{S}^2(V(b)),W(b)) \tag{3.2}$$

$$- \int_{t=0}^{b} (\overline{g}(V',W') - \overline{g}(\overline{R}(V,\beta')\beta',W))\, dt.$$

Here we use the quotient metric $\overline{g}$, covariant derivative $\overline{D}$ and curvature tensor $\overline{R}$ as defined in Section 2. Also if $X \in V^{\perp}(\beta,K_1,K_2)$ with $V = \pi(X)$, then

$$\overline{S}^1(V(0)) := S_{\beta'(0)}X(0) + [\beta'(0)] = \pi(S_{\beta'(0)}X(0))$$

and

$$\overline{S}^2(V(b)) := S_{\beta'(b)}X(b) + [\beta'(b)] = \pi(S_{\beta'(b)}X(b)).$$

Notice that the nullity of the index form (3.2) is exactly the space of K_1-Jacobi classes that are simultaneously K_2-Jacobi classes along the geodesic $\tilde{\beta}(t) = \beta(b-t)$. Also, with our choice of sign convention on the index form $\bar{I}_{(b,K_1)}$ is negative definite on $\chi_0(\beta,K_1)$ up to the first K_1-focal point. Thus it makes sense to define

Definition 3.1: The index $\bar{I}(\beta,K_1,K_2)$ of the quotient index form $\bar{I}_{(b,K_1,K_2)}$ on $\chi(\beta,K_1,K_2)$ is defined to be

$$\bar{I}(\beta,K_1,K_2) = \sup\{\dim A;\ A \subset \chi(\beta,K_1,K_2) \text{ is an } \mathbf{R}\text{-vector subspace on which } \bar{I}_{(b,K_1,K_2)} \text{ is positive definite}\}.$$

As in [4, section 9.3], one could also define an augmented index $\bar{I}_0(b,K_1,K_2)$ by considering subspaces A of $\chi(\beta,K_1,K_2)$ on which $\bar{I}_{(b,K_1,K_2)}$ is positive semi-definite. If we considered a corresponding augmented index $I_{(b,K_1,K_2)}$ for the ordinary index form (3.1) on $V^{\perp}(\beta,K_1,K_2)$, we would always obtain an infinite augmented index by considering vector fields of the form $X = f\beta'$ where $f: [0,b] \to \mathbf{R}$ is a smooth function with $f(0) = f(b) = 0$, cf. section 2 or [4, p. 293]. This is, of course, a second motivation for using the quotient index form (3.2) rather than the usual index form (3.1) in the case of a null geodesic.

In order to obtain the finiteness of $\bar{I}(\beta,K_1,K_2)$ for arbitrary β,K_1,K_2, via the classical Jacobi field approximation technique, we need to obtain the maximality of the quotient index form with respect to Jacobi classes up to the first K_1-focal point. Thus as in Section 2, we let $K = K_1$ and restrict attention again to $\bar{I}_{(b,K)}: \chi(\beta,K) \times \chi(\beta,K) \to R$. A crucial step in the proof is the following argument: we need to find (n-2) - K-Jacobi classes $V_1,V_2,\ldots,V_{n-2}$ in $\chi(\beta,K)$ so that if $X \in \chi(\beta,K)$ is arbitrary, then $X = \sum_{j=1}^{n-2} f_jV_j$ for piecewise smooth functions $f_j: [0,b] \to \mathbf{R}$.

The technical difficulty occurs for $t = 0$ since $\dim \pi(T_pK) = \dim K = k$ may be less than $(n-2)$ in which case the Jacobi classes V_j do not have linearly independent values at $t = 0$.

To deal with this problem, we use the following construction suggested by that of Ambrose [1] for the Riemannian case. Let $n \in (T_pK)^\perp$ be a null vector with $g(n,\beta'(0)) = -1$. Since g is a Lorentzian (nondegenerate) metric on $(T_pK)^\perp$, we may find orthonormal spacelike vectors $e_{k+1}, e_{k+2}, \ldots, e_{n-2}$ in $(T_pK)^\perp$ such that $\mathrm{Span}\{e_{k+1}, \ldots, e_{n-2}\} \cap \mathrm{Span}\{n,\beta'(0)\} = \{0\}$. Let $e_1, \ldots, e_k \in T_pK$ be orthonormal spacelike vectors. Then the set $\{e_1, \ldots, e_k, e_{k+1}, \ldots, e_{n-2}\}$ of spacelike vectors in T_pM has the property that $g(e_i,e_j) = \delta_{ij}$, $g(e_i,\beta'(0)) = g(e_i,n) = 0$ for all i,j. For $1 \leq j \leq k$ let J_i be the unique Jacobi field along β satisfying the initial condition

$$J_j(0) = e_j,\ J_j'(0) = -S_{\beta'(0)}e_j \tag{3.3}$$

and for $k + 1 \leq j \leq n-2$, let J_i be the unique Jacobi field along β satisfying the initial conditions

$$J_j(0) = 0,\ J_j'(0) = e_j. \tag{3.4}$$

Then the J_j's are K-Jacobi fields along β and if we set $V_j := \pi(J_j) \in \chi(\beta,K)$, then the V_j's are K-Jacobi classes in $\chi(\beta,K)$. It is easily checked that if $\beta|_{[0,b]}$ has no K-focal points, then the J_j's are spacelike vector fields and also for $t > 0$ we have $G(\beta(t)) = \mathrm{Span}\{V_1(t),V_2(t),\ldots,V_{n-2}(t)\}$, cf. equation (2.3). Thus for all $t > 0$ we may expand $X \in \chi(\beta,K)$ as

$$X(t) = \sum_{j=1}^{n-2} f_j(t)V_j(t)$$

for piecewise smooth functions $f_j\colon (0,b] \to \mathbf{R}$. However, as $V_j(0) = [\beta'(0)]$ for $k + 1 \leq j \leq n - 2$, it is not apparent that this decomposition extends to $t = 0$. To achieve this extension, we note the following two lemmas (cf. [10, Lemmas 3.4, 3.6]). The first lemma is immediate from Cramer's Rule.

Lemma 3.2: Let ε be arbitrary with $0 < \varepsilon < b$ and let $\{W_i\}$, $i = 1,2,\ldots,q$, be smooth vector classes in $\chi(\beta)$ such that $\{W_i(0)\}$, $i = 1,2,\ldots,q$ are $\bar{g}$-orthonormal in $G(\beta(0)) = (\beta'(0))^{\perp}/[\beta'(0)]$. Suppose Z is a smooth vector class along $\beta|_{(0,\varepsilon]}$ with $Z(t) = \sum_{i=1}^{q} f_i(t)W_i(t)$ for all $t \in (0,\varepsilon]$ for smooth functions $f_i\colon (0,\varepsilon] \to \mathbf{R}$. Then if $\lim_{t\to 0} Z(t)$ exists, $\lim_{t\to 0} f_i(t)$ exists for $1 \leqslant i \leqslant q$ and

$$Z(0) = \sum_{i=1}^{q} (\lim_{t\to 0} f_i(t))W_i(0).$$

Using this first result, we obtain

Lemma 3.3: Let $\varepsilon > 0$ be arbitrary with $0 < \varepsilon \leqslant b$ and let $\{U_i\}$, $i = 1,2,\ldots,q$, be smooth vector classes in $\chi(\beta)$ with $U_i(0) = [\beta'(0)]$ for $1 \leqslant i \leqslant q$ and with $\{U_i'(0)\}$, $i = 1, \ldots, q$, $\bar{g}$-orthonormal in $G(\beta(0))$. Suppose $W \in \chi(\beta)$ with $W(0) = [\beta'(0)]$ and

$$W(t) = \sum_{i=1}^{q} f_i(t)U_i(t) \text{ for smooth functions } f_i\colon (0,\varepsilon] \to \mathbf{R}.$$

Then $\lim_{t\to 0} f_i(t)$ exists for $1 \leqslant i \leqslant q$ and $W'(0) = \sum_{i=1}^{q} (\lim_{t\to 0} f_i(t))U_i'(0)$.

Proof. Note first that $W'(0) = \lim_{t\to 0} \frac{W(t)}{t}$ in $\chi(\beta)$ since $W(0) = [\beta'(0)]$. Also the vector fields

$$W_i(t) = \begin{cases} U_i'(0) & \text{if } t = 0 \\ \dfrac{U_i(t)}{t} & \text{if } t > 0 \end{cases}$$

and

$$Z(t) = \begin{cases} W'(0) & \text{if } t = 0 \\ \dfrac{W(t)}{t} & \text{if } t > 0 \end{cases}$$

satisfy the hypothesis of Lemma 3.2 and for $t > 0$

$$Z(t) = \frac{1}{t} \sum_{i=1}^{q} f_i(t)U_i(t) = \sum_{i=1}^{q} f_i(t)W_i(t).$$

Hence

$$W'(0) = Z(0) = \sum_{i=1}^{q} (\lim_{t\to 0} f_i(t))W_i(0)$$

$$= \sum_{i=1}^{q} (\lim_{t\to 0} f_i(t))U_i'(0)$$

as required. □

We are now ready to show that the K-Jacobi classes $V_1,\ldots,V_{n-2}$ constructed according to (3.3) and (3.4) form a basis for $\chi(\beta,K)$ if β has no K-focal points.

Theorem 3.4: Suppose $\beta\colon [0,b] \to (M,g)$ has no K-focal points and let $V_1,\ldots,V_{n-2}$ be K-Jacobi classes in $\chi(\beta,K)$ satisfying the initial conditions (3.3) and (3.4). Then give any piecewise smooth vector class $X \in \chi(\beta,K)$, there exist piecewise smooth functions

$f_j\colon [0,b] \to \mathbf{R}$ such that $X = \sum_{j=1}^{n-2} f_j V_j$.

Proof: If β has no K-focal points, then $\{V_1(t),\ldots,V_{n-2}(t)\}$ forms a basis for $G(\beta(t))$ for all $t \in (0,b]$. Thus $X(t)$

$= \sum_{j=1}^{n-2} f_j(t)V_j(t)$ for piecewise smooth functions $f_j\colon (0,b] \to \mathbf{R}$. We

must show that the f_j's may be extended to $t = 0$.

Let $A(t) = \sum_{j=1}^{k} f_j(t)V_j(t)$ and $B(t) = \sum_{j=k+1}^{n-2} f_j(t)V_j(t)$ for all

$t \in (0,b]$. Thus $X(t) = A(t) + B(t)$ for all $t \in (0,b]$.

Set

$$P_j(t) = \begin{cases} V_j'(0) & \text{if } t = 0 \\ \dfrac{V_j(t)}{t} & \text{if } t > 0 \end{cases}$$

for $k + 1 \leqslant j \leqslant n-2$ and define two smooth subbundles ξ,η of $G(\beta)$ by $\xi|_{\beta(t)}$ = Span $\{V_1(t),\ldots,V_k(t)\}$ and $\eta|_{\beta(t)}$ = Span$\{P_{k+1}(t),P_{k+2}(t),\ldots,P_{n-2}(t)\}$. Since β has no K-focal points, it may be checked that $\xi|_{\beta(t)} \cap \eta|_{\beta(t)} = [\beta'(t)]$ for all $t \in [0,b]$ and thus $G(\beta) = \xi \oplus \eta$. Let p: $G(\beta) \to \xi$ denote the smooth projection map.

Given the arbitrary $X \in \chi(\beta,K)$, then $X(0) \in \pi(T_pK)$ may be written as $X(0) = \sum_{j=1}^{k} a_jV_j(0)$ for some constants $a_j \in \mathbf{R}$. Moreover $Y: = p(X)$ is a piecewise smooth section of ξ with $Y(0) = X(0)$ and $Y(t) = A(t)$ for all $t > 0$. Since $\lim_{t\to 0} Y(t) = p(X(0)) = Y(0)$ exists and $\{V_1(0),\ldots,V_k(0)\}$ are $\bar{g}$-orthonormal, Lemma 3.2 implies that $\lim_{t\to 0} f_j(t)$ exists for $1 \leqslant j \leqslant k$ and moreover, $a_j = \lim_{t\to 0} f_j(t)$. Thus we set $f_j(0) = a_j$ for $1 \leqslant j \leqslant k$ and with this extension, $Y(t) = A(t)$ on $[0,b]$ and $X(0) = A(0)$.

Now let $W: = X - p(X) = X - A$ for all $t \in [0,b]$. Then $W \in \chi(\beta,K)$, $W(0) = [\beta'(0)]$ and $W(t) = B(t)$ for all $t \in (0,b]$.

Expand $W'(0) = \sum_{j=k+1}^{n-2} b_jV_j'(0)$ for some $b_j \in \mathbf{R}$. Applying Lemma 3.3 to

W and $\{V_{k+1},V_{k+2},\ldots,V_{n-2}\}$, we obtain that $\lim_{t\to 0} f_j(t)$ exists for $k+1 \leqslant j \leqslant n-2$ and that $\lim_{t\to 0} f_j(t) = b_j$. Thus setting $f_j(0) = b_j$ for $k-1 \leqslant j \leqslant n-2$ extends f_j to $t = 0$ for $k+1 \leqslant j \leqslant n-2$.

With Theorem 3.4 in hand, it may be checked that the proof given in Bishop and Crittenden [5, p. 229] for the basic index inequality may be adapted to the quotient bundle setting, cf. [10, chapter 4.4, Theorem 4.2] for the proof.

Theorem 3.5: (Maximality of K-Jacobi Classes) Let $\beta: [0,b] \to (M,g)$ be a null geodesic segment with no K-focal points and let $V \in \chi(\beta,K)$ be arbitrary. Then there exists a unique K-Jacobi class $J \in \chi(\beta,K)$ with $J(b) = V(b)$. Further

$$\bar{I}_{(b,K)}(V,V) \leq \bar{I}_{(b,K)}(J,J) \tag{3.5}$$

and equality holds iff $J = V$.

Recall that $\chi_0(\beta,K) = \{V \in \chi(\beta,K): \ V(b) = [\beta'(b)]\}$.

Corollary 3.6: Suppose $\beta: [0,b] \to (M,g)$ has no K-focal points. Then the quotient bundle index form $\bar{I}_{(b,K)}: \chi_0(\beta,K) \times \chi_0(\beta,K) \to \mathbf{R}$ is negative definite.

With this last result in hand, we may now establish the finiteness of $\bar{I}(\beta,K_1,K_2)$ for arbitrary null geodesic segments. If $K_1 = \{p\}$, choose $t_1 > 0$ such that $\beta|_{[0,t_1]}$ has no conjugate points to $t = 0$. Otherwise, choose $t_1 > 0$ such that $\beta|_{[0,t_1]}$ has no K_1-focal points to $t = 0$. Let $\sigma(t): = \gamma(b-t)$ so that $\sigma(0) = q$, $\sigma(b) = p$. If $K_2 = \{q\}$, choose $\delta > 0$ such that $\sigma|_{[0,\delta]}$ has no conjugate points and $t_1 < b-\delta$. Otherwise, choose $\delta > 0$ such that $\sigma|_{[0,\delta]}$ has no K_2-focal points and $t_1 < b-\delta$. Set $\varepsilon = b-\delta$. By compactness, choose $t_2,\ldots,t_{m-1}$ with $t_1 < t_2 < \ldots < t_{m-1} < \varepsilon$ such that $\beta|_{[t_j,t_{j+1}]}$ has no conjugate points to $t = t_j$ for $1 \leq j \leq m-1$ and $\beta|_{[t_{m-1},\varepsilon]}$ has no conjugate points to $t = t_{m-1}$. Now relabel $t_m: = \varepsilon$. Thus we have a partition $t_0 = 0 < t_1 <$

$t_2 < \ldots < t_m < t_{m+1} = b$ with the following property. Given any $v \in G(\beta(t_1))$, there exists a unique K_1-Jacobi class J on $\beta|_{[t_0,t_1]}$ with $J(t_1) = v$. For $1 \leqslant j \leqslant m-1$, given any $v_1 \in G(\beta(t_j))$ and $v_2 \in G(\beta(t_{j+1}))$, there exists a unique Jacobi class J on $\beta|_{[t_j,t_{j+1}]}$ with $J(t_j) = v_1$ and $J(t_{j+1}) = v_2$. Given any $v \in G(\beta(t_m))$, there exists a unique K_2-Jacobi class J along $\beta|_{[t_m,b]}$ (really $\sigma|_{[0,b-t_m]}$) with $J(t_m) = v$. We may thus define the usual approximation map

$$\Phi: \ \chi(\beta,K_1,K_2) \to J\{t_j\}$$

where $J\{t_j\}$ denotes the **R**-vector space of piecewise smooth Jacobi classes V such that $V|_{[0,t_1]}$ is a smooth K_1-Jacobi class, $V|_{[t_j,t_{j+1}]}$ is a smooth Jacobi class for $1 \leqslant j \leqslant n$, and $V|_{[t_m,b]}$ is a K_2-Jacobi class. Given $W \in \chi(\beta,K_1,K_2)$, $\Phi(W) \in J\{t_j\}$ is defined to be the K_1-Jacobi class with $\Phi(W)(t_1) = W(t_1)$ on $\beta|_{[0,t_1]}$, the Jacobi class with $\Phi(W)(t_j) = W(t_j)$ and $\Phi(W)(t_{j+1}) = W(t_{j+1})$ on $\beta|_{[t_j,t_{j+1}]}$ for $1 \leqslant j \leqslant m-1$, and the K_2-Jacobi class with $\Phi(W)(t_m) = W(t_m)$ for $\beta|_{[t_m,b]}$. Especially $\Phi|_{J\{t_j\}} = \text{Id}$. Using inequality (3.5) and its conjugate point analogue ([4, p. 316]), it is readily checked that if $W \in \chi(\beta,K_1,K_2)$ and $W \notin J\{t_j\}$, then

$$\overline{I}_{(b,K_1,K_2)}(W,W) < \overline{I}_{(b,K_1,K_2)}(\Phi(W),\Phi(W)) \qquad (3.6)$$

From (3.6) it is immediate that if A is a subspace of $\chi(\beta,K_1,K_2)$ on which $\overline{I}_{(b,K_1,K_2)}$ is positive (semi)definite, then $\Phi|_A: A \to J\{t_j\}$ is injective. Since $\dim J\{t_j\} < +\infty$, it follows that the index

$\bar{I}(\beta,K_1,K_2)$ is finite, cf. [4, p. 324]. We have established the following

Proposition 3.7: Let β: $[0,b] \to (M,g)$ be an arbitrary null geodesic segment which is perpendicular at $p = \beta(0)$ to the spacelike submanifold K_1 and at $q = \beta(b)$ to the spacelike submanifold K_2. Then the index $\bar{I}(\beta,K_1,K_2)$ of the quotient bundle index form

$$\bar{I}_{(b,K_1,K_2)}: \chi(\beta,K_1,K_2) \times \chi(\beta,K_1,K_2) \to \mathbf{R}$$

is finite.

Notice that it is <u>not</u> assumed in Proposition 3.7 that b is not a K_1-focal point to $t = 0$ along β in obtaining the finiteness. Also in the case $K_2 = \{q\}$ and hence $\bar{I}_{(b,\, K_1,K_2)} = \bar{I}_{(b,K_1)}$, it may be checked that the function $f(t)$ = Index of $\bar{I}_{(t,K_1)}$ on $\beta|_{[0,t]}$ is a nondecreasing function of t, regardless of whether or not b is a K_1-focal point. Thus the conjugate point proof of the null Morse index theorem given in [4, section 9.3] for $K_1 = \{p\}$ and $K_2 = \{q\}$ extends to the case that K_1 is arbitrary and $K_2 = \{q\}$, cf. [5, pp. 231-236] for the Riemannian case). Thus we have (remembering Corollary 2.7)

Proposition 3.8: Let β: $[0,b] \to (M,g)$ be an arbitrary null geodesic segment which is perpendicular at $p = \beta(0)$ to the spacelike submanifold K. Let

$$J_t(\beta) = \{K \text{ - Jacobi fields } J \in V^{\perp}(\beta,K) \text{ with } J(t) = 0\}.$$

Then, for the quotient bundle index form

$$\bar{I}_{(b,K)}: \chi_0(\beta,K) \times \chi_0(\beta,K) \to \mathbf{R},$$

$$\text{Index}\,(\bar{I}_{(b,K)}) = \sum_{t\,\in\,(0,b)} \dim J_t(\beta)$$

and

$$\text{Augmented Index}\,(\bar{I}_{(b,K)}) = \sum_{t\,\in\,(0,b]} \dim J_t(\beta).$$

When $K_2 \neq \{q\}$, difficulties arise in studying the Morse approximation of $V^{\perp}(\beta,K_1,K_2)$ by Jacobi fields, especially when $t = b$ is a K_1-focal point to $t = 0$ along β, cf. Ambrose [2], Bolton [6] for results concerning the Riemannian case. Thus in Section 4, we will assume that $t = b$ is <u>not</u> a K_1-focal point in the case K_2 is an arbitrary spacelike submanifold in order to bring Kalish's methods [9] to bear on the index problem.

SECTION 4. THE MORSE INDEX THEOREM.

Let (M,g) be an arbitrary Lorentzian manifold of dimension $n \geq 3$ and let $\beta\colon [0,b] \to (M,g)$ be a null geodesic segment which is orthogonal at $p = \beta(0)$ to a spacelike submanifold K_1 of dimension k_1 and orthogonal at $q = \beta(b)$ to a spacelike submanifold K_2 of dimension k_2. As the K_i are spacelike and $g(\beta'(0),\beta'(0)) = g(\beta'(b),\beta'(b)) = 0$, we have $\beta'(0) \notin T_pK_1$ and $\beta'(b) \notin T_pK_2$. Also, $k_1,k_2 \leq n-2$. Finally we may reparametrize the null geodesic β and thus take $b = 1$. Further, in order to adapt Kalish's proof method to our quotient bundle setting, we need to assume that $t_0 = b = 1$ is not a K_1-focal point to $t = 0$.

Recalling that $V^{\perp}(\beta,K_1,K_2) = \{Y \in V^{\perp}(\beta);\ Y(0) \in T_pK_1$ and $Y(1) \in T_qK_2\}$ and $\chi(\beta, K_1,K_2) = \{V \in \chi(\beta);\ V(0) \in \pi(T_pK_1)$ and $V(1) \in \pi(T_qK_2)\}$, consider the $\mathbf{R}$-vector subspace $\mathcal{R}$ of $\chi(\beta,K_1,K_2)$ given by

$$\mathcal{R} = \{K_1\text{-Jacobi classes } V \in \chi(\beta,K_1);\ V(1) \in \pi(T_qK_2)\}.$$

In view of (2.3), since $t_0 = 1$ is not a K_1- focal point, $\mathcal{R} \neq 0$. Also, motivated by (3.2), consider the following functional S: $\mathcal{R} \times \mathcal{R} \to \mathbf{R}$ given for $V,W \in \mathcal{R}$ by

$$S(V,W) = \bar{g}(V'(1) + \bar{S}^2(V(1)),W(1)). \qquad (4.1)$$

Note that for Jacobi classes V, W $\in$ R we have

$\bar{I}_{(b,K_1,K_2)}(V,W) = \bar{I}_{(b,K)}(V,W) - S(V,W) + \bar{g}(V'(1),W(1)).$

Fix $t_1,t_2,\ldots,t_k$ with $0 < t_1 < t_2 < \ldots < t_k = 1$ such that $\beta(t_1),\beta(t_2),\ldots,\beta(t_k)$ are the K_1-focal points to t = 0 along β. (In view of Prop. 3.7 with $K_2 = \{q\}$, there are only finitely many K_1-focal points to t = 0 along β). Let $m_i := \dim J_{t_i}(\beta)$ = the multiplicity of the K_1-focal point $\beta(t_i)$, i.e., the dimension of the space of smooth K_1-Jacobi fields along β which vanish at t_i, (recall Corollary 2.7 here). Then the index theorem we wish to obtain may now be stated.

Theorem 4.1: Let β: [0,1] → (M,g) be a null geodesic segment in an arbitrary space-time of dimension n ⩾ 3 and let K_1,K_2 be spacelike submanifolds of dimension ⩽ n-2 which are perpendicular to β at p = β(0) and q = β(1) respectively. Suppose further that t = 1 is not a K_1-focal point to t = 0 along β. Then

$$\mathrm{Ind}(\bar{I}_{(\beta,K_1,K_2)}) = \sum_{t\in(0,1)} \dim J_t(\beta) + \mathrm{Ind}(S)$$

Here Ind(S) = $\sup\{\dim \mathcal{B}$; $\mathcal{B}$ is a **R**-vector subspace of R on which $S|_{\mathcal{B}\times\mathcal{B}}$ is negative definite$\}$.

As in [3], in view of Corollary 2.7, it suffices to work with Jacobi vector classes and show that $\mathrm{Ind}(\bar{I}) = \sum_{t\in(0,1)} \dim \bar{J}_t(\beta) + \mathrm{Ind}(S)$. On $\chi(\beta,K_1,K_2)$, we have the Sobelev norm

$$(V,W) = \int_{t=0}^{1} (\bar{g}(V(t),W(t)) + \bar{g}(V'(t),W'(t)))\, dt.$$

Let M denote the completion of $\chi(\beta,K_1,K_2)$ in the Sobelev norm, cf. [9]. In order for Kalish's proof method to be valid for null geodesics, we must check for any $t_0 \in [0,1]$ that if $v \in G(\beta(t_0))$ is an arbitrary vector coset, then the evaluation map

$$L_v : \chi(\beta, K_1, K_2) \to \mathbf{R}$$

given for $W \in \chi(\beta, K_1, K_2)$ by

$$L_v(W) = \bar{g}(v, W(t_0)) \tag{4.2}$$

is bounded in the sup norm and hence extends to a continuous linear functional L_v: $M \to \mathbf{R}$. But assuming $\bar{g}(W(t),W(t)) \leq 1$ for all $t \in [0,1]$ and recalling that $\bar{g}$ is a positive definite metric on the quotient bundle $G(\beta(t))$ so that the Cauchy-Schwartz inequality is valid, we have $|L_v(W)| \leq \bar{g}(v,v)$ and L_v is thus a bounded functional. It is interesting to note by way of contrast that if γ is a spacelike geodesic, then $(\gamma'(t))^\perp = \{v \in T_{\gamma(t)}M;\ g(v,\gamma'(t)) = 0\}$ contains a null vector w and for such a vector L_w: $V^\perp(\gamma) \to \mathbf{R}$ is not continuous in the sup norm.

In view of decomposition (2.3), let

$A := \{V \in \chi(\beta, K_1, K_2);\ V(1) = [\beta'(1)]$ and $V(t_i) \in A_{\beta(t_i)}$ for $1 \leq i \leq k\}$.

Because of decomposition (2.3), it makes sense to utilize projection functionals L_v of the above form, for which the v's span the spaces $B_{\beta(t_i)}$, cf. [9]. At each K_1-focal point t_i of multiplicity m_i, let vector cosets $\{e_1, \ldots, e_{m_i}\} \subset G(\beta(t_i))$ be a $\bar{g}$-orthonormal basis for $B_{\beta(t_i)}$. Define p_{i,j_i}: $M \to \mathbf{R}$ by

$$p_{i,j_i}(V) := \bar{g}(V(t_i), e_{j_i}),\ 1 \leq i \leq k,\ 1 \leq j_i \leq m_i.$$

At $q = \beta(1)$, let $\{e_1, \ldots, e_r\} \subset G(\beta(1))$ be a $\bar{g}$-orthonormal basis for $\pi(T_q K_2)$, $r = \dim K_2$. Define p_i: $M \to \mathbf{R}$ by

$$p_i(V) := \bar{g}(V(1), e_i),\quad 1 \leq i \leq r.$$

One easily checks that these two sets of functionals are linearly independent and hence since M is complete (as a Hilbert space) that

$$M = W \oplus W^{\perp} \tag{4.3}$$

where

$$W = (\bigcap_{i,j_i} \ker(p_{i,j_i})) \cap (\bigcap_{i} \ker(p_i)) \tag{4.4}$$

with $\dim (W^{\perp}) = \sum_{i=1}^{k} m_i + \dim K_2 < +\infty$, cf. [9].

We now wish to replace $W^{\perp}$ with a more "geometric" complement to W in M utilizing the following Lemma from [9].

Lemma 4.2: Let X be any vector space and let V be an n-dimensional subspace. Let W be a subspace of X with $X = V \oplus W$. Suppose V_1 is a subspace of X with $V_1 \cap W = \{0\}$. Then $\dim V_1 \leqslant n$. Also if $\dim V_1 = \dim V = n$ and $V_1 \cap W = \{0\}$, then $X = W \oplus V_1$.

In constructing this complement, we use the assumption "t = 1 is not a K_1-focal point" to obtain $A_{\beta(1)} = G(\beta(1))$ so that the quadratic form S: $R \times R \to \mathbf{R}$ may be "diagonalized" with respect to the positive definite inner product $\bar{g}$ on $R \times R$ in the following sense. We may choose $r = \dim K_2$ linearly independent K_1-Jacobi classes $J_1,\ldots,J_r$ in $\chi(\beta,K_1)$ so that

(i) $J_1(1),\ldots,J_r(1)$ span $\pi(T_qK_2)$,

(ii) S is negative definite on the span of $J_1,\ldots,J_N$, where $N = \mathrm{Ind}(S)$, $N \leqslant r$, and

(iii) S is positive semi-definite on the span of $J_{N+1},\ldots,J_r$.

(Note the change needed in (3) of [9, p. 8] from "positive definite" to "positive semi-definite"; otherwise, the construction in [9] is unaltered.) The span of the vector classes $J_1,\ldots,J_r$ will form part of the "geometric complement" A_1 to W. We need to construct additional fields which span the spaces $B_{\beta(t_i)}$ at the focal points

t_i, again utilizing decomposition (2.3) for the existence. Specifically, let $Y_{i,j_i} \in \chi(\beta,K_1)$ be K_1-Jacobi classes with

$Y_{i,j_i}(t_i) = [\beta'(t_i)]$ for $j_i = 1,2,\ldots,m_i = \dim \bar{J}_{t_i}(\beta)$ and

$\{Y'_{i,j_i}(t_i)\}_{i=1}^{m_i}$ a $\bar{g}$-orthonormal basis for $B_{\beta(t_i)}$ for

$i = 1,2,\ldots,k$. Let Z_{i,j_i} be the smooth parallel vector class in $\chi(\beta)$ with $Z_{i,j_i}(t_i) := -Y'_{i,j_i}(t_i)$ for $i = 1,2,\ldots,k$ and $j_i = 1,2,\ldots,m_i$. Further construct for $i = 1,\ldots,k$ smooth bump functions $\phi_i\colon [0,1] \to [0,1]$ with mutually disjoint support such that $\phi'_j(t_i) = 0$, $\phi_i(t_i) = 1$, $\phi_1(0) > 0$, and $\phi_k(1) < 1$. Then define vector classes $V_{i,j_i} \in \chi(\beta,K_1)$ for a positive parameter $\lambda > 0$ to be chosen later to make the index positive definite on the span of the V_{i,j_i}'s as follows:

$$V_{i,j_i}(t) = \begin{cases} Y_{i,j_i}(t) + \lambda\phi_i(t)Z_{i,j_i}(t) & \text{if } 0 \leqslant t \leqslant t_i \\ \lambda\phi_i(t)Z_{i,j_i}(t) & \text{if } t_i \leqslant t \leqslant 1. \end{cases}$$

Now let $A_1 \subset \chi(\beta,K_1)$ be given by

$$A_1 := \mathrm{Span}_R\{V_{i,j_i}, J_q;\ i = 1,2,\ldots,k,\ j = 1,2,\ldots,m_i,\ q = 1,2,\ldots,r\}$$

Just as in [9], one checks readily using decomposition (2.3) that $\dim A_1 = \sum_{i=1}^{k} m_i + r = \dim(W^{\perp})$ and $A_1 \cap W = \{0\}$. Hence, by Lemma 4.2 applied with $X = \mathcal{M}$, we obtain $\mathcal{M} = W \oplus A_1$. Recall that $A := \{V \in \chi(\beta,K_1,K_2);\ V(1) = [\beta'(1)]$ and $V(t_i) \in A_{\beta(t_i)}$ for $1 \leqslant i \leqslant k\}$. We want to use the decomposition $\mathcal{M} = W \oplus A_1$ to show that $\chi(\beta,K_1,K_2) = A \oplus A_1$ as in [9]. First one checks that $A \subset W$ whence $A \cap A_1 = \{0\}$. Now let $V \in \chi(\beta,K_1,K_2)$. As $V \in \mathcal{M}$ as well, we may write $V = Y + Z$ where $Y \in W$ and $Z \in A_1$. Thus $Y = V - Z \in \chi(\beta,K_1,K_2)$, i.e., Y is a piecewise smooth vector class. Hence $Y \in W$ piecewise smooth and decomposition (2.3) implies that $Y \in A$ as required. Operationally, this means that if $J_1,\ldots,J_p$, $p = \dim M - 2$, are p linearly independent K_1-Jacobi classes in $\chi(\beta,K_1)$, then Y as above may be written as $Y = \sum_{j=1}^{p} f_j J_j$ for piecewise smooth functions $f_j: [0,1] \to \mathbf{R}$, $1 \leq j \leq p$. It is then well known that for $Y \in A$ decomposed as above,

$$\bar{I}_{(\beta,K_1,K_2)}(Y,Y) = -\int_0^1 \bar{g}\left(\sum_{j=1}^{p} f_j'(t)J_j(t), \sum_{j=1}^{p} f_j'(t)J_j(t)\right) dt$$

whence $\bar{I}_{(\beta,K_1,K_2)}(Y,Y) \leq 0$ for any $Y \in A$ since the metric $\bar{g}$ is positive definite on the quotient bundle $G(\beta)$. Recall also, that if V is a smooth Jacobi class and $W \in \chi(\beta,K_1,K_2)$ is arbitrary, then formula (3.1) for the index form $\bar{I}_{(\beta,K_1,K_2)}$ simplifies to

$$\begin{aligned}\bar{I}_{(\beta,K_1,K_2)}(V,W) &= \bar{g}(V'(0) + S^1(V(0)),W(0)) \\ &\quad - \bar{g}(V'(1) + S^2(V(1)),W(1)) \\ &= \bar{g}(V'(0) + S^1(V(0)),W(0)) - S(V,W).\end{aligned} \tag{4.5}$$

With this formula, it is immediate that $\bar{I}_{(\beta,K_1,K_2)} \leq 0$ in $\mathcal{R}_2 = \mathrm{Span}_R\{Y_{N+1}, Y_{N+2},\ldots,Y_r\}$. Now let $V \in A_1$ be written as $V = Z + W$ where $Z = \sum_{i,j_i} \alpha_{i,j_i} V_{i,j_i}$ and $W = \sum_{k=1}^{N} \beta_k J_k$. Since each J_k is a smooth K_1-Jacobi class and each $V_{i,j_i}(1) = [\beta'(1)]$, it is immediate that $\bar{I}_{(\beta,K_1,K_2)}(V,W) = 0$ and $\bar{I}(W,W) \geq 0$, and if some $\beta_k \neq 0$, then $\bar{I}(W,W) > 0$ for any choice of $\lambda > 0$. Now let I denote the matrix

$$\bar{I}_{(\beta,K_1,K_2)}(\phi_i(t)Z_{i,j_i}(t), \quad \phi_h(t)Z_{h,l_h}(t))$$

One checks via (4.5) that

$$\bar{I}_{(\beta,K_1,K_2)}(Z,Z)=2\lambda\sum_i (a_{i,j_i})^2+\lambda^2\sum_{i,h} a_{i,j_i}a_{h,l_h}\bar{I}_{(\beta,K_1,K_2)}(\phi_i Z_{i,j_i},\phi_h Z_{h,l_h}).$$

This may be reinterpreted in terms of the usual Euclidian inner product $\langle , \rangle$ as

$$\bar{I}_{(\beta,K_1,K_2)}(Z,Z) = 2\lambda \langle Z,Z \rangle + \lambda^2 \langle IZ,Z \rangle$$

$$\geqslant 2\lambda \langle Z,Z \rangle - \lambda^2 ||I|| \langle Z,Z \rangle$$

Hence in the unlikely event that $I = 0$, we have $\bar{I}_{(\beta,K_1,K_2)}(Z,Z) > 0$ unless $Z = 0$ for any $\lambda > 0$. If $I \neq 0$, choose $\lambda > 0$ with $0 < \lambda < 2/||I||$. Then one also obtains from the above inequality that $\bar{I}_{(\beta,K_1,K_2)}(Z,Z) > 0$ if $Z \neq 0$. Combining these two cases yields $\bar{I}_{(\beta,K_1,K_2)}(V,V) > 0$ if $V \in \mathcal{R}_1$, $V \neq 0$. The above calculations show that

$$\mathrm{Ind}(\bar{I}_{(\beta,K_1,K_2)}|A_1 \times A_1) = \sum_{i=1}^{k} m_i + \mathrm{Ind}(S).$$

The fact that $\bar{I}_{(\beta,K_1,K_2)}$ is negative semi-definite on A essentially means that $\bar{I}_{(\beta,K_1,K_2)}$ restricts to A_1 in calculating the index, whence the desired formula follows.

More exactly, the following Lemma may be established as in [9].

Lemma 4.3: Let W_1 be a subspace of $\chi(\beta,K_1,K_2)$ with $\chi(\beta,K_1,K_2) = A \oplus W_1$. Then

$$\mathrm{Ind}(\bar{I}_{(\beta,K_1,K_2)}|W_1 \times W_1) \leqslant \mathrm{Ind}\,(\bar{I}_{(\beta,K_1,K_2)}|A_1 \times A_1)$$

$$= \sum_{j=1}^{k} m_i + \mathrm{Ind}(S).$$

Proof: Note first that because A_1 is finite dimensional, and $W_1 \cap A = \{0\}$, we have $\dim W \leqslant \dim A_1$ by Lemma 4.2. Hence if there exists a subspace $V \subset W_1$ with $\bar{I}_{(\beta,K_1,K_2)}$ positive definite on V and $\dim V \geqslant \dim \mathcal{R}_1 + 1$, then the dimension of the subspace of W on which $\bar{I}_{(\beta,K_1,K_2)}$ is negative semi-definite is of dimension less than r - N.

On the other hand, a subspace of W of dimension (r-N) on which $\bar{I}$ is negative semi-definite may be constructed as follows. Using the decomposition $\chi(\beta,K_1,K_2) = A \oplus W_1$, write each $Y_{N+j} \in \mathcal{R}_2$ as $Y_{N+j} = V_j + Z_j$ with $V_j \in A$ and $Z_j \in W$, for $j = 1,2,\ldots,r-N$. Then the vector classes $\{Z_1,Z_2,\ldots,Z_{r-N}\}$ are easily seen to be linearly independent over $\mathbf{R}$ and calculating from (4.5) using $Z_j = J_{N+j} - V_j$, one checks easily that $\bar{I}_{(\beta,K_1,K_2)}$ is negative semi-definite on $\mathrm{Span}_{\mathbf{R}}\{Z_1,Z_2,\ldots,Z_{r-N}\}$.

The proof of the Index Theorem 4.1 is immediate from Lemmas 4.2 and 4.3.

SECTION 5: CONCLUDING REMARKS.

If γ: $[a,b] \to (M,g)$ is a unit speed timelike geodesic segment, then for each $t \in [a,b]$ the set of tangent vectors

$$(\gamma'(t))^{\perp} := \{v \in T_{\gamma(t)}M;\ \ g(v,\gamma'(t)) = 0\}$$

is a vector subspace of $T_{\gamma(t)}M$ of dimension (n-1) consisting of spacelike tangent vectors, i.e., the Lorentzian metric g is positive definite on the vector space $(\gamma'(t))^{\perp}$. Thus the Cauchy-Schwartz inequality argument of [9] may be applied just as in Section 4 to the evaluation functionals L_v: $(\gamma'(t))^{\perp} \to \mathbf{R}$ given by $L_v(w) := g(v,w)$ to obtain the boundedness of these functionals. Also, one has a direct sum decomposition of $(\gamma'(t))^{\perp}$ using the values and derivatives of K_1-Jacobi fields analogous to (2.3) for the quotient bundle $G(\beta)$. Using the analogue for timelike geodesics of Theorem 3.5 (cf. [10]), the timelike analogues of Propositions 3.7 and 3.8 may be obtained just as in Section 3 for null geodesics. Finally, the timelike analogue of the Morse Index Theorem 4.1 for spacelike endmanifolds K_1, K_2 to a timelike geodesic segment γ: $[a,b] \to (M,g)$ where $t = b$ is not a K_1-focal point for $t = a$ along γ may be obtained by arguments similar to those of Section 4. Here we may allow the hypersurface case, i.e., $\dim K_1$, $\dim K_2 \leq n-1$ since $\gamma'(t) \notin (\gamma'(t))^{\perp}$ for a timelike geodesic.

REFERENCES

1. Ambrose, W., "The Cartan structural equations in classical Riemannian geometry", J. Indian Math. Soc. 24, 23-76(1960).

2. Ambrose, W., "The index theorem in Riemannian geometry", Annals of Math. 73, 49-86(1961).

3. Beem, J. and Ehrlich, P., "A Morse index theorem for null geodesics", Duke Math. J. 46 561-569(1979).

4. Beem, J. and Ehrlich, P., Global Lorentzian Geometry, Marcel Dekker Pure and Applied Mathematics 67 (1981).

5. Bishop, R. and Crittenden, R., Geometry on Manifolds, Academic Press, Pure and Applied Mathematics 15, (1964).

6. Bolton, J., "The Morse index theorem in the case of two variable endpoints", J. Differential Geometry 12 567-581(1977).

7. Bölts, G., "Existenz und Bedeutung von konjugierten Werten in der Raum-Zeit", Diplomarbeit, Bonn University (1977).

8. Hawking, S. and Ellis, G., The Large Scale Structure of Space-Time, Cambridge University Press (1973).

9. Kalish, Diana, "Aspects of Morse Theory", Thesis, CUNY, (1984).

10. Kim, S., "K-Jacobi fields and Lorentzian comparison theorems", Thesis, University of Missouri-Columbia, (1986).

11. Warner, F., "Extension of the Rauch comparison theorems to submanifolds", Trans. Amer. Math. Soc. 122, 341-356(1966).

12. Warner, F., "The conjugate locus of a Riemannian manifold", Amer. J. Math. 87, 575-604(1965).

Paul E. Ehrlich*
Department of Mathematics
University of Missouri-Columbia
Columbia, Missouri 65211
U.S.A.

and

Department of Mathematics
201 Walker Hall
University of Florida
Gainesville, Florida 32611

Seon - Bu Kim
Department of Mathematics
Chonnam National University
Kwangju City
500 Korea
South Korea

*Supported in part by the National Science Foundation under grant number DMS 8601342 and DMS 8802672.

GEOMETRY AND TOPOLOGY (pp. 134–151)
edited by G. M. Rassias & G. M. Stratopoulos

GEOMETRY OF UNIFORM BODIES

Marek Elżanowski
Department of Mechanical Engineering and
Department of Mathematics and Statistics
The University of Calgary
Calgary, Alberta, Canada T2N 1N4

Marcelo Epstein
Department of Mechanical Engineering
The University of Calgary
Calgary, Alberta, Canada T2N 1N4

Jędrzej Śniatycki
Department of Mathematics and Statistics
The University of Calgary
Calgary, Alberta, Canada T2N 1N4

ABSTRACT

The paper discusses the uniformity and homogeneity of elastic bodies. After introducing some basic definitions such as configuration space, virtual displacement, forces, global elasticity and several aspects of local elasticity the idea of global uniformity is formulated and compared with the standard definition of (local) uniformity. Then, the existence of the material G-structure for a uniform jet-local elastic body is proved. Observing that homogeneity is equivalent to the flatness of such a G-structure, new explicit conditions for local homogeneity of certain classes of materials are also presented.

INTRODUCTION

The material continuous body, as understood in Continuum Mechanics, is probably the simplest concrete example of a differentiable manifold. And since a material body manifests itself through embeddings into another differentiable manifold, namely physical space, it is reasonable to expect a natural geometrical setting for Continuum Mechanics. Although this fact has been recognized early in the modern revival of this discipline (see, e.g., [1,2]), as attested by the careful definitions of the basic notions, it appears that a full utilization of geometrical and topological methods is still wanting. In previous papers [3,4,5] we have attempted to fill this gap and the present paper contains a review of our previous work as well as new results. Specifically, in Section 1, we briefly state the main facts about G-structures with particular attention to the notion of characteristic fields. It appears that it is the G-structure language (rather than that of affine connections) which is best suited to dealing with questions of material uniformity and homogeneity. In Section 2, we review the basic definitions of configuration space, virtual displacements, forces and global elasticity and, in Section 3, several aspects of locality, understood as certain types of symmetry of the elastic constitutive law [5], are discussed.

Section 4 and 5 are entirely devoted to uniformity and homogeneity. The geometrical theory of inhomogeneities of elastic bodies was formulated two decades ago by Noll [6] and Wang [7]. They have introduced a motion of a material connection consistent with the given response functional of the body. This material connection becomes the geometrical tool for the analysis and classification of different types of inhomogeneity. In particular the existence of a flat material connection was shown to be equivalent to the local homogeneity of the body. If however the group of material symmetries happens to be continuous, it is rather difficult to prove the existence of such a flat connection. In this paper, by observing that homogeneity is equivalent to the flatness of an appropriately defined G-structure, new explicit conditions for local homogeneity are obtained for certain classes of materials.

1. MATHEMATICAL FRAMEWORK

Let M be a differentiable n- manifold and let L(M) denote its bundle of linear frames (see [8]). The group GL(n,R) (general linear group in n- variables) acts on L(M) to the right as follows: if $A \in GL(n,R)$ and $p \in L(M)$, where $p=(\underline{u}_1,\ldots,\underline{u}_n)$ is a linear frame at $x \in M$, then $pA \in L(M)$ is a frame at $x \in M$ given by

$$pA = (\underline{u}_i\, \beta^i{}_1,\ldots,\, \underline{u}_i\, \beta^i{}_n) \tag{1.1}$$

where $A = (\beta^i{}_j)$ and the summation convention is used. This action is free and transitive making L(M) into a right principal GL(n,R)-bundle. Let G be a closed (Lie) subgroup of GL(n,R). A G-structure on M is a submanifold (a G-subbundle) B_G of L(M) stable under the right action of G on L(M), i.e., for any $p\epsilon B_G$ and any $A\epsilon GL(n,R)$ the point pA belongs to B_G if, and only if, $A\epsilon G$. The induced action of G on B_G is obviously free and transitive.

Naturally, not every G-structure is admissible on a particular manifold M.

If U is now a neighborhood of a point in M (an open submanifold of M) then the set $B_G|U \equiv \pi_G^{-1}(U)$ is a G-structure over U, where π_G is the natural projection of B_G onto M. We shall call this "local" G-structure, which consists of all the frames in B_G attached at points in U, <u>the restriction of B_G to U</u>. Since principal bundles are locally trivial, i.e., for each $x\epsilon M$ there exists an open neighborhood $U\subset M$ such that $B_G|U$ is isomorphic to U x G, there exists always a <u>smooth frame field</u>

$p(x) = (\underline{u}_1(x),\ldots,\underline{u}_n(x))$ on U such that

$$\pi_G^{-1}(x) = \{p(x)A \mid A\epsilon G\} \tag{1.2}$$

for every $x\epsilon U$.

<u>Definition 1</u>. The "local" G-structure $B_G|U$ is said to be flat if there exists a smooth section $(\underline{u}_1,\ldots,\underline{u}_n)$: $U\to B_G|U$ such that the coordinate vector fields $\underline{u}_j$, $j = 1,\ldots,n$, commute with each other.

A G-structure B_G on M is said to be <u>locally flat (integrable)</u> at $x\epsilon M$ if there exists a neighborhood V of x in M such that $B_G|V$ is flat.

In the light of the above definitions the following is the obvious consequence of Frobenius' theorem.

<u>Proposition 1</u>. The G-structure B_G on M is locally flat if and only if there exists an atlas $\{U_\alpha,\psi_\alpha\}$ on M with $\psi_\alpha=(x_\alpha^1,\ldots,x_\alpha^n)$ such that $(\frac{\partial}{\partial x_\alpha^1}(x),\ldots,\frac{\partial}{\partial x_\alpha^n}(x))$ is a cross-section of $B_G|U_\alpha$ for any $U_\alpha\subset M$.

For a certain class of closed subgroups G of GL(n,R), namely groups of finite type, there exists a general theory based on the notion of prolongations of G-structures, [8][9], which enables one to determine if a given G-structure is locally flat (integrable) or not. However, in this paper we show that in some special cases one can resolve the integrability problem in a more elementary manner, using, introduced here, notion of a characteristic field of the G-structure.

Let W be a finite dimensional vector space with the differentiable action of GL(n,R) to the left, and let ξ be the space of all equivalence classes in L(M) x W under the following equivalence relation: $(p,f) \in L(M) \times W$ is equivalent to $(p',f') \in L(M) \times W$, say $(p,f) \sim (p',f')$, if and only if there is $A \in GL(n,R)$ such that $p' = pA$ and $f' = A^{-1}f$. The space ξ defined this way is an <u>associated bundle of L(M) with a typical fibre W</u> corresponding to the given action of GL(n,R) on W. Each equivalence class $\{(p,f)\}$ defines a geometric quantity of type W, and for a given frame p the corresponding element $f \in W$ can be understood as the components of the quantity in the frame p. Geometric object bundles and in particular tensor bundles and bundles of tensor densities on M are useful examples of associated bundles of L(M).

A section $s\colon M \to \xi$ of the associated bundle ξ (a field s of quantities of type W) with a typical fibre W corresponds to a function $s^{\#}\colon L(M) \to W$ such that

$$s^{\#}(pA) = A^{-1}s^{\#}(p) \tag{1.3}$$

for every $p \in L(M)$ and $A \in GL(n,R)$.

The value $s^{\#}(p)$ of $s^{\#}$ on a frame $p \in L(M)$ gives the components of the field s relative to this frame.

Let now B_G be a G-structure over a manifold M.

<u>Definition 2</u>. A <u>local characteristic field of B_G</u> is a local section $s\colon U \subset M \to \xi$ of some associated bundle ξ of L(M) such that

i. $$s^{\#-1}\{As^{\#}(p) \mid A \in G\} = B_G \mid U \tag{1.4}$$

for any $p \in B_G|U$.

ii. there exists a section $r\colon U \to B_G|U$ such that $s\# \circ r\colon U \to W$ is constant.

The name "characteristic" is justified by the following observation: Let the orbit $\{As^{\#}(p) | A \varepsilon G\}$ be a singleton, and let $G_1 \subset G$ be a subgroup of G, then

$$s^{\#-1}\{As^{\#}(p) \mid A \varepsilon G_1\} = B_G \mid U. \qquad (1.5)$$

In other words, G is the maximal group for which the given object enjoys the property (1.4). It should be noted here that a globally defined characteristic field of any G-structure B_G need not necessarily exist and even if it does exist need not be unique. If however, the orbit $\{As^{\#}(p) \mid A \varepsilon G\}$ is a singleton then s is globally defined. This way, a characteristic field of an $\mathcal{O}(n)$ - structure (the orthogonal structure) is a Riemamian metric on the manifold.

The importance of characteristic fields of G-structures lies in the fact that they provide convenient criteria for checking the local flatness (integrability) of these structures, by comparing the characteristic field of a given structure with the characteristic field of the standard flat G-structure. In particular, one can easily prove [10] that,

<u>Proposition 2</u>. Let $B_G|U$ be a "local" G-structure over $U \subset M$. Let s be a characteristic field of $B_G|U$ such that the orbit $\{A\, s^{\#}(p(x)) \mid A \varepsilon G\}$ is a singleton at each $x \varepsilon U$. Then, $B_G|U$ is integrable if and only if there exists an embedding ψ_U of U (a coordinate chart) such that $s^{\#} \circ \psi^{-1}_{U*}$: $U \to W$ is constant, where with some abuse of notation ψ^{-1}_{U*} is understood as a section of $L(U) \equiv L(M)|U$.

2. PHYSICAL CONTEXT

We define a <u>body</u> B as a 3-dimensional compact and orientable differentiable manifold (possibly with boundary) which manifests itself through configurations in a given <u>physical space</u> S, which is an n-dimensional (n≧3) differentiable manifold, without boundary. Any such configuration κ is, by definition, an embedding

$$\kappa: \quad B \to S \tag{2.1}$$

of class $C^k (0 < k < +\infty)$.
The set

$$Q_B = \{\kappa: \quad B \to S\} \tag{2.2}$$

of all possible configurations of a given body B is an infinite dimensional manifold [11] which we shall call the <u>configuration space</u> (in the sequel, where there is no ambiguity, we shall use the notion Q for the configuration space).

An element $\delta\kappa$ of the tangent bundle TQ of the configuration space has the physical meaning of a <u>virtual displacement</u> measured away from the configuration

$$\kappa = \pi_Q \; (\delta\kappa), \tag{2.3}$$

where π_Q: $TQ \to Q$ is the tangent bundle projection. Indeed, such an element can be reinterpreted (see for example [12]) as a mapping

$$\delta\kappa: \quad B \to TS \tag{2.4}$$

of the body into the tangent bundle TS of the physical space such that the diagram

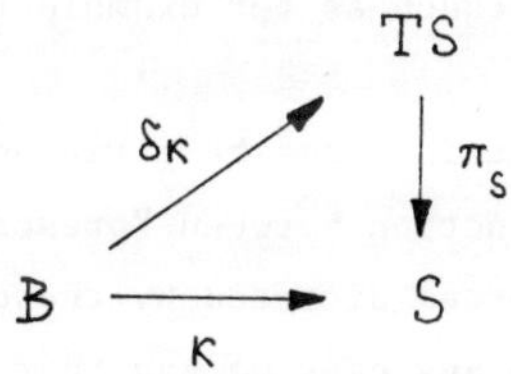

commutes, where π_S is the natural projection of TS. In other words, at a given configuration κ, to each material point $X \in B$, $\delta\kappa$ assigns a vector $\delta\kappa(\underset{\sim}{X}) \in TS$ at a point $\kappa(\underset{\sim}{X})$ in the physical space.

Intuitively, we conceive of a force as an entity which performs work linearly on virtual displacements. From the mathematical viewpoint such a <u>force</u> f is then an element of the cotangent bundle T*Q of the configuration space, that is a co-vector on Q.
A force f is associated thus with the configuration

$$\kappa = \tau_Q(f) \tag{2.6}$$

where τ_Q is the natural projection of T*Q. Given a force f and a virtual displacement $\delta\kappa$ at the same base configuration, i.e. such that

$$\tau_Q(f) = \pi_Q(\delta\kappa), \tag{2.7}$$

the virtual work W of f on $\delta\kappa$ is naturally given by

$$W = \langle f, \delta\kappa \rangle \tag{2.8}$$

where $\langle\ ,\ \rangle$ denotes the evaluation map.
It is worth mentioning that although, as remarked above, a virtual displacement $\delta\kappa$ can be associated with a vector field on a submanifold $\kappa(B)$ S, there is no such natural representation of a force f as a one-form on $\kappa(B)$.

A global <u>elastic constitutive law</u>, completely defining the material behaviour of the body B, is a one form ζ on Q, viz

$$\zeta: \quad Q \to T^*Q\ . \tag{2.9}$$

The use of the term "global" is meant to stress the generality of this constitutive law in which not only the operator ζ depends on the entire configuration κ but also its action on a virtual displacement is non-local in the sense that it involves the entire $\delta\kappa$ rather than any particular characteristics of it (such as for example the value of its jet map at a material point).

As a force f may in certain cases be given as a smooth function of the configuration, the distinction between "prescribed" (external) forces and the "internal" forces dictated by the constitutive laws becomes rather artificial. In any case we say that a configuration κ_o is a state of <u>equilibrium</u> if and only if

$$\zeta(\kappa_o) + f = 0 \tag{2.10}$$

where the force f satisfies also

$$\tau_Q(f) = \kappa_o\ . \tag{2.11}$$

If f is given as a smooth function of configuration, the equilibrium equation may be conceived as a boundary value problem to be solved for the state of equilibrium.

3. CONSTITUTIVE SYMMETRIES AND LOCALITY

A diffeomorphism ψ of the configuration space Q is, by definition, a (global) symmetry of the constitutive law ζ if

$$\psi^* \zeta = \zeta , \tag{3.1}$$

namely, if ζ as a one-form on Q is invariant under ψ. In equation (3.1) the star denotes the pull-back operator, so that more explicitly we can write

$$< \zeta(\psi(\kappa)), \quad T\psi(\delta\kappa) \quad > = < \zeta(\kappa) \ , \quad \delta\kappa > \tag{3.2}$$

for all $\delta\kappa$ and for corresponding κ, where $T\psi$: TQ -> TQ is the tangent map of ψ.
It is not hard to see that the set

$$G_Q = \{\psi \varepsilon \text{ Diff. } Q | \ \psi^* \zeta = \zeta\} \tag{3.3}$$

forms a group under composition law, which we call the global symmetry group of a constitutive law ζ. Two particular sub-groups of G_Q are of interest. Let β be a diffeomorphism of the body B. It obviously induces by composition, a unique diffeomorphism ψ_β on Q. Similarly, a diffeomorphism σ of the physical space S induces a diffeomorphism ψ_σ on Q. The sub-groups G_B and G_S of G_Q defined by

$$G_B = \{\psi \ \varepsilon \text{ Diff.} Q| \ \psi \ \varepsilon \ G_Q \text{ and } \psi = \psi_\beta \text{ for some } \beta\varepsilon \text{ Diff. } B\} \tag{3.4}$$

and

$$G_S = \{\psi \ \varepsilon \text{ Diff.} Q| \ \psi \ \varepsilon \ G_Q \text{ and } \psi = \psi_\sigma \text{ for some } \sigma\varepsilon \text{ Diff. } S\} \tag{3.5}$$

will be called, respectively, the material and spatial global symmetry groups of the constitutive law.

Let VcB be an open set in B (a subbody of B) and denote by $\delta\kappa_V$ any virtual displacement with compact support in V. If we now should restrict the "test functions" in equation (3.2) to such $\delta\kappa_V$, then the group G_Q will in general be expanded to a group $G_Q(V)$ say, which will include more diffeomorphisms ψ preserving the constitutive law under the smaller class of virtual displacements. We may write

$$G_Q(V) = \{\chi \in \text{Diff.}Q \mid < \psi^* \zeta, \ \delta\kappa_V > = < \zeta, \ \delta\kappa_V > \text{ for each } \delta\kappa_V\} \tag{3.6}$$

and we call $G_Q(V)$ the <u>global symmetry group of ζ relative to V B</u>. Let $V_{\underset{\sim}{X}}$ denote now the family of neighborhoods of a material point $\underset{\sim}{X}\varepsilon B$. We define the <u>global symmetry group $G_Q(X)$ of ζ relative to X</u> as the union

$$G_Q(\underset{\sim}{X}) = \bigcup_{V\varepsilon V_{\underset{\sim}{X}}} G_Q(V) \tag{3.7}$$

Thus, $G_Q(\underset{\sim}{X})$ provides us with information about symmetries of a global constitutive law at a fixed material point $\underset{\sim}{X}$. We show now that if this group is large enough the material can be considered to be <u>germ-local</u> in the precise sense that its response to virtual displacements with compact support in some neighborhood is unaffected by the configuration outside that neighborhood. To make precise the notion of just how "big"$G_Q(\underset{\sim}{X})$ must be, we define the global identity group $L_Q(V)$ relative to the open set $V \subset B$ as

$$L_Q(V) = \{\psi \ \varepsilon \ \text{Diff.}Q \mid \psi(\kappa)_V = \kappa|_V \text{ for each } \kappa \ \varepsilon \ Q\} \tag{3.8}$$

where $|_V$ denotes the restriction to V.
It is important to note the existence of non-trivial elements of $L_Q(V)$, such as the ones induced by composition by any diffeomorphism of B which reduces to the identity on V.
By analogy to (3.9) we now define the local identity group $L_Q(\underset{\sim}{X})$ at the point $\underset{\sim}{X}\varepsilon B$ as the union

$$L_Q(\underset{\sim}{X}) = \bigcup_{V\varepsilon V_{\underset{\sim}{X}}} L_Q(V) \tag{3.9}$$

where $V_{\underset{\sim}{X}}$ is again the family of neighborhoods of the point $\underset{\sim}{X}$.

We are now in a position to formalize our notion of locality. We state that a body is <u>germ-local at X</u> if its local identity group at $\underset{\sim}{X}$ is contained in its global symmetry group relative to $\underset{\sim}{X}$.

Germ locality seems to be the least restrictive form of locality and, in a different framework, was introduced by Noll [13]. The stronger and more common notion of locality is what may be called <u>jet-locality of order k</u>.

We shall say that an elastic material as defined by a constitutive law ζ is of <u>local action</u> if there exists a linear mapping P from the space TQ of virtual displacements to the space $\Omega(B)$ of m-forms on B with the property

$$\text{supp } P(\delta\kappa) \subset \text{supp } \delta\kappa \tag{3.10}$$

and such that for any given configuration κ and any corresponding virtual displacement $\delta\kappa$ the virtual work of "internal forces" $\zeta(\kappa)$ on $\delta\kappa$ is given by

$$< \zeta(\kappa), \delta\kappa > = \int_B P(\delta\kappa) \tag{3.11}$$

Here we have ignored a possible contribution from the boundary of B.

Note, that P represents a "density" of work and the evaluation $P(\delta\kappa)$ at a material point represents the contribution of that point towards the total virtual work. The condition (3.10) is, however, crucial to ensure that work is not assigned to a point unless there is a non-vanishing virtual displacement on a neighborhood of it. It is also interesting to realize that due to the assumption (3.10) the linear operator P: TQ -> $\Omega(B)$ becomes a linear differential operator on B. Consequently, P is locally generated by the derivatives of virtual displacement, [14].

The local action P in (3.11) may still depend at any given material point $\underset{\sim}{X} \in B$ on the configuration of points away from it. If, however, the local action is such that there exists a mapping σ: $J^k TQ \to \Lambda^* B$, called a <u>local response functional</u>, such that for each virtual displacement $\delta\kappa$ and each material point $\underset{\sim}{X} \in B$

$$P(\delta\kappa)(\underset{\sim}{X}) = \sigma(j^k\, \delta\kappa(\underset{\sim}{X})) \tag{3.12}$$

and where $j^k\, \delta\kappa$ is the section of the k-th jet fibre bundle $J^k(B,TS)$ induced by $\delta\kappa \in C^k(B,TS)$, then we say that the material is <u>jet-local of order k</u>.

As an example we consider now a jet-local hyperelastic materials, i.e., materials with an elastic potential θ such that $\zeta = d\theta$. In general, for a global hyperelastic material, we have

$$< \zeta, \delta\kappa > = < d\theta, \delta\kappa > = \delta\kappa(\theta) \tag{3.13}$$

which means that the virtual work is given by the Fréchet differential of θ in the direction $\delta\kappa$. We say that a hyperelastic material has a <u>local potential</u> if there exists a scalar function

$$\Phi: B \times Q -> R \tag{3.14}$$

such that

$$\theta = \int_B \Phi \, v_B \tag{3.15}$$

where v_B is a volume element on B.
Therefore, the hyperelastic local action P is at each point $\underset{\sim}{X} \in B$ given by

$$P(\delta\kappa)\ (\underset{\sim}{X}) = \delta\kappa(\Phi(\underset{\sim}{X},\kappa),\underset{\sim}{X}) \tag{3.16}$$

If the local potential Φ is such that

$$\Phi(\underset{\sim}{X},\kappa) = \Phi(\underset{\sim}{X},\ j^k\ \kappa(\underset{\sim}{X})) \tag{3.17}$$

then we say that the material is jet-local of order k as the response functional σ defined previously in (3.12) has a form

$$\sigma(j^k\ \delta\kappa(\underset{\sim}{X})) = \delta\kappa(\Phi(\underset{\sim}{X},j^k\ \kappa(\underset{\sim}{X})),\underset{\sim}{X}) \tag{3.18}$$

for a given material point $\underset{\sim}{X} \in B$.
The virtual work is now given by a first variation

$$< \zeta(\kappa)\delta\kappa > = \int_B \sum_{j=0}^{k} \frac{\partial\phi}{\partial\kappa^i},\ \alpha_1\ldots\alpha_j\ \ \delta(\kappa^i,\ \alpha_1\ldots\alpha_j)v_B\ . \tag{3.19}$$

4. MATERIAL UNIFORMITY

A body B is understood to be materially uniform if its points are made of the same material. In the context of a completely global theory, this way of formulating the idea of uniformity is problematic since it presupposes locality. The key idea, however, is that of moving one piece of the body into another and then checking for the local response. We make this idea more precise by suggesting the following definition: we say that the point $\underset{\sim}{Y} \varepsilon B$ is materially isomorphic with the point $X \varepsilon B$, if there exists a diffeomorphism β of B such that

$$\underset{\sim}{X} = \beta(\underset{\sim}{Y}) \tag{4.1}$$

and

$$\psi_B \varepsilon\, G_Q(Y) \tag{4.2}$$

where ψ_β is a diffeomorphism of Q generated by β. It is not difficult to show reflexivity and transitivity of the above definition.

We are now in a position to adopt the standard [6] definition of uniformity: a body is materially uniform if and only if all its points are pairwise materially isomorphic. Note that because of the required compactness of B, this definition may imply some physically unreasonable behaviour in an "boundary" layer of a truly global body, particularly when the dimensions of B and S are the same. At this level of generality, however, no precise statement can be made to this effect. For a jet-local material of order 1, on the other hand, we can rephrase the standard definition of material isomorphism as follows. We say that two material points, $\underset{\sim}{X}$ and $\underset{\sim}{Y}$ say, of an elastic body B with the local response functional σ are materially isomorphic if and only if there exists a linear isomorphism

$$L_{\underset{\sim}{X}\underset{\sim}{Y}}: \; T_{\underset{\sim}{Y}}B \rightarrow T_{\underset{\sim}{X}}B \tag{4.3}$$

such that

$$\sigma(\delta(j^1\, \kappa o L_{\underset{\sim}{X}\underset{\sim}{Y}})) = L^*_{XY}\, \sigma(j^1 \delta\kappa(\underset{\sim}{Y})) \tag{4.4}$$

for all $\kappa \varepsilon Q$, where $L^*_{\underset{\sim}{X}\underset{\sim}{Y}}$ is induced by $L_{\underset{\sim}{X}\underset{\sim}{Y}}$.

If in addition, $L_{\underset{\sim}{X}\underset{\sim}{Y}}$ is smoothly distributed over B the body is called smoothly materially uniform. In particular, the material isomorphisms $L_{\underset{\sim}{X}\underset{\sim}{X}}$ are called material symmetries at the point $\underset{\sim}{X} \varepsilon B$ and are assumed for physical reasons to be elements of the special linear group SL(3). It can easily be shown that the set of material symmetries at the arbitrary material point $\underset{\sim}{X} \in B$ form a subgroup of SL(3) and that the <u>intrinsic material symmetry groups</u> $G_{\underset{\sim}{X}} = \{L_{\underset{\sim}{X}\underset{\sim}{X}}\}$ of a uniform body are all isomorphic. Indeed, let $i_{\underset{\sim}{X}}: T_{\underset{\sim}{X}}B \to V_3$ be an isomorphism of the fibre $T_{\underset{\sim}{X}}B$ with V_3 (the translation space of R^3). For a uniform body B we can define at each point $\underset{\sim}{Y} \in B$, the isomorphism $i_{\underset{\sim}{Y}} = i_{\underset{\sim}{X}} \circ L_{\underset{\sim}{Y}\underset{\sim}{X}}: T_{\underset{\sim}{Y}}B \to V_3$. Then the group

$$G \equiv i_{\underset{\sim}{X}} \circ G_{\underset{\sim}{X}} \circ i_{\underset{\sim}{X}}^{-1} \tag{4.5}$$

will be called a <u>typical isotropy group relative to i</u> and, as defined, is obviously independent of position in the body. One can now show [10],

<u>Proposition 3</u>. For a uniform material body B the set $Q_G = \{L_{\underset{\sim}{Y}\underset{\sim}{X}}p \mid \underset{\sim}{Y} \in B\}$, of all linear frames on B obtained from a given frame $p(\underset{\sim}{X})$ at a given $\underset{\sim}{X} \in B$ by the action of all material isomorphisms $L_{\underset{\sim}{Y}\underset{\sim}{X}}$, is a G-structure over B whose structure group G is a typical isotropy group of B.

Let us now take an open, simply connected subbody $U \subset B$ and a trivialization of its tangent bundle TB relative to U, i.e., a smooth map

$$\mu_U: T_UB \to U \times V_3. \tag{4.6}$$

Such a trivialization is called a <u>material chart</u> of the uniform material body if for any $\underset{\sim}{X}, \underset{\sim}{Y} \in U$

$$\mu_U(\underset{\sim}{X}) = \mu_U(\underset{\sim}{Y}) \circ L_{\underset{\sim}{Y}\underset{\sim}{X}} \tag{4.7}$$

where $L_{\underset{\sim}{Y}\underset{\sim}{X}}$ is a smooth field of material isomorphisms on U and $\mu_U(\underset{\sim}{Z}): T_{\underset{\sim}{Z}}B \to V_3$ is a vector space isomorphism for all $\underset{\sim}{Z} \in U$. We can recall after [7] that a (smooth) uniform elastic body U is said to be <u>homogeneous</u> if there exists a material chart μ_U and a configuration $\kappa_U: U \to S$ such that

$$\mu_U(\underset{\sim}{X}) = \kappa_{U*}^{-1}(\underset{\sim}{X}) \tag{4.8}$$

for every $\underset{\sim}{X} \in U$.

Accordingly, an elastic uniform body B is called locally homogeneous if for every material point $\underset{\sim}{X}$ there exists a neighborhood (a subbody) U of $\underset{\sim}{X}$ which is homogeneous. Any configuration κ_U of U such that κ_{U*}^{-1} is a material chart on U is called a homogeneous configuration of U. Taking into account the above definitions as well as Propositions 1 & 3 it follows immediately that

Theorem 1. A uniform elastic body B with the typical isotropy group G is locally homogeneous if and only if the material G-structure Q_G is locally flat (integrable)*.

In the light of the above theorem it is interesting to realize as pointed out in [5] that for a jet-local material global uniformity is an intermediate stage between local uniformity and local homogeneity. Indeed, if $p = (u_1 \ldots, u_n): U \to B_G|U$ is a local field of frames (a material chart) and if the structure functions C^k_{ij} of p are defined on B by

$$[\underline{u}_i, \underline{u}_j](\underset{\sim}{X}) = C^k_{ij}(\underset{\sim}{X})\, \underline{u}_k(\underset{\sim}{X}) \qquad (4.9)$$

then it can be shown [8] that the local uniformity implies global uniformity if there exists a local material chart such that its structure functions are constant on B. If in addition, one can find a material chart such that $C^k_{ij} \equiv 0$ then according to Definition 1 the body is locally homogeneous.

5. HOMOGENEITY CRITERIA

To illustrate the application of these ideas to the actual solution of a problem in Continuum Mechanics we show explicit homogeneity criteria for two important classes of materials.

We start with the isotropic solid.

---

* As the typical isotropy groups are material chart dependent (see(4.5)) the respective material structures Q_G are not unique. They are however conjugate one to each other.

<u>Definition 3</u>. An <u>Isotropic Elastic Solid</u> is a uniform elastic body for which there exists a material chart such that a typical isotropy group relative to this chart is the group of all rotations in R_3, that is, the proper orthogonal group SO(3). This special material chart is called <u>undistorted</u>.

Any $\mathcal{O}(3)$ - structure is just a choice of a Riemann metric $\tilde{g}$ on B. Specifically, $\tilde{g}$ is the pull-back of the cartesian inner product through an undistorted chart. On the other hand, the Riemann metric chooses the bundle of orthogonal frames. The characteristic field of any $S\mathcal{O}(3)$ - structure is a Riemann matric on B and the Riemannian volume element associated with it.

As it is shown in [15] an $\mathcal{O}(3)$ - structure is locally flat if and only if the Riemann curvature tensor of the corresponding Riemannian metric vanishes. In 3-dimensions, however, the Riemann curvature tensor is proportional to the Ricci tensor. Hence, in order to be able to determine whether a given $\mathcal{O}(3)$ - structure is locally flat it suffices to verify whether the Ricci tensor of the characteristic Riemannian metric vanishes. In the light of the fact that $S\mathcal{O}(3) = \mathcal{O}(3) \cap SL(3)$, that any SL(3) - structure is locally integrable, [5,16] and that the Riemannian volume element is parallel with respect to any Riemannian parallel transport, we can claim that the $S\mathcal{O}(3)$ - structure is locally flat if and only if the respective $\mathcal{O}(3)$ - structure is locally flat. Therefore,

<u>Proposition 3</u>. A uniform isotropic elastic solid body is locally homogeneous if and only if the Ricci tensor of its intrinsic (material) metric, that is the one generated by its material $S\mathcal{O}(3)$ - structure, vanishes.

A non-trivial reduction of the orthogonal structure is the so-called transversely isotropic elastic solid.

<u>Definition 4</u>. A <u>Transversely Isotropic Elastic Solid</u> is a uniform isotropic elastic body with an intrinsic field of unit vectors $\tilde{e}$ in the tangent space of the body manifold.

The typical isotropy group of this material is the $S\Theta(1) \times S\Theta(2)$ group, that is, the group of all rotations about the fixed axis $\underset{\sim}{e}$. Such a G - - structure defines a Riemannian structure and an orientation (Riemannian volume element) on B. This Riemannian structure (called sometimes the associated Riemannian structure) can also be viewed as the embedded $S\Theta(2)$ - structure. The characteristic fields of the $S\Theta(1) \times S\Theta(2)$ - structure are given by the associated metric $\underset{\sim}{H}$, its volume element, and the distribution e. It can now be proven that

<u>Proposition 4</u>. The transversely isotropic uniform elastic structure is locally homogeneous (integrable) if and only if its associated Riemannian structure is locally flat, and the intrinsic vector field $\underset{\sim}{e}$ is covariantly constant (holonomic structure [17]) with respect to the associated Levi-Civita connection.

<u>Proof</u>. As it is evident from [16], the orientation of the structure causes no trouble as it gives no new compatibility criteria for the local flatness. Now, if the given $S\Theta(1) \times S\Theta(2)$ - structure is locally flat then there exists a trivialization of Q_G such that the vector field $\underset{\sim}{e}$ as well as the associated metric $\underset{\sim}{H}$ have constant components. But then the Riemann curvature tensor of $\underset{\sim}{H}$ vanishes and the associated structure $S\Theta(2)$ is integrable. On the other hand, if the associated Riemannian structure is locally flat, then there exists a local chart with respect to which $\underset{\sim}{H}$ has constant components. Since e is parallel with respect to the Levi-Civita connection of $\underset{\sim}{H}$, then it has also constant components in this chart. Hence, according to proposition 1, the material $S\Theta(1) \times S\Theta(2)$ - structure is locally flat.

ACKNOWLEDGEMENT

This work has been supported through the operating grants from the Natural Sciences and Engineering Research Council of Canada.

REFERENCES

1. Truesdell, C. and Noll, W., "The Nonlinear Field Theories of Mechanics", Encyclopedia of Physics, Vol. 111/3, Springer-Verlag, Berlin Heidelberg New York (1965).

2. Wang, C.C. and Truesdell, C., "Introduction to Rational Elasticity", Noordhoff International Publishing, Leyden (1973).

3. Epstein, M. and Segev, R., "Differentiable Manifolds and the Principle of Virtual Work in Continuum Mechanics", J. Math. Phys. 21(5), 1243-1245 (1980).

4. Elżanowski, M., and Epstein, M., "Geometric Characterization of Hyperelastic Uniformity", Arch. Rational Mech. Anal., 88, 347-357 (1985).

5. Epstein, M., Elżanowski, M. and Śniatycki, J., "Locality and Uniformity in Global Elasticity", Lecture Notes in Mathematics No. 1139, 300-310, Springer-Verlag (1985).

6. Noll, W., "Materially Uniform Simple Bodies with Inhomogeneities", Arch. Rational Mech. Anal., 27, 1-32 (1967).

7. Wang, C.C., "On the Geometric Structure of Simple Bodies, a Mathematical Foundation for the Theory of Continuous Distribution of Dislocations", Arch. Rational Mech. Anal., 27, 33-94 (1967).

8. Sternberg, S., "Lectures on Differential Geometry", Chelsea Publishing Company, N.Y. (1983).

9. Hirsch, M.W., "Differential Topology", Springer-Verlag, New York, Heidelberg, Berlin (1976).

10. Elżanowski, M., Epstein, M. and Śniatycki, J., "G-Structures and Material Homogeneity", Departmental Report No. 386, Dept. of Mechanical Engineering, University of Calgary, March (1987).

11. Michor, R.W., "Manifolds of Differentiable Mappings", Shiva, London (1980).

12. Ebin, D.G. and Marsden, J., "Groups of Diffeomorphisms and the Motion of an Incompressible Fluid", Annals. Math., 92, 102-163 (1970).

13. Noll, W., "A Mathematical Theory of the Mechanical Behaviour of Continuous Media", Arch. Rat. Mech. Anal. 2, 197-226 (1958).

14. Kahn, D.W., "Introduction to Global Analysis", Academic Press, New York (1980).

15. Guillemin, V., "The Integrability Problem for G-Structures", Trans. Amerc. Math. Soc., 116, 544-560 (1965).

16. Moser, J., "On the Volume Elements of a Manifold", Trans. Amer. Math. Soc. 120, 286-294 (1965).

17. Chern, S.S., "The Geometry of G-Structures", Bull. Amer. Math. Soc. 72, 167-219 (1966).

GEOMETRY AND TOPOLOGY (pp. 152–158)
edited by G. M. Rassias & G. M. Stratopoulos

An analytic computation of the additive structure of $MU_*(BZ_4)$

Peter B. Gilkey *
University of Oregon
Eugene Oregon 97403 (USA)

The eta invariant of Atiyah et al [1] is an R/Z valued invariant of equivariant unitary bordism which completely detects $MU_*(BG)$ for spherical space form groups. We use the eta invariant to compute the additive structure of $MU_*(BZ_4)$.

A finite group G is called a spherical space form group if it admits a fixed point free unitary representation; we restrict to such G henceforth. These groups have been classified by Wolf [13]. Let $MU_*(BG)$ be the reduced equivariant unitary bordism groups; this is often denoted by $\tilde{\Omega}_*^U(BG)$ and is a module over the coefficient ring MU_*. Let R(G) be the representation ring of G, let $R_0(G)$ be the augmentation ideal, and let R(U) be the representation ring of the unitary group. If a manifold M defines an element of $MU_*(BG)$ and if $\theta \in R_0(G) \otimes R(U)$, let $\eta(\theta,M) \in R/Z$ be the eta invariant corresponding to the tangential operator of the Dolbeault complex on M with coefficients in the bundle defined by θ. See [1,8] for further details. The map $(\theta,M) \rightarrow \eta(\theta,M)$ extends to a map in bordism:

$$\eta: R_0(G) \otimes R(U) \otimes MU_*(BG) \rightarrow R/Z.$$

Lemma 1: Let G be a spherical space form group. If $M \in MU_*(BG)$ and if $\eta(\theta,M)=0$ for all $\theta \in R_0(G) \otimes R(U)$, then $M=0$.

This follows from the arguments given in [3,7] for $MSpin_*^c(BG)$; the argument for $MU_*(BG)$ is simpler than the argument for $MSpin_*^c(BG)$ since the coefficient ring MU_* is without torsion. See also Wilson [12].

The Hattori-Stong theorem states that all relations among characteristic classes follow from the index theorem. If $\psi \in R(U)$ and $M \in MU_*$, let $\text{index}(\psi,M) \in Z$ be the index of the Dolbeault complex with coefficients in the bundle defined by ψ. If $n>0$ and $M \in MU_*$, then $M \in n \cdot MU_*$ if and only if $\text{index}(\psi,M) \equiv 0 \bmod n$ for all $\psi \in R(U)$ so $\text{index}(\psi,M)$ completely detects $MU_* \otimes Z_n$. The ordinary index and the eta invariant are closely related. If $M=M_1 \times M_2$ for $M_1 \in MU_*(BG)$ and $M_2 \in MU_*$ then $\eta(-,M)$ can be computed in terms of $\eta(-,M_1)$ and $\text{index}(-,M_2)$ (see [3,7]). Lemma 1 can be thought of as the generalization of

---

1980 Mathematics Subject Classification (1985 Revision). 58G12
*Research partially supported by NSF grant DMS8614715

the Hattori-Stong theorem to equivariant unitary bordism. The Hattori-Stong theorem is an essential ingredient in the proof of Lemma 1, but can also be deduced from Lemma 1.

The Anderson-Brown-Peterson splitting expresses $MSpin_*^c(BG)$ in terms of the connective K-theory bu_* and ordinary homology at the prime 2:

$$MSpin_*^c(BG)_{(2)}=\{bu_*(BG)\otimes Z[x_4,x_6,\ldots]\oplus\tilde{H}_*(BG;Tor(MSpin^c))\}_{(2)}$$

(see [3,4,7]). The corresponding splitting of the spectrum $\underline{MU}$ or $\underline{MSpin^c}$ odd primes, is in terms of Brown-Peterson homology BP_* and not bu_*. We conjectured several months ago that never-the-less there is an <u>additive</u> splitting of the form:

$$MU_*(BG)\approx bu_*(BG)\otimes Z[x_4,x_6,\ldots].$$

A topological proof of this conjecture has been given recently by Bahri et al [2]; we also refer to B.Flynn [5] for cyclic case. The additive structure of $bu_*(BG)$ is known explicitly (see [2] for details) so this gives the exact additive structure of these groups. For example, we refer to Fujii et al [6] for $bu_*(BZ_n)$.

In this note, we will use the eta invariant to give an analytic computation of the special case $G=Z_4$; a similar approach computes $MU_*(BQ_8)$ where $Q_8=\{\pm1,\pm i,\pm j,\pm k\}$ (see [10]). Our approach here is totally different from [2] and we hope will lead to new insights into these matters. This approach was what lead us to the conjecture in the first instance and provided motivation and insight into the general problem. Eventually we hope to generalize these methods to give a new proof of the additive splitting based on analytic methods instead of the topology of [2].

Let $Z_n=\{\lambda\in C:\lambda^n=1\}$ be the cyclic group of order n and let $\rho_s(\lambda)=\lambda^s$ be the irreducible representations of Z_n for $0\leq s\leq n$. Let $\vec{q}=(q_1,\cdots,q_k)$ be integers coprime to n. Let $\tau_{\vec{q}}(\lambda)=\mathrm{diag}(\lambda^{q_1},\cdots,\lambda^{q_k})$ be a fixed point free representation of Z_n to $U(k)$. Let $L(n;\vec{q})=S^{2k-1}/\tau_{\vec{q}}(Z_n)$ be the corresponding lens space. Let V_s be the complex line bundle over $L(n;\vec{q})$ corresponding to the representation ρ_s. Let $K(L(n;\vec{q}))$ and $\tilde{K}(L(n;\vec{q}))$ be the complex K-theory and reduced complex K-theory groups. Extend the map $\rho_s\rightarrow V_s$ by linearity to define ring homomorphisms

$$\pi:R(Z_n)\rightarrow K(L(n;\vec{q}))\quad\text{and}\quad\pi:R_0(Z_n)\rightarrow\tilde{K}(L(n;\vec{q})).$$

Then it is well known that:

$$bu_{2k-3}(BZ_n)\approx R_0(Z_n)/R_0(Z_n)^k\approx\tilde{K}(L(n;\vec{q}))$$

which relates the K-theory, representation theory, and the connective K-theory for the cyclic groups; there are similar formulas for general spherical space form groups. Thus the conjecture expresses the equivariant bordism in terms of the representation theory.

We give $L(n;\vec{q})$ a stable complex structure by means of the standard isomorphism of the underlying real vector bundle of $\psi(\tau)$ with the stable tangent bundle $T(L(n;\vec{q}))\oplus 1$. Since $L(n;\vec{q})$ is odd dimensional, it bounds in MU_{2k-1} so $L(n;\vec{q})$ belongs to the reduced bordism group $MU_{2k-1}(BZ_n)$. The $\{L(n;\vec{q})\}$ generate $MU_*(BZ_n)$ as an MU_* module.

Lemma 2: Let $\rho\in R_0(Z_n)$ and let $\vec{q}=(q_1,\ldots,q_k)$. Then

$$\eta(\rho,L(n;\vec{q}))=n^{-1}\cdot\Sigma_{\lambda^n=1,\lambda\neq 1}\mathrm{Tr}(\rho(\lambda)\cdot\det(\tau_{\vec{q}}(\lambda))/\det(\tau_{\vec{q}}(\lambda)-I).$$

Proof: The corresponding calculation for the tangential operator of the signature complex appears in Atiyah et al [1,II-(2.9)]; the calculations for the tangential operator of the Dolbeault complex are the same; this expresses the eta invariant in terms of Dedekind sums.

The eta invariant is closely related to the K-theory groups.

Lemma 3: Let $\rho\in R_0(Z_n)$ and let $\vec{q}=(q_1,\cdots,q_k)$.
(a) $\rho\in R_0(Z_n)^k$ if and only if $\eta(\rho\otimes\sigma,L(n;\vec{q}))=0$ for all $\sigma\in R_0(Z_n)$.
(b) The map $\rho\rightarrow V_\rho$ is an isomorphism from $R_0(Z_n)/R_0(Z_n)^k$ to $\tilde{K}(L(n;\vec{q}))$.
(c) The map $\rho\otimes\tau\rightarrow\eta(\rho\otimes\tau,L(n,\vec{q}))$ extends to a non-singular bilinear R/Z valued form $\eta:\tilde{K}(L(n;\vec{q}))\otimes\tilde{K}(L(n;\vec{q}))\rightarrow R/Z$.

Proof: This lemma provides the link between K-theory and equivariant bordism we will need. We refer to Gilkey [9].

We embed a copy of $bu_*(BZ_4)$ in $MU_*(BZ_4)$ as follows. Let $i:Z_2\rightarrow Z_4$ be the natural inclusion and let $i\cdot L(2;1,1,\ldots,1)$ have the induced Z_4 structure. Define $N_i(k)\in MU_{2k-1}(BZ_4)$ and $x_i\in R_0(Z_4)$ for $i=1,2,3$ by:

$$x_1=(\rho_0-\rho_1)=(\rho_3-\rho_0)/\rho_3\ ,\ x_2=(\rho_0-\rho_3)=(\rho_1-\rho_0)/\rho_1\ ,\ x_3=\rho_0+\rho_2-\rho_1-\rho_3$$

$$N_1(k)=L(4;1,1,\ldots,1)\quad,\quad N_2(k)=L(4;3,1,\ldots,1)\quad,\quad N_3(k)=i\cdot L(2;1,1,\ldots,1)$$

The $\{x_1,x_2,x_3\}$ form a basis for $R_0(Z_4)$. Let $A_{2k-1}=\mathrm{span}_Z\{N_i(k)\}\subseteq MU_{2k-1}(BZ_4)$. If $\rho=\Sigma_i n_i\cdot x_i$, let $f_k(\rho)=\Sigma_i n_i N_i(k)$ define $f_k:R_0(Z_4)\rightarrow A_{2k-1}$.

Lemma 4: Let ρ and σ belong to $R_0(Z_4)$.

(a) $\eta(\sigma, f_k(\rho)) = \eta(\sigma\rho, N_2(k+1))$.

(b) $\rho \in \ker(f_k)$ **if and only if** $\rho \in R_0(Z_4)^{k+1}$. **Thus** $A_{2k-1} \approx R_0(Z_4)/R_0(Z_4)^{k+1} \approx bu_{2k-1}(BZ_4)$.

Proof: Fix an integer k. By Lemma 2:

$$\eta(\sigma, N_1(k)) = \eta(\sigma\cdot(\rho_3 - \rho_0)/\rho_3, N_2(k+1)) = \eta(\sigma \cdot x_1, N_2(k+1))$$

and similarly $\eta(\rho, N_i(k)) = \eta(\rho\cdot x_i, N_2(k+1))$ for $i=2,3$. This proves (a). To prove (b), suppose first $f_k(\rho)=0$. Then $\eta(\tau\cdot\rho, N_2(k+1)) = \eta(\tau, f_k(\rho)) = 0$ for all $\tau \in R_0(Z_4)$ and by Lemma 3, $\rho \in R_0(Z_4)^{k+1}$.

To establish the converse is more difficult. Let

$$\ker(\eta) = \{M \in MU_*(BZ_4) : \eta(\theta, M) = 0 \forall \theta \in R_0(Z_4) \otimes R(U)\}.$$

We show $f_k(\rho)=0$ by showing $f_k(\rho) \in \ker(\eta) = 0$. Let $\theta = \sigma \otimes \psi$ for $\sigma \in R_0(Z_4)$ and $\psi \in R(U)$. If $\psi = 1$,

$$\eta(\sigma, f_k(\rho)) = \eta(\sigma\rho, N_2(k+1)) = 0$$

by Lemma 3(a). To deal with the general case, we must get rid of the dependence on $\psi \in R(U)$. Suppose for the sake of illustration that $\psi = (\Lambda^k)^a \in R(U)$. Then

$$\psi(N_1(k)) = (\rho_{ak}\cdot\rho_a)(N_1(k)) \quad , \quad \psi(N_2(k)) = (\rho_{ak}\cdot\rho_{3a})(N_2(k)) \quad ,$$
$$\psi(N_3(k)) = (\rho_{ak}\cdot\rho_a)(N_3(k)).$$

More generally, we can find $\alpha(a,k) \in R(Z_4)$ so that

$$\psi(N_1(k)) = (\Sigma_a \alpha(a,k)\rho_a)(N_1(k)), \psi(N_2(k)) = (\Sigma_a \alpha(a,k)\rho_{3a})(N_2(k)), \psi(N_3(k))$$
$$= (\Sigma_a \alpha(a,k)\rho_a)(N_3(k)).$$

By replacing σ by $\sigma\cdot\alpha(a,k)$ and studying each term separately, we may assume without loss of generality that

$$\theta(N_1(k)) = (\sigma\cdot\rho_a)(N_1(k)), \theta(N_2(k)) = (\sigma\cdot\rho_{3a})(N_1(k)), \theta(N_3(k)) = (\sigma\cdot\rho_a)(N_3(k)).$$

for $\sigma \in R_0(Z_4)$. To apply Lemma 3 to this setting, let $x_1(a) = \rho_a \cdot x_1$, $x_2(a) = \rho_{3a}\cdot x_2$, and $x_3(a) = \rho_a \cdot x_3$. Decompose $\rho = \Sigma_i n_i x_i$ and define $x(a) = \Sigma_i n_i \cdot x_i(a)$ so

$$\eta(\theta, f_k(\rho)) = \eta(\sigma\cdot x(a), N_2(k+1)).$$

If we can show $x(a) \in R_0(Z_4)^{k+1}$, then $\eta(\theta, f_k(\rho)) = 0$ as before which will complete the proof. Since $x(a) = \rho_a\cdot\rho + \rho_a\cdot(\rho_{2a} - \rho_0)\cdot n_2 \cdot x_2$ we must show $(\rho_{2a} - \rho_0)\cdot n_2 \cdot x_2 \in R_0(Z_4)^{k+1}$. The case $a=0,2$ is immediate so we may assume $a=1,3$. $R_0(Z_4)$ is invariant under the involution sending $\rho_s \to \rho_s^* = \rho_{-s}$. Thus

$$\rho=n_1(\rho_0-\rho_1)+n_2(\rho_0-\rho_3)+n_3(\rho_0+\rho_2-\rho_1-\rho_3)\in R_0(Z_4)^{k+1}$$

$$\rho^*=n_1(\rho_0-\rho_3)+n_2(\rho_0-\rho_1)+n_3(\rho_0+\rho_2-\rho_1-\rho_3)\in R_0(Z_4)^{k+1}$$

$$\rho_1\cdot\rho^*=n_1(\rho_1-\rho_0)+n_2(\rho_1-\rho_2)+n_3(\rho_1+\rho_3-\rho_0-\rho_2)\in R_0(Z_4)^{k+1}$$

$$\rho+\rho_1\rho^*=n_2(\rho_0+\rho_1-\rho_2-\rho_3)=n_2(\rho_0-\rho_2)(\rho_0-\rho_3)=n_2(\rho_0-\rho_2)x_2\in R_0(Z_4)^{k+1}.$$

which completes the proof.

The Smith homomorphism is an MU_* module morphism

$$\Delta:MU_{2k-1}(BZ_n)\rightarrow MU_{2k-3}(BZ_n)\rightarrow 0.$$

with $\Delta(N_i(k))=N_{i-1}(k)$ and $\ker(\Delta)=L^1(4;1)\cdot MU_*$ (see [3]). The coefficient ring $MU_*=Z[x_2,x_4,...]$ is a polynomial ring with generators in even dimensions. Let $x_2=CP^1$ and choose x_{2i} for $i>1$ so the arithmetic genus or Todd genus $Td(x_{2i})=0$. The major result of this note is the following

Theorem 5: Let $P_*=Z[x_4,x_6,...]$ so $MU_*=P_*[x_2]$. Let S_* be the P_* submodule of $MU_*(BZ_4)$ generated by $A_*=\mathrm{span}_Z\{N_i(-))\}=bu_*(BZ_4)$.
(a) $L(4;1)\times(CP^1)^k\in S_{2k+1}$.
(b) $MU_*(BZ_4)=S_*(BZ_4)$.
(c) $|A_{2k-1}|=4^k$.
(d) Cartesian product gives an isomorphism $A_*\otimes P_*\approx MU_*(BZ_4)$.

Proof:: We prove (a) by induction; it is immediate if $k=0$. Let $\tau=k\cdot\rho_1\oplus\rho_3$ and let $x=\det(\tau-I)/\det(\tau)\in R_0(Z_4)^{k+1}$. Set $M=f_{k+1}(x)\in A_{2k+1}$. Then

$$\eta(\rho_1-\rho_0,M)=\eta((\rho_1-\rho_0)\cdot x,N_2(k+2))=\eta(\rho_1-\rho_0,L(4;1))=-1/4$$

by Lemmas 2, 3, and 4. Since $x\in R_0(Z_4)^{k+1}$ and since $\Delta(f_{k+l}(x))=f_k(x)$, $\Delta(M)=0$. Since $\ker(\Delta)=L^1(4;1)\times MU_*)$, we may decompose

$$M=L(4;1)\{a\cdot X_2^k+\Sigma_{v>0}X_2^{k-v}\cdot Y_v\}$$

for $Y_v\in P_v$. Using the multiplicative properties of the eta invariant with respect to such products (see [8] for example), we compute

$$\eta(\rho_1-\rho_0,M)=\eta(\rho_1-\rho_0,L(4;1))\{a\cdot Td(X_2)^k+\Sigma_{v>0}Td(X_2)^{k-v}\cdot Td(Y_v)\}$$

so $\eta(\rho_1-\rho_0,M)=-a/4$ since $Td(X_2)=1$ and $Td(Y_v)=0$. This shows $a\equiv 1(4)$. By induction, $L(4;1)\cdot X_2^{k-v}\in S_*$ so the remaining terms belong to S_* and $a\cdot L(4;1)\times X_2^k\in S_*$. Since $a\equiv 1(4)$ and since $L(4;1)$ is an element of order 4 in $MU_1(BZ_4)=Z_4$, this proves (a). Since $\Delta(N_i(k))=N_i(k-1)$, we see $\Delta:S_{2k-1}\rightarrow S_{2k-3}\rightarrow 0$. By (a) $\ker(\Delta)=L^1(4;1)\cdot MU_*\subseteq S_*$. We use induction and the five-Lemma to conclude $S_*=MU_*(BZ_4)$ which proves (b). (c) follows from the identification of $A_{2k-1}=R_0(Z_4)/R_0(Z_4)^{k+1}$ and from the calculation $|R_0(Z_4)/R_0(Z_4)^{k+1}|=4^k$ given in [9,Lemma 4.2].

Cartesian product gives a surjective map $A_*\otimes P_*\to MU_*(BZ_4)$. We show this is an isomorphism and complete our determination of the additive structure of $MU_*(BZ_4)$ by comparing the orders of the groups involved. As the bordism spectral sequence for $MU_*(BZ_4)$ collapses by Landweber [11].

$$\begin{aligned}
&|MU_{2k-1}(BZ_4)|=|\oplus_{c\le k}\tilde{H}_{2c-1}(BZ_4;MU_{2k-2c})|\\
&=|\oplus_{c\le k}\tilde{H}_{2c-1}(BZ_4;\oplus_{d\le k-c}P_{2d})|\\
&=|\oplus_{c+d\le k}\tilde{H}_{2c-1}(BZ_4;P_{2d})|=|\oplus_{c\le a}\tilde{H}_{2c-1}(BZ_4;\oplus P_{2k-2a})|\\
&=|\oplus_a A_{2a-1}\otimes P_{2k-2a}|=|\{A_*\otimes P_*\}_{2k-1}|
\end{aligned}$$

which completes the proof.

References

[1] M.F.Atiyah, V.K.Patodi, and I.M.Singer, Spectral asymmetry and Riemannian geometry I. Math. Proc. Camp. Phil. Soc. , vol. 77 (1975), 43-69; II, vol. 78 (1975), 405-432; III, vol. 79 (1976), 71-99.

[2] A.Bahri, M.Bendersky, and D.Davis, The complex bordism of groups with periodic cohomology (to appear).

[3] A.Bahri and P.Gilkey, The eta invariant, Pin^c bordism, and equivariant Spin^c bordism for cyclic 2-groups. Pacific J Math vol. 128 (1987), 1-24.

[4] A.Bahri and P.Gilkey, Pin^c bordism and equivariant Spin^c bordism of cyclic 2-groups, Proceedings of the AMS, vol. 99 (1987), 380-382.

[5] B.Flynn, The complex bordism of cyclic groups, Osaka J. Math 11 (1974), 503-516.

[6] K.Fujii, T.Kobayashi, K.Shimomura, M.Sugawara, KO groups of lens spaces modulo powers of two. Hiroshima Math J 8 (1978), 469-489.

[7] P.Gilkey , The eta invariant and equivariant Spin^c bordism for spherical space form groups (to appear).

[8] P.Gilkey, Invariance Theory, The Heat Equation, and the Atiyah-Singer Index Theorem. Publish or Perish (1984).

[9] P.Gilkey, The eta invariant and the K-theory of spherical space forms, Inventiones Math. 76 (1984), 421-453.

[10] P.Gilkey, The eta invariant and $MU_*(BQ_8)$ (to appear).

[11] P.Landweber, Complex bordism of classifying spaces, Proceedings of AMS V2 (1971), 175-179.

[12] G.Wilson, K-theory invariants for unitary bordism, Quarterly J Math V2 (1973), 499-526.

[13] J.Wolf, Spaces of Constant Curvature, 5th ed, Publish or Perish (1985).

GEOMETRY AND TOPOLOGY (pp. 159–207)
edited by G. M. Rassias & G. M. Stratopoulos

PORTRAITS OF HARMONIC FUNCTIONS

H.B. Griffiths

University of Southampton, U.K.

ABSTRACT. If M is a smooth compact surface with boundary, with Riemannian metric, then a generic smooth function f has a 'portrait' consisting of the level curves of f and the trajectories orthogonal to them. Isomorphisms between portraits are defined in terms of homeomorphisms of M, and we show that each portrait is essentially determined by a linear graph determined by the stable manifolds of f. The classification of portraits is reduced to a combinatorial problem in graph theory. We consider also a 'generic' version of the Dirichlet problem, classifying the set of solutions by their portraits.

1. INTRODUCTION

Let M be a smooth compact surface with a Riemannian metric μ. We are concerned with the qualitative description of a typical smooth function $f : M \to \mathbb{R}$, in terms of its 'portrait' $\Pi(f)$. This consists of the family of level curves of f together with their orthogonal trajectories (defined in terms of μ). The object of this paper is to begin a topological classification of the different types of portrait.

Here, we concentrate on those portraits $\Pi(f)$ whose structure is stable under small changes in f. They are the portraits of 'generic' functions f on M, which have a finite set crit (f) of critical points in the interior of M and behave well on the boundary

∂M . If ∂M is empty and we remove certain small discs surrounding the maxima and minima of f on M , we obtain a restriction $g : N \to \mathbb{R}$ of f on a sub-surface N ; and now crit (g) consists only of saddle-points, and g takes a constant value on each component of ∂N . We call such a function g an 'elliptic measure' on N because of its resemblance to certain types of harmonic function. Indeed, this theory was suggested by contemplation of such functions on a planar domain bounded by three circles. In that case (see Griffiths [3]) we were able to show that there are just twelve essentially different types of portrait $\Pi(g)$ when g is a harmonic elliptic measure; these types include only six stable ones in the sense of the present paper, because the others describe the cases when g is either degenerate in the sense of Morse Theory, or possesses a critical point on the boundary. With more general surfaces, we have to exclude such 'unstable' cases for simplicity; but they should not be forgotten, as they yield the structures that allow one stable type of portrait to bifurcate into another.

Returning to the function f and its restriction g on N , we observe that $\Pi(f)$ is easily reconstructed from $\Pi(g)$, so it is the latter that we study most. From $\Pi(g)$ we extract a certain linear graph $K(g)$ which is defined by the stable manifolds of g , and it turns out (Theorem 10.3) that the embedding of $K(g)$ in M determines $\Pi(g)$ - to within equivalences defined in terms of the group of homeomorphisms $N \approx N$. We also prove (Theorem 8.5) that there is only a finite number of different types of embedded linear graph $K(g)$, although a finer classification gives a countable set. Thus the classification of portraits is reduced to a combinatorial problem, which is as yet unsolved except in special cases. Linear graphs arising from stable manifolds have also been considered by Peixoto [10], in connection with a classification of vector fields on surfaces, but there is little overlap between the work of his paper and the present one; they have different purposes.

When g is harmonic, the type of graph $K(g)$ is a conformal invariant; and we show in a 'realisation theorem' (11.1 below) that every possible graph occurs in some conformal structure on N . (For technical simplicity, we confine ourselves in this theorem, as in most of the later ones, to orientable surfaces.)

These considerations suggest a formulation of a 'generic' version of the Dirichlet problem when ∂M is non-empty. We think of it in terms of classifying all possible stable types of portrait $\Pi(f)$ whose restriction to ∂M has a given form, provided that crit $(f) \cap \mathrm{Int}(M)$ consists only of saddle-points. By eliminating the elliptic operator that occurs in the classical form of the problem, and preferring genericity, we lose uniqueness. The enumeration of the various portraits is reduced, in Theorem 14.7, to the problem of enumerating a certain set of linear graphs; but we do not attempt that problem here.

The plan of the paper is therefore as follows. In Sections 2-4 we introduce notation and motivate certain of our problems in a planar setting. We pass in Sections 5-8 to the study of abstract graphs on M like K(g) , mentioned above, and their equivalence classes. This theory leads via Sections 9-12 to the finiteness theorem 12.3 for smooth elliptic measures, although the proofs depend on the use of non-smooth homeomorphs of smooth functions. In the remaining Sections, we formulate the generic extension problem on M in terms of the double 2M , to conclude with Theorem 14.7.

2. NOTATION

If M is a compact surface, then ∂M will denote its boundary, and Int M denotes $M - \partial M$. A ∂-automorphism of M is a homeomorphism $\phi : M \approx M$ which maps each component of ∂M to itself. In particular if $\phi : M \approx M$ and ϕ is isotopic to the identity we write $\phi \simeq 1_M$; and ϕ is then a ∂- automorphism. We call M 'smooth' if it is C^∞ or real-analytic, and similarly for functions defined on M . If $\partial M \neq \emptyset$ we can form the double of M , denoted by 2M , which contains M as a sub-surface. Moreover, there is an involution $\zeta : 2M \to 2M$, such that

$$2M = M \cap \zeta(M) , \qquad M \cap \zeta(M) = \partial M . \qquad (2.1)$$

When M is smooth, its structure can be extended to $2M$ so that ζ becomes a smooth diffeomorphism: see Munkres [9] p.52.

If $f : M \to \mathbb{R}$ is a smooth real-valued function, we are interested in its critical points; such a point will be one which is interior to M and at which grad f is zero. We denote the set of critical points of f by

$$\text{crit}\ (f) \subseteq \text{Int}\ M\ .$$

The levels of f are the sets $\{f = \text{const}\}$; a level is critical or ordinary according as it contains a critical point of f, or not. A critical point c of f is non-degenerate provided the Hessian determinant of f at c is non-zero; and f is non-degenerate provided that all its critical points are non-degenerate, in which case they are isolated. A critical point is then of index 0, 1, or 2 according as it is a minimum, a saddle-point, or a maximum of f.

Suppose M is smooth and has a smooth Riemannian metric μ. Then (M,μ) is a Riemannian surface. The metric allows us to define the trajectories of $f : M \to \mathbb{R}$, orthogonal (in the sense of μ) to the level curves (see, e.g. Morse [6] p.150). Unless the contrary is specifically stated, we shall assume a fixed metric on M and assume that 'trajectories' are defined with respect to this metric. If M happens to lie in the plane, we also assume that it has the Euclidean metric. A trajectory which ends at a critical point is called a critical trajectory.

For brevity we say that $f : M \to \mathbb{R}$ is 'ND' provided (i) crit (f) is finite and (ii) no trajectory joints a pair of saddle-points. If, moreover, each level of f contains at most one critical point, we say that f is monic.

Homeomorphs

Given a continuous function $f : M \to \mathbb{R}$, a homeomorph of f is a function $g : M \to \mathbb{R}$ of the form

$$g = \rho\circ f\circ\theta\ , \tag{2.3}$$

where $\rho : \mathbb{R} \approx \mathbb{R}$ is a strictly increasing continuous function, and $\theta : M \approx M$ is a ∂-automorphism. We write $f \sim g$, and this is clearly an equivalence relation. Now, if f is smooth, then in Int M, its levels are locally arcs except at a critical point c, at which c is recognizable topologically as a singularity of the level $\{f = f(c)\}$. Hence $\theta^{-1}(c)$ is similarly recognizable on the level $\{g = \rho(f(c))\}$. We therefore call $\theta^{-1}(c)$ a critical point of g; and then

$$\text{crit}\,(g) = \theta^{-1}\,\text{crit}\,(f) \subseteq \text{Int}\,M\,.$$

The trajectories of g (through $x \in M$) are the images by θ^{-1} of the trajectories of f through $\theta(x)$. Thus we translate, without comment, valid topological statements about the trajectories of f into analogues about g. It should therefore be clear what we mean by saying that g is ND or monic, and that if g has these properties, so has any homeomorph. For brevity we say that g is portrayable when f in (2.3) is smooth. Thus a portrayable function g has a portrait $\Pi(g)$, which consists of level curves and trajectories of g. Clearly if $g \sim h$, then h is also portrayable; and if $h = \rho'\circ g\circ\theta'$ analogously to (2.3) then θ' carries levels and trajectories of h to those of g and preserves the sense of trajectories: if h increases along T then g increases along $\theta(T)$. Hence θ' induces an isomorphism between $\Pi(h)$ and $\Pi(g)$: we write $\Pi(h) \sim \Pi(g)$. Conversely, if a ∂-automorphism $\theta' : M \approx M$ induces such an isomorphism, then there exists ρ' such that $h = \rho'\circ g\circ\theta'$. Therefore we have:-

2.1 PROPOSITION. *Suppose $g,h : M \to \mathbb{R}$ are portrayable functions. Then $g \sim h$ iff $\Pi(g) \sim \Pi(h)$.*

2.2 FUNCTION-SPACES

Let $C^2(M)$ denote the set of all ND smooth functions on M, with the C^2-topology (i.e. the distance between $f, g \in C^2(M)$ is the maximum of the maximum values of $f - g$ and its first and second derivatives on M). It is well-known that the set C^2_{mon} of monic functions is an open dense subset of $C^2(M)$ (see, e.g. Morse-Van Schaak [8] p.547). More precisely, if $f \in C^2(M)$ there exists $\varepsilon > 0$ such that if crit $f = \{c_1,\ldots,c_p\}$ and $g \in C^2_{mon}(M)$ is within ε of

f , then the ε-neighbourhoods V_i , centred at c_i , lie in Int M ; they are mutually disjoint, while each V_i contains exactly one critical point d_i of g , and index c_i = index d_i . The construction shows that f and g agree except in sets V_i for which the critical point c_i is not the only one on its f-level. We shall call g a monic modification of f .

3. JORDAN REGIONS

In this Section we shall indicate the features of the problems considered in this paper, by considering the special case when M is planar. Thus let D be a plane region bounded by $h + 1$ smooth Jordan curves $C_0, C_1, \ldots, C_h$, where $h \geq 2$. We refer to C_0 as the 'outer boundary' of D , and $C_1, \ldots, C_h$ as the boundaries of its h 'holes'. We write

$$\partial D = C_0 \cup C_1 \cup \ldots \cup C_h , \quad \text{Int } D = D - \partial D .$$

It is well-known that D can be conformally mapped onto a planar region bounded by $h + 1$ circles, but we need to consider the more general shape.

Given $p = (p_0, p_1, \ldots, p_h) \in \mathbb{R}^{h+1}$, there is a unique continuous function $f_p : D \to \mathbb{R}$ which is harmonic on Int D and such that for each $i = 0,1,\ldots,h$, $f_p | C_i = p_i$. We call f_p the harmonic measure levelled at p . Denote by $e(0),\ldots,e(h)$ the natural basis of $\mathbb{R}^{h+1}$ for which the ith co-ordinate of $e(i)$ is 1 and the rest zero. Then by uniqueness of harmonic functions

$$f_p = \sum_{i=0}^{h} p_i f_e(i) \tag{3.1}$$

so the harmonic measures form a vector space H ; and the correspondence $p \mapsto f_p$ is an isomorphism $\mathbb{R}^{h+1} \approx H$. By the maximum property of harmonic functions, if all the co-ordinates of p are constant (say k), then f_p is the constant function k . In particular,

$$1 = \sum f_{e(i)} , \quad \text{so} \quad 1 = f_u \quad \text{(say)}$$

where $u = (1,1,\ldots,1)$.

We are interested in the portrait $\Pi(f_p)$ of each f_p where, as in Section 1, we mean by $\Pi(f_p)$ the pair consisting of the set of levels of f_p, and its set of trajectories orthogonal to these levels. These trajectories are considered as sensed, in the direction of f_p increasing. It is then clear that $\Pi(f_p + \text{const}) = \Pi(f_p)$. In particular, if $q = p - tu$ where $t = \sum p_i/(h+1)$, then $f_q = f_p - t$ so $\Pi(f_q) = \Pi(f_p)$ and $\sum q_i = 0$. Moreover, if $k \in \mathbb{R}$ then $f_{kp} = kf_p$, so if $k > 0$ then $\Pi(f_p) = \Pi(kf_p) = \Pi(f_{kp})$. Hence we may restrict attention to investigating $\Pi(f_p)$ when

$$p \in \textstyle\sum^{h-1} = S^h \cap L \tag{3.2}$$

where S^h is the unit sphere in $\mathbb{R}^{h+1}$, and L the hyperplane $\sum x_i = 0$. Observe that $\sum^{h-1}$ is a great $(h-1)$-sphere on S^h, and it is invariant under the antipodal map $\alpha : x \mapsto -x$ of $\mathbb{R}^{h+1}$. However, since $f_{-p} = -f_p$ then $\Pi(f_{-p})$ differs from $\Pi(f_p)$ in that the sense of each trajectory in the former is opposed to the sense of the same trajectory in the latter. If we were to neglect this sense, then we could assume that the 'control' parameter p runs through real projective $(h-1)$-space $\mathbb{P}^{h-1}$. For our purposes, however, it is better to take p in the $(h-1)$-sphere $\sum^{h-1}$, on which the involution α acts without fixed points; so $\sum^{h-1}$ is our 'control space'.

As mentioned in the Introduction, a detailed discussion of the case when $h = 2$ is given in Griffiths [3]. When $h > 2$, we would like to make a complete description of the classes X of homeomorphs of the harmonic measures f_p, by representing each class X by a portrait $\Pi(f_p)$, and by describing the structure and situational properties of $P(X)$ in $\sum^{h-1}$ in (3.2), where

$$P(X) = \{p \in \textstyle\sum^{h-1} \mid f_p \in X\} \tag{3.3}$$

(For example: is each $P(X)$ a topological simplex?) We are here unable to formulate such a precise result as Theorem 4.1 of [3], and instead confine ourselves to a specific subset P_h of $\sum^{h-1}$ given by:-

3.1 Definition. In $\sum^{h-1}$, P_h consists of all p such that f_p is ND and the normal derivative of f_p never vanishes on ∂D . As explained in Griffiths [3], P_h is open in $\sum^{h-1}$, but unfortunately we cannot assert that it is dense in $\sum^{h-1}$, - contrary to Theorem 6.1 of that paper. The 'proof' of that theorem contains a gross error (line 4, p.520), and the theorem now seems to the author to be unlikely. In this paper, however, we solve only the problem of enumerating the classes X mentioned above, in that by Prop. 8.5 we reduce it to the enumeration of a finite set of planar graphs (and that graphical problem is as yet unsolved in general).

If $p \in P_h$, then Morse Theory tells us that if $h > 2$ then f_p has exactly $h - 1$ critical points, all in Int D and each a non-degenerate saddle.

The equation grad $f_p(x) = 0$ near c_i can be solved using an Implicit Function theorem, since the Jacobian determinant is the Hessian of f_p which is non-zero by non-degeneracy. Hence there is a neighbourhood Q of p in $P_h \subseteq \sum^{h-1}$, and $h - 1$ real-analytic functions

$$\phi_i : Q \to \text{Int } D \text{ , } \quad \phi_i(p) = c_i \tag{3.4}$$

such that for all $q \in Q$, the critical points of f_q are precisely the points $\phi_i(q)$, $1 \leq i < h$. If f_p is monic, then so is f_q if Q is sufficiently small. Our interest lies in describing the equivalence classes of portraits and, as we shall see following Cor. 6.8, we have:-

3.2 PROPOSITION. Suppose f_p is monic and $p \in P_h$. If $q \in Q$ in (3.4), then $\Pi(f_q) \sim \Pi(f_p)$.

This proposition tells us that the structure of $\Pi(f_p)$ is stable, and hence that the control-space P_h is divided into components, each of which correspond to a single equivalence class of portraits. The points of $\sum^{h-1} - P_h$ correspond to portraits with degeneracies of various kinds, that allow a portrait to change drastically as we pass from one component on P_h to another. By Theorem 12.3 below, the number of components is finite.

4. GENERAL SURFACES

If the plane region D is replaced by any smooth compact surface M with h + 1 boundary curves and the Laplace operator by any linear elliptic operator L , then (since the maximum principle still applies), everything we have said in Section 3 applies equally well to M , except that the number of critical points is changed from h - 1 to

$$\sigma = -\chi(M) , \tag{4.1}$$

where χ denotes the Euler characteristic. (This holds, by Morse Theory, since the critical points are all saddle points.) It is well-known that L defines upon M a Riemannian metric μ for which L is the Laplace-Beltrami operator, so local co-ordinates exist in which L becomes the (planar) Laplace operator (see e.g. Courant-Hilbert [2] p.160).

If M has only two boundary components then a measure has no critical points if M is planar (an annulus), but this is not so if genus(M) > 0 .

Until Section 11, however, we shall not need the operator L , and instead we discuss a class of functions that we shall call <u>elliptic measures</u> on M , defined as follows. (Cf. Def. 3.1.)

4.1 <u>Definition</u>. A <u>smooth elliptic measure</u> on M is a smooth function $g : M \to \mathbb{R}$ which is ND , and whose critical points are all saddle points. Moreover, for each component C_i of ∂M , $g|C_i$ is constant; and if $x \in C_i$ exactly one trajectory T of g has x as an end-point, and T depends continuously on x . An <u>elliptic measure</u> on M is a function $f : M \to \mathbb{R}$ which is a homeomorph (see (2.2)) of a smooth elliptic measure on M .

The condition concerning trajectories is a restriction, even for a harmonic measure g , because it is possible for the normal derivative of g to vanish - as is shown in Griffiths [3]. However, the measures $f_{e(i)}$ in (3.1) are known to have non-vanishing normal derivative. Whether they are necessarily ND is a difficult question that we hope to discuss elsewhere.

5. THE LINEAR GRAPH $K(f)$

As in the last Section, let $f:M \to \mathbb{R}$ denote an elliptic measure where M is a smooth compact connected surface (e.g. a Jordan domain) with $h+1$ boundary curves. Then f has σ non-degenerate critical points c_i, with σ as in (4.1). Through each point c_i there pass exactly four (oriented) trajectories, two of which form the stable manifold A_i at c_i (along which f increases up to c_i) while the remaining two form the unstable manifold B_i at c_i (along which f increases away from c_i). No two critical points are joined by a trajectory (since f is ND.) Hence the two end-points of A_i lie on boundary curves $C_{\alpha(i)}$, $C_{\beta(i)}$ where $\alpha(i)$ may equal $\beta(i)$. Moreover, every boundary curve C_i of ∂ is a 'source' or 'sink', in that either at least one A_j starts from C_i, or at least one B_k ends on C_i using an obvious terminology. Clearly, no C_i can be both a source and a sink, and not all can be sources, nor all sinks. Thus, if there are π sources and ν sinks,

$$1 \leqslant \pi \leqslant h\,, \quad 1 \leqslant \nu \leqslant h\,, \quad 1+h=\pi+\nu\,. \tag{5.1}$$

Let X denote the union of all sources, and let $K = K(f)$ denote the 1-dimensional complex formed from the union of X and all the stable manifolds of f. By Morse Theory, K is connected, and a deformation retract of M. Hence for any homology coefficients, the inclusion induces an isomorphism of singular homology:-

$$H_1(K) \approx H_1(M)\,. \tag{5.2}$$

Before describing K further, we proceed rather more abstractly, in order to deal with sets of arcs on M that need not arise initially as the stable manifolds of a function.

Let M be the surface previously considered, with $h+1$ boundary curves. If α is an arc in M, let $\dot{\alpha}$ denote its pair of end-points. (Here, an arc is considered as a set of points, for convenience.) We say that α spans M if $\dot{\alpha} \subseteq \partial M$ and $\alpha - \dot{\alpha} \subseteq \operatorname{Int} M$. Let $\mathcal{B}$ denote the set of boundary components of M.

5.1 <u>Definition</u>. <u>Suppose</u> $\emptyset \neq \mathcal{D} \subseteq \mathcal{B}$. A $\mathcal{D}$-<u>set</u> U on M is a set of mutually disjoint spanning arcs α such that

(i) each end-point of α lies on some member of $\mathcal{D}$ and

(ii) if K(U) denotes the complex formed by all the arcs of U and all the circles in $\mathcal{D}$, then the inclusion $K(U) \subseteq M$ induces isomorphisms

$$H_i K(U) \approx H_i M \qquad (i = 0,1) .$$

The isomorphism with $i = 0$ shows that U cannot be empty if $2 \leq \text{card } \mathcal{D}$, since M is connected. We shall see later (Cor. 5.6) that card $U = -\chi(M)$, and this is zero only when M is an annulus. It will turn out also that, always, $\mathcal{D}$ is never the whole of $\mathcal{B}$.

For example, if $\mathcal{D}$ is the set of sources of f as in (5.1) then the set $S(f)$ of stable manifolds of f is a $\mathcal{D}$-set and the complex K in (5.2) is $K(S(f))$. Similarly the set $U(f)$ of unstable manifolds is a $(\mathcal{B}-\mathcal{D})$ -set.

Now K(U) is a 1-dimensional complex, and we need to consider certain neighbourhoods of it in M .

5.2 <u>Definition</u>. A <u>thickening</u> ΘL of a 1-dimensional complex L in M is a compact surface in M , and consists of a union of topological discs D_v and strips $S_e \approx e \times I$ in M , as v and e respectively run through the vertices and edges of L . Also D_v and S_e are compact, and:-

(i) if $v \in \partial M$, then a neighbourhood of v in ∂M lies in ∂D_v , and if $e \subseteq \partial M$ then $e \subseteq \partial S_e$;

(ii) all the discs are mutually disjoint;

(iii) D_v meets S_e if and only if v is an end-point of e ;

(iv) $D_v \cap S_e$ is then the four-sided subset of D_e bounded by the arcs $\{v\} \times I$, $D_v \cap (e\times 0)$, $D_v \cap (e\times 1)$ and an arc $\bar{e}$ of ∂D_v ;

(v) the arcs $\bar{e}$, for different e's , are mutually disjoint;

(vi) if a boundary curve C of M is disjoint from L , then $C \cap \Theta L = \emptyset$.

Clearly L has such a thickening in M - indeed it has many. If we ignore the product structure in each S_α, ΘL is a union of discs and forms a compact surface with boundary; this surface we call the 'underlying set' of ΘL. The boundary $\partial\Theta L$ of ΘL is a linear graph, and it contains any edge of L that lies in ∂M. Also L is a deformation retract of ΘL; so the inclusion induces an isomorphism of homology groups:-

$$H_1 L \approx H_1 \Theta L . \tag{5.3}$$

As to mappings of thickenings we formulate:-

5.3 Definition. If L' is a second 1-dimensional complex in M, and $\Theta L'$ is a thickening of L', a thickened isomorphism $\phi : \Theta L \to \Theta L'$ is a homeomorphism such that (i) $\phi(L) = L'$ (ii) for each disc D_v and strip S_e of ΘL, $\phi(D_v)$ and $\phi(S_e)$ are the discs $D_{\phi(v)}$, $S_{\phi(e)}$ respectively in $\Theta L'$, (iii) $\phi|\partial\Theta L$ is an isomorphism between the linear graphs $\partial\Theta L'$.

Any simplical isomorphism $L \to L'$ induces a thickened isomorphism $\Theta L \to \Theta L'$, but not necessarily in a canonical way.

With U a $\mathcal{D}$-set as in Def. 5.1, we now investigate the set $M - K(U)$. We use the following terminology. An annulus A is homeomorph of the Cartesian product of a circle and a closed interval. An open annulus is a homeomorph of $\operatorname{Int} A$, and a semi-annulus B is a homeomorph of $A - C$ where C is a boundary curve of A. Thus B has a single boundary curve, corresponding to $\partial A - C$.

5.4 PROPOSITION. *Let U be a $\mathcal{D}$-set, and P a thickening of $K(U)$. For each curve $C \in \mathcal{B} - \mathcal{D}$, C lies in a unique component F_C of $M - P$, and F_C is a semi-annulus with boundary C. Its closure in M is an annulus, with boundary curves C, C' where $C' \subseteq \partial P$.*

Also F_C lies in a unique component G_C of $M - K(U)$, and G_C is a semi-annulus with boundary C. (The correspondences $C \to F_C$, $C \to G_C$ are bijections.)

Proof. Let $2M$ denote the double of M (see (2.1)), with the automorphism ζ of $2M$ there described. The given thickening P is a compact surface with boundary, as is $Q = P \cup \zeta(P)$. By (ii) of Def. 5.1, and by (5.3).

$$H_j(P) \approx H_j(M) , \quad j \geqslant 0 , \tag{i}$$

so in particular P is connected. Hence so is Q since $P \cap \zeta(P)$ contains all curves in $\mathcal{D}$, and $\mathcal{D} \neq \emptyset$. Using (vi) of Def. 5.2, we see that the Euler characteristic of Q is

$$\chi(Q) = 2\chi(P) = 2(1-\beta_P)$$

where $\quad \beta_P = \dim H_1(P)$

and we regard all homology groups as vector spaces over $\mathbb{Z}_2$. Hence since $Q \neq 2M$, then $H_2(Q) = 0$, so

$$\beta_Q = 2\beta_P - 1 , \quad \beta_{2M} = 2\beta_M \tag{ii}$$

Since $2M$ is a closed surface and coefficients are in $\mathbb{Z}_2$, we have the exact Poincaré-Lefschetz cohomology sequence

$$0 \to H^0(2M) \to H^0(Q) \to H_1(2M-Q) \to H^1(2M) \to H^1(Q) \to H_0(2M-Q) \to H^2(2M) \to 0$$

in which $\quad H^0(2M) \approx H^0(Q) \approx H^2(2M) \approx \mathbb{Z}_2$,

- whether or not M is orientable. Taking the Euler characteristic of the exact sequence therefore gives

$$p_1 - \beta_{2M} + \beta_Q - p_0 + 1 = 0$$

where $p_i = H_i(2M-Q)$; so by (ii) we have $p_0 = p_1$. Each component G of $2M - Q$ is of the form $F \cup \zeta(F)$ where $F \subseteq M - P$; so, if $\mathcal{D}' = \mathcal{B} - \mathcal{D}$, each curve $C \in \mathcal{D}'$ lies in such a set F . Since C does not bound in M , it cannot bound in F and hence not in G ; so $\beta_G \gtrsim 1$. Therefore, since there are p_0 components G and $p_0 = p_1$ then $\beta_G = 1$ for each G . But G is of the form $F \cup \zeta(F)$, so F is a semi-annulus with a boundary curve in $\mathcal{D}'$. Each member C of $\mathcal{D}'$ must lie in such a component F_C of $M - P$ so the correspondence $C \to F_C$ is a bijection from $\mathcal{D}'$ to the set of components of $M - P$.

Now P is a thickening of $K(U)$, so the closure of F_C in M is an annulus with boundary curves C, C' where C' lies in the boundary of the surface P. This proves the first assertion of the Proposition.

Hence, if $C \in \mathcal{D}'$, then the component F of $M - K(U)$ that contains C is the union of all semi-annuli F_C, as P runs through all thickenings of $K(U)$ in M. Hence F is a semi-annulus, with boundary C. This completes the proof.

Three consequences follow. First, as each semi-annulus mentioned in the Proposition can be retracted along 'radii' into the complex $K(U)$, then $K(U)$ is a deformation retract of M and we have at once an analogue of (4.1):-

5.5 <u>COROLLARY</u>. <u>The inclusion</u> $K = K(U) \subseteq M$ <u>induces isomorphisms</u>

$$\pi_1(K) \approx \pi_1(M) \ ; \ H_1(K) \approx H_1(M) .$$

A more algebraic argument gives:-

5.6 <u>COROLLARY</u>. <u>A</u> $\mathcal{D}$-set U <u>contains precisely</u> $-\chi(M)$ <u>members, and the number of components of</u> $M - K(U)$ <u>is the number of members of</u> ∂M <u>not in</u> $\mathcal{D}$.

For, since $\partial M \neq \emptyset$, then $\dim H_1(M) = 1 - \chi(M)$; but if $\mathcal{D}$ has n elements while U has m, then $n - 1$ arcs of U are needed to connect all circles of $\mathcal{D}$, so

$$\dim H_1(K) = (m - (n-1)) + n = m + 1 .$$

By (ii) of Def. 5.1, $\dim H_1(K) = \dim H_1(M)$ so $1 + m = 1 - \chi(M)$ and $m = -\chi(M)$ as required. The number of components F in $M - K$ is card $\mathcal{D}'$, where $\mathcal{D}' = \mathcal{B} - \mathcal{D}$; for by Prop. 5.4 each F has its boundary curve in $\mathcal{D}'$, and each curve in $\mathcal{D}'$ must lie in some F. This proves the Corollary.

Since $M - K(U)$ is non-empty, we may therefore assert:-

5.7 <u>COROLLARY</u>. $\mathcal{D}$ <u>cannot equal</u> $\mathcal{B}$. <u>Also</u>, $U = \emptyset$ <u>iff</u> M <u>is an annulus</u>.

Consider again the elliptic measure f in (5.1); we apply Corollary 5.6 taking $\mathcal{D}$ to be the set of sources of f, and U to be the set $S(f)$ of stable manifolds, of which there is one for each critical point of f. Therefore we have at once, on writing K for $K(S(f))$:-

5.8 <u>COROLLARY</u>. <u>If</u> f <u>has</u> π <u>sources and</u> ν <u>sinks, then</u> M - K <u>has exactly</u> ν <u>components, each a semi-annulus, of which the boundary component is a sink of</u> ∂M. <u>Also</u> f <u>has</u> $-\chi(M)$ <u>critical points</u>.

6. RECIPROCAL SETS

Let f be an elliptic measure on M, as before, with its sets $S(f)$, $U(f)$ of stable and unstable manifolds. The relationship between arcs of $S(f)$ and thos of $U(f)$ can make sense for arbitrary $\mathcal{D}$-sets, as in the following definition.

6.1 <u>Definition</u>. Let U be a $\mathcal{D}$-set of M, and let V be a $\mathcal{D}'$-set, where $\mathcal{D}' = \mathcal{B} - \mathcal{D}$. Suppose there is a bijection $\omega : U \to V$ such that each $\alpha \in U$ is crossed exactly once by $\omega(\alpha)$, while $\alpha \cap \beta = \phi$ if $\beta \in V - \omega(\alpha)$. Then we call V a <u>reciprocal</u> of U, and write $U \perp V$. Clearly, $V \perp U$.

The force of the word 'crosses' is this. If $\alpha \in U$ then it is a common edge of at most two components R_1, R_2 of $M - K(U)$, each of which has respectively a boundary curve A_1, A_2 in $\mathcal{B} - \mathcal{D}'$ (see Fig. 1); perhaps $R_1 = R_2$, in which case $A_1 = A_2$, as for α' in Fig. 1.

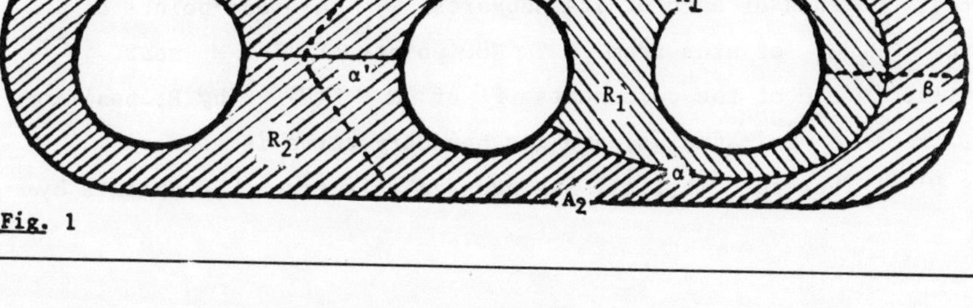

Fig. 1

If α is crossed by $\beta \in V$, and exactly once, then the ends of β must lie on $A_1 \cup A_2$ since β spans M. In an obvious sense, the point $\alpha \cap \beta$ divides α into two 'halves', as it does for β. Such a point is called the '(U,V)-point' on α or β.

Thus, for the elliptic measure $f : M \to \mathbb{R}$, we have

$$S(f) \perp (f) \tag{6.1}$$

and the $S(f)$-points are the critical points. More generally we prove:-

6.2 <u>PROPOSITION</u>. <u>Every $\mathcal{D}$-set U has a reciprocal.</u>

<u>Proof</u>. We use the above notation A, R for arcs and components. If $\alpha \in U$ is in the boundary of two annuli R_1, R_2, we can choose any arc β_α on $R_1 \cup R_2$ that spans M, crosses α exactly once and has its ends on $A_1 \cup A_2$. We can choose the arcs β_α, as α runs through U, so that $\gamma \cap \beta = \emptyset$ if $\gamma \neq \alpha$, while W is a $\mathcal{D}'$-set. Hence $U \perp W$, as required.

A pair U, V of reciprocal sets will divide M into a <u>reciprocal subdivision</u> of 'special regions' which we describe in the following Proposition.

6.3 <u>PROPOSITION</u>. <u>Corresponding to the reciprocal pair</u> (U,V) <u>there is a family</u> $\mathcal{R}$ <u>of compact annuli and discs that covers</u> M, <u>and such that each member of</u> $\mathcal{R}$ <u>has a boundary composed of halves of arcs in</u> $U \cup V$, <u>or portions of boundary curves of</u> ∂M. <u>Different members of</u> $\mathcal{R}$ <u>intersect, if at all, along common boundary arcs</u>. <u>An annulus in</u> $\mathcal{R}$ <u>contains just two</u> (U,V)-<u>points in its boundary, and a disc just one.</u>

<u>Proof</u>. We first note that if C_i is a boundary curve of M, and we remove from it the end-points of all arcs in $U \cup V$, then there remain mutually disjoint arcs $\gamma_{i1},\ldots,\gamma_{i,r(i)}$ (say). If C_i is a source, the ends of each of its sub-arcs γ_{ip} are end-points of halves α_s, α_t of arcs from U. Now points of Int M near γ_{ip} must lie in one of the components F of $M - K(U)$; by Proposition 5.4, F is a semi-annulus, and its boundary curve is some sink $C_j \in \mathcal{D}'$. Therefore α_s, α_t are crossed at their (U,V)-points by

halves β_s, β_t of arcs in V , and β_s, β_t lie in F with an end on C_j . There are now two possibilities:-

(a) $\beta_s = \beta_t$, so α_s, α_t are the two halves of the same $\alpha \varepsilon U$. Here $\alpha \cup \gamma_{ip}$ is a simple closed curve, that bounds with C_j an annulus spanned by β_s . (E.g. R_1 in Fig. 1.)

(b) $\beta_s \neq \beta_t$, so their two ends are joined by one of the sub-arcs γ_{jq} of C_j . If $\alpha_s \neq \alpha_t$ then $J = \gamma_{ip} \cup \alpha_s \cup \beta_s \cup \gamma_{jq} \cup \beta_t \cup \alpha_t$ is a Jordan curve, of which the interior lies in the semi-annulus F . Therefore J bounds a disc in this case (Fig. 1 contains two such discs.) But, if $\alpha_s = \alpha_t$, then β_s and β_t are two halves of the same $\beta \varepsilon V$, and (dually to (a)) $\beta \cup \gamma_{jq}$ is a Jordan curve that bounds with C_i an annulus spanned by α_s ; one such is seen on the left of Fig. 1.

Every point of M lies either on some source, or some $\alpha \varepsilon U$, or some component of $M - K(U)$, so it must lie on an annulus or disc of the type specified in (a) or (b). These sets, then, are the 'special regions' that cover M , and any two of these intersect, if at all, along halves of arcs of U or V .

From (a) and (b) we see that if a special region is an annulus then it has just one (U,V)-point in its boundary, while if it is a disc then it has two. This completes the proof of the Proposition.

We use reciprocal subdivision in the next proof. Recall the notions of monic and ∂-automorphism, from Section 2. Note that if f, g in the following Proposition are smooth, their trajectories may be defined relative to different metrics.

6.4 <u>PROPOSITION</u>. <u>Let</u> $f, g : M \to \mathbb{R}$ <u>be two monic elliptic measures with the same critical points, stable manifolds, and unstable manifolds. There is a</u> ∂<u>-automorphism</u> $\phi : M \to M$ <u>which carries levels and trajectories of</u> f <u>to those of</u> g . <u>Hence</u> f <u>and</u> g <u>are homeomorphs, with isomorphic portraits.</u>

<u>Proof</u>. The hypotheses ensure that the reciprocal subdivisions of M , corresponding to the reciprocal pairs $(\mathcal{S}(f),\mathcal{U}(f))$, $(\mathcal{S}(g),\mathcal{U}(g))$,

have exactly the same 'special regions'. If X is such a 'special region', and $x \in X$, then there passes through x exactly one level curve L and one trajectory S of f; and L is $\{f = f(x)\}$. Now S cannot cross an arc of $\mathcal{S}(f)$ or $\mathcal{U}(f)$ so S is an arc which is either a boundary arc of X, or else the two ends of S lie on distinct boundary curves C_i, C_j of ∂M, where $C_i \in \mathcal{D}$, and parts of these curves lie in ∂X. We are going to define a homeomorphism $\theta : X \to X$ and begin by setting $\theta(x) = x$ if $x \in C_i \cup C_j$.

Now X is either an annulus or a disc (see Prop. 6.3); suppose first that X is an annulus. Then it has just one critical point u in its boundary, and if $x \in X - (C_i \cup C_j)$, then $f(C_i) < f(u) < f(C_j)$; with similar inequalities for g since the unstable manifold of f through u is also unstable for g, by hypothesis. Hence there is a monotonic increasing continuous function $\sigma : \mathbb{R} \to \mathbb{R}$ that maps $f(C_i), f(u), f(C_j)$ to $g(C_i), g(u), g(C_j)$. Also, there is exactly one critical level K of f in X, with a similar contour K' for g. Consider x, L, S as above; and if x lies on one of the critical trajectories of f in ∂X, define $\theta(x)$ to be the point y on that same trajectory for which $g(y) = \sigma(f(x))$. If $x \in X - \partial X$, let S hit C_i in z, and let T be the trajectory of g through z. Define $\theta(x)$ to be that point $y \in S$ such that $g(y) = \sigma(f(x))$. In all cases then, we have a homeomorphism $\theta : X \approx X$ such that $g(\theta(x)) = \sigma(f(x))$, so θ maps trajectories and levels of f to those of g.

Suppose instead that X is a disc, so that it has just two critical points c, d. Then we must have $f(C_i) < f(c) < f(d) < f(C_j)$ since f is monic, with similar inequalities for g for the same reason as before. Hence there is a monotonic increasing continuous function $\tau : \mathbb{R} \to \mathbb{R}$ which maps $f(C_i), f(c), f(d), f(C_j)$ to $g(C_i), g(c), g(d), g(C_j)$ respectively. By the same process as before, but using τ instead of σ, we construct a homeomorphism $\theta : X \approx X$ that maps trajectories and levels of f to those of g. Note that on any half α of a stable manifold in ∂X, we have $\theta|\alpha = \mathrm{id}$, and if $x \in X$, then $g(\theta(x)) = \tau(f(x))$.

For each such special region X we obtain thus a homeomorphism $\theta_X : X \approx X$ with the given properties. Moreover, if X, Y are two such regions that have a half, β, of an unstable manifold as common edge, then the monotonic functions σ or τ described above can be chosen so as to agree on the interval $f(\beta)$: if θ_X is constructed before θ_Y, then the monotonic function for Y is already determined on β by that for X. Hence θ_X and θ_Y agree on β and (noted above) they agree on any common half of a stable manifold. Therefore there is a ∂-automorphism $\phi : M \to M$ such that, for each X, $\phi|X = \theta_X$; so ϕ carries levels and trajectories of f to those of g, as required.

Note that the proof explicitly constructs a monotonic increasing, continuous function $\rho : \mathbb{R} \to \mathbb{R}$ such that $g \circ \phi = \rho \circ f$. Hence f and g are homeomorphs of each other. This completes the proof of Proposition 6.4.

7. RECIPROCAL THICKENINGS

Besides reciprocal sets, we need to consider also reciprocal thickenings, which we now describe. We use them to prove Prop. 3.6, and to show (Prop. 7.6) that portraits are independent of the Riemannian metric used to define them.

Let U be a $\mathcal{D}$-set, with a thickening $P = \Theta K(U)$, and let U' be a second $\mathcal{D}$-set with a bijection $\alpha \leftrightarrow \alpha'$ between U and U'. We say that P thickens U' provided each α' in U meets ∂P only at the end-points of α'.

7.1 LEMMA. If P thickens U', then it is the underlying surface of a thickening of U'.

Proof. Using the notation of Def. 5.2, each α' in U lies in the strip S_α and joins the ends $p \times I$, $q \times I$ – where $\{p,q\} = \dot{\alpha}$ – without touching the two sides $\alpha \times 0$, $\alpha \times 1$. Therefore there is a homeomorphism $\psi : S_\alpha \approx S_\alpha$ that carries α to α', leaving the sides fixed and the ends invariant. This induces a new product structure $S_{\alpha'}$ on S_α to make it $\alpha' \times I$, and leaves the discs D_p, D_q

invariant. There results a thickening of U' with the same discs and strips as P , (considered as point-sets) and the Lemma follows.

Next let (U,V) be a reciprocal pair with bijection $\omega : U \to V$ (as in Def. 6.1) and thickenings P, Q respectively. We say that P and Q form a reciprocal pair iff, for each α in U the strips S_α, $S_{\omega(\alpha)}$ have the property that in the product structures on S_α, $S_{\omega(\alpha)}$, $S_\alpha \cap S_{\omega(\alpha)}$ is a disc of the form $\gamma \times \delta$ where γ and δ are sub-arcs of α, $\omega(\alpha) = \beta$. Hence $S_\alpha \cup S_\beta$ is a homeomorph of a cross-shaped disc (see Fig. 2(a)); for brevity we omit here some straightforward details of the proof.

Fig. 2.

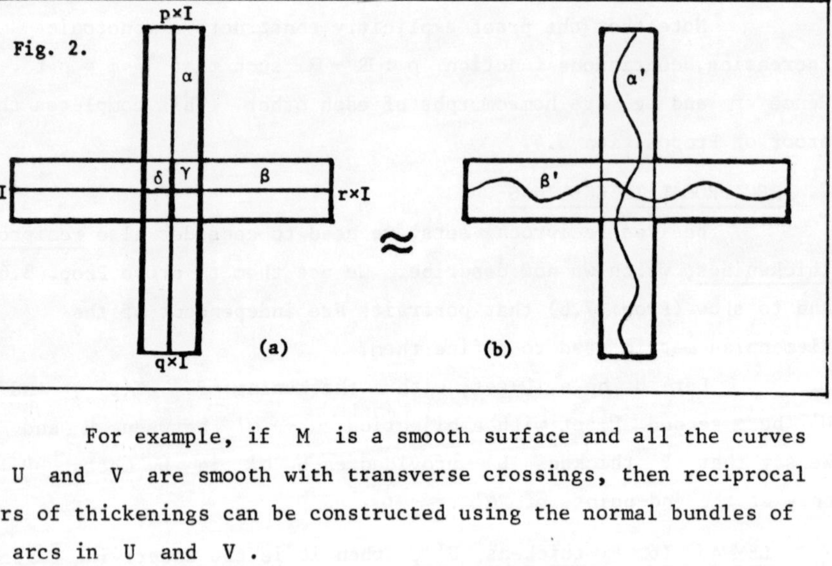

For example, if M is a smooth surface and all the curves in U and V are smooth with transverse crossings, then reciprocal pairs of thickenings can be constructed using the normal bundles of the arcs in U and V .

7.2 LEMMA. Suppose that (U,V), (U',V) are reciprocal pairs and (P,Q) a reciprocal thickening of (U,V) such that P, Q thicken U', V' respectively. Then there is a ∂-automorphism $\eta : M \to M$ that carries the curves of U, V respectively to those of U', V' , and η is the identity outside $P \cup Q$.

Proof. There are bijections $\alpha \leftrightarrow \alpha'$, $\beta \leftrightarrow \beta'$ between the elements of U, U' and V, V' respectively, such that if α in U is crossed by $\beta = \omega(\alpha) \in V$, then α' in U' is crossed by $\beta' \in V'$. Since P, Q thicken U', V' , then α', β' lie in the disc $R_\alpha = S_\alpha \cup S_\beta$ as shown in Fig. 2(b), and there is then a homeomorphism $\psi_\alpha : R_\alpha \approx R_\alpha$ which carries α to α' , β to β' and is the identity on all arcs of ∂R_α except $p \times I, q \times I, r \times I, s \times I$ (these lying on ∂M). We can therefore extend the maps $\psi_\alpha(\alpha \in U)$ to one homeomorphism $\eta : M \approx M$ for which η is the identity except on each R_α , and the lemma follows.

7.3 <u>COROLLARY</u> (of proof). $\eta \simeq 1_M$.
(For ψ_α , in the proof, is isotopic to the identity on R_α .)

If U, U' are two $\mathcal{D}$-sets, we measure their 'nearness' by writing

$$\mathrm{dist}(U,U') < \varepsilon \tag{7.1}$$

iff each arc of U is within ε of some arc of U' and <u>vice-versa</u>. It is then clear that there is a thickening of U that thickens U' .

Similarly, if (U,V) is a reciprocal pair, then there exists some $\varepsilon > 0$ such that if (U',V') is a reciprocal pair and

$$\mathrm{dist}(U,U') < \varepsilon , \quad \mathrm{dist}(V,V') < \varepsilon \tag{7.2}$$

then there is a reciprocal thickening (P,Q) of (U,V) such that P thickens U' and Q thickens V' . Hence from Lemma 7.2 and Cor. 7.3 we have a result that can be expressed as:-

7.4 <u>COROLLARY</u>. <u>If</u> $(U,V), (U',V')$ <u>are reciprocal pairs which are sufficiently near</u>, <u>then there is a</u> ∂-<u>automorphism</u> $\eta : M \approx M$ <u>which carries</u> (U,V) <u>to</u> (U',V') ; <u>and</u> $\eta \simeq 1_M$.

By arguing similarly from the proof of Lemma 7.1 we also have:-

7.5 <u>COROLLARY</u>. <u>If</u> U, U' <u>are</u> $\mathcal{D}$-<u>sets which are sufficiently close</u>, <u>then there is a</u> ∂-<u>automorphism</u> $\eta : M \approx M$ <u>which carries</u> U <u>to</u> U' ; <u>and</u> $\eta \simeq 1_M$.

Two applications of Cor. 7.4 now follow, concerning the portraits of elliptic measures. Consider first the situation of Prop. 3.6. Starting with the given elliptic measure $f_p : M \to \mathbb{R}$, we take

(U,V) to be its sets $S(f)$, $U(f)$ of stable and unstable manifolds. We then determine $\varepsilon > 0$ as in (7.1). By the continuous dependance of trajectories on parameters, we can then restrict the neighbourhood Q of P in (3.4) to be so small that for all $q \in Q$

$$\mathrm{dist}(S(f_p), U(f_q)) < \varepsilon ,$$

and similarly for the unstable manifolds. Hence by Cor. 6.8, there is a ∂-automorphism $\eta_q : M \to M$ that carries the stable and unstable manifolds of f_p to those of f_q. Therefore the functions f_p, $h = f_q \circ h$ satisfy the conditions of Prop. 6.4, so

$$f_p \approx h \approx f_q . \tag{7.2}$$

Hence by Prop. 2.1, $\Pi(f_p) \sim \Pi(f_q)$, and Prop. 3.6 is established.

Secondly, we show that the portrait of an elliptic measure is independent of the particular metric μ used for defining the trajectories. Call such a portrait a 'μ-portrait'.

7.5 THEOREM. <u>Let μ_1, μ_2 be two Riemannian metrics on M, and let $\Pi_i(f)$ denote the μ_i-portrait of a smooth monic elliptic measure $f : M \to \mathbb{R}$ (i=1,2). Then there is a ∂-automorphism θ of M, such that θ maps $\Pi_1(f)$ to $\Pi_2(f)$ and θ is isotopic on M to the identity. Indeed, $f \circ \theta = f$, and θ fixes crit f.</u>

<u>Proof</u>. Let $\mathcal{D}$ denote the set of sources of f. These, as well as the critical points and level curves of f, are independent of any metric. Now for all t in the unit interval I we obtain a Riemannian metric $\nu_t = t\mu_2 + (1-t)\mu_1$, with a corresponding ν_t-portrait $\Pi_t(f)$; and $\nu_0 = \mu_1$, $\nu_1 = \mu_2$. Let (U_t, V_t) denote the reciprocal pairs of stable and unstable sets $S(f_t)$, $U(f_t)$ in $\Pi_t(f)$. Since trajectories of f with respect to ν_t are defined by differential equations whose coefficients (written in terms of a coordinate system) depend smoothly on ν_t, so do U_t and V_t.

As in (7.1), there is for each $t \in I$ an $\varepsilon > 0$ such that for any reciprocal pair (U',V') with $\mathrm{dist}(U_t, U')$, $\mathrm{dist}(V_t, V')$ each $< \varepsilon$, Cor. 7.4 applies to (U_t, V_t) and (U',V'). Hence, by the continuous dependance on ν_t noted above, there exists $\delta > 0$ such that for all $s \varepsilon I$ within distance δ of t, the pair (U_s,V_s) satisfies the conditions on (U',V'). By compactness of I, therefore, there is a finite set

$$0 = t(0) < t(1) < \ldots < t(n) = 1$$

in I such that for each $i = 0,\ldots,n-1$, Cor. 7.4 applies to the pairs $(U_{t(i)},V_{t(i)})$, $(U_{t(i+1)},V_{t(i+1)})$. Thus, there exists a ∂-automorphism $\eta_i : M \to M$ which carries $(U_{t(i)},V_{t(i)})$ to $(u_{t(i+1)},V_{t(i+1)})$ and $\eta_i \simeq 1_M$. Therefore $\theta = \eta_{n-1} \circ \eta_{n-2} \circ \cdots \circ \eta_0$ is a ∂-automorphism of M, that carries (U_0,V_0) to (U_1,V_1); and $\theta \simeq 1_M$. Just as we argued above for (7.2), it now follows from Prop. 6.4 that $\Pi_0(f) \sim \Pi_1(f)$, and the proof is complete.

8. EQUIVALENCE BETWEEN $\mathcal{D}$-sets on M

To discuss possible ways of regarding two $\mathcal{D}$-sets U, V on M as 'equivalent' we use subgroups of the group $\underset{\sim}{H}$ of all automorphisms of M. If $\theta \varepsilon \underset{\sim}{H}$ then θ induces a permutation θ_∂ of the set $\mathcal{B}$ of all components of ∂M. Now let $H_\mathcal{D}$ denote the subgroup

$$H_\mathcal{D} = \{\theta \varepsilon \underset{\sim}{H} | \theta_\partial \text{ leaves } \mathcal{D} \text{ fixed}\} \tag{8.1}$$

so if $\theta \varepsilon H_\mathcal{D}$ then θ_∂ permutes the curves of $\mathcal{D}' = \mathcal{B} - \mathcal{D}$. Hence we have a homomorphism

$$\psi : H_\mathcal{D} \to \text{Perm } \mathcal{D}' , \quad \psi(\theta) = \theta_\partial | \mathcal{D}' ; \tag{8.2}$$

and its kernel is $H_\mathcal{B}$, the set of ∂-automorphisms of M. We shall shortly prove:-

8.1 <u>PROPOSITION</u>. <u>The group</u> $H_\mathcal{D}/H_\mathcal{B}$ <u>is isomorphic to</u> Perm $\mathcal{D}'$.

First, however, suppose that $\mathcal{D}$ has π elements, and that $\mathcal{B}$ is labelled in a particular way as $\mathcal{B} = \{C_0,C_1,\ldots,C_h\}$. We prove:-

8.2 <u>LEMMA</u>. <u>Let</u> $\alpha, \beta : A \to \mathcal{B}$ <u>be two injections</u>. <u>Then there is an automorphism</u> $\phi \varepsilon \underset{\sim}{H}$ <u>such that for each</u> $i \varepsilon A$, $\phi_\partial(\alpha(i)) = \beta(i)$.

<u>Proof</u>. It suffices to be able to interchange any two boundary components C_i, $C_j \varepsilon \mathcal{B}$. If $h = 1$ we use reflection across a central

curve, so assume $h > 1$. Then there is a planar domain $D \subseteq M$ as in Figure 3, whose boundary components are C_i, C_j, and E; so we simply rotate D through a half-turn about E, leaving $M - D$ fixed. This interchanges C_i and C_j as required.

8.3 COROLLARY (of proof). If $\alpha(A) = \beta(A) = \mathcal{D}$, then $\phi_\partial|(\mathcal{B}-\mathcal{D})$ is the identity.

We can now prove Proposition 8.1 as follows. The homomorphism ψ in (8.2) was seen there to have kernel $H_\mathcal{B}$ so we need only prove that ψ is onto. Let then $\omega \in \text{Perm}\,\mathcal{D}'$, and let $B = \{1,2,\ldots,\nu\}$ where $\nu = \text{card}\,\mathcal{D}'$. Choose a bijection $\alpha : B \approx \mathcal{D}'$, and let $\beta = \omega\circ\alpha$. Then by Lemma 8.2, there exists $\phi \in \underset{\sim}{H}$ such that for each $i \in B$, $\phi_\partial(\alpha(i)) = \omega\circ\alpha(i)$, whence $\phi_\partial|\mathcal{D}' = \omega$. But by Cor. 8.3, $\phi_\partial|\mathcal{D}$ is the identity, so $\phi \in H_\mathcal{D}$ by (8.1). Hence by (8.2), $\omega = \psi(\phi)$ and ψ is onto. This proves Proposition 8.1.

8.4 REMARK. The last proof works even if $\mathcal{D} = \emptyset$ since $\mathcal{D}' \neq \emptyset$. By (8.1) we may interpret $H_\emptyset$ as $\underset{\sim}{H}$ itself, and then Proposition 8.1 yields

$$\underset{\sim}{H}/H_\mathcal{B} \approx \text{Perm}\,\mathcal{B}\,.$$

Let $\underset{\sim}{I}$ denote the (normal) subgroup of $\underset{\sim}{H}$ that consists of all elements θ isotopic to the identity (in brief, $\theta \simeq 1$). Then for all non-empty $\mathcal{D} \subseteq \mathcal{B}$ we have

$$\underset{\sim}{I} < H \leqslant H_\mathcal{B} \leqslant H_\mathcal{D} \quad (\text{and } H_\mathcal{B} \lhd \underset{\sim}{H})\,. \tag{8.3}$$

The group $\underset{\sim}{D} = \underset{\sim}{H}/\underset{\sim}{I}$ is the homeotopy group of M, and $\underset{\sim}{D}$ contains the (normal) subgroup $D_\mathcal{B} = H_\mathcal{B}/\underset{\sim}{I}$ that we call the ∂-homeotopy group M. Then $\underset{\sim}{D}/D_\mathcal{B}$ is finite; for by an isomorphism theorem and Remark 8.4 we have

$$\text{Perm}\,\mathcal{B} \approx \underset{\sim}{H}/H_\mathcal{B} \approx \underset{\sim}{D}/D_\mathcal{B}\,. \tag{8.4}$$

Let $\mathcal{Y} = \mathcal{Y}(\mathcal{D})$ denote the family of all $\mathcal{D}$-sets on M. Then $\underset{\sim}{H}$ acts on $\mathcal{Y}$ by the following rule: if $\theta \in \underset{\sim}{H}$ and $U = \{\alpha_1,\ldots,\alpha_0\} \in \mathcal{Y}$, then $\theta.U$ is the $\mathcal{D}$-set $\{\theta(\alpha_i),\ 1 \leqslant i \leqslant \sigma\}$. We then have various notions of equivalence of $\mathcal{D}$-sets U, V, according as they are equivalent under the groups $H_{\mathcal{B}}$, $\underset{\sim}{I}$, $\underset{\sim}{H}(=H_{\emptyset})$, or H_C where $\emptyset \subseteq C \subseteq \mathcal{D}$. It is therefore of interest to study the orbit-sets in the sequence

$$\mathcal{Y}/I \to \mathcal{Y}/H_{\mathcal{B}} \to \mathcal{Y}/H_{\mathcal{D}} \to \mathcal{Y}/H_C \qquad (\emptyset \subseteq C \subseteq \mathcal{D})$$

of surjections, and our main theorem in this section is as follows.

8.5 THEOREM. $\mathcal{Y}/H_{\mathcal{B}}$ is finite, the quotient of $\mathcal{Y}/\underset{\sim}{I}$ by an action of the ∂-homeotopy group $\underset{\sim}{D}_{\mathcal{B}}$ of M . Also, $\mathcal{Y}/\underset{\sim}{H}_{\mathcal{D}}$ is the quotient of $\mathcal{Y}/H_{\mathcal{B}}$ by an action of Perm C' $(C'=\mathcal{B}-C)$.

Proof. If R, S are any two equivalence relations on a set X and xSy implies xRy , then (using an obvious notation) $X/R \approx (X/S)/(R/S)$. Hence

(i) $\mathcal{Y}/H_{\mathcal{B}} \approx (\mathcal{Y}/\underset{\sim}{I})/(H_{\mathcal{B}}/\underset{\sim}{I}) = (\mathcal{Y}/\underset{\sim}{I})/D_{\mathcal{B}}$,

(ii) $\mathcal{Y}/H_C \approx (\mathcal{Y}/H_{\mathcal{B}})/(H_C/H_{\mathcal{B}})$.

Therefore, by (i) and Proposition 8.1, the elements of $\mathcal{Y}/\underset{\sim}{H}_{\mathcal{B}}$, $\mathcal{Y}/\underset{\sim}{H}_C$ are, respectively orbits of $\mathcal{Y}/\underset{\sim}{I}$, $\mathcal{Y}/\underset{\sim}{H}_{\mathcal{B}}$ under the action of $\underset{\sim}{D}_{\mathcal{B}}$, and Perm C' . Hence it suffices to prove that $\mathcal{Y}/\underset{\sim}{H}_{\mathcal{D}}$ is finite.

We shall therefore construct a mapping

(iii) $\kappa : \mathcal{Y}/\underset{\sim}{H}_{\mathcal{D}} \to F$

where F is a finite set, and then prove that κ is an injection.

Suppose then that $U \in \mathcal{Y}$ and $h \in H_{\mathcal{D}}$. Then h induces a simplicial isomorphism $K(U) \approx K(hU)$; hence using the terminology of Def. 5.3, h also induces a thickened isomorphism between thickenings ΘK(U) and ΘK(hU) . Observe that ∂ΘK(U) contains the circles of $\mathcal{D}$, which are invariant under h . But h preserves further structure, as we now explain.

A Jordan-chain c in ΘK(U) will consist of a collection of edges in the linear graph ∂ΘK(U) whose union |c| forms a Jordan curve; let J(U) denote the family of Jordan-chains of ΘK(U) . If c ε J(U) we call c annular if |c| is one of the two boundaries of a component (annulus) of M − ΘKU (see Prop. 5.4). Let A(U) denote the family of annular Jordan chains of $\Theta K(U)$. Then card $A(U)$ is the number of components of $M - \Theta KU$, and given by Cor. 5.6. Also, the homeomorphism $\theta \in \underset{\sim}{H}$ preserves the pair $(K(U),A(U))$, in that

(iv) $\theta : K(U) \approx K(\theta U)$, $\theta(A(U) = A(\theta(U))$.

If the circles of $\mathcal{D}$ are labelled $C_1,\ldots,C_\pi$ (π = card($\mathcal{D}$)) let $C'_1,\ldots,C'_\pi$ be a collection of mutually disjoint copies of the C_i , in the plane $z = 0$ of $\mathbb{R}^3$. Let $J_{\pi\sigma}$ denote a minimal collection of representatives of all isomorphism classes of connected graphs formed from the circles C'_i and σ arcs, where $\sigma = -\chi(M)$. It is easily shown that $J_{\pi\sigma}$ is finite; and we may assume that each graph in $J_{\pi\sigma}$ lies in the upper half-space $z \geqslant 0$ of $\mathbb{R}^3$. Now choose a thickening ΘT (relative to $z \geqslant 0$) of each $T \in J_{\pi\sigma}$, and let the set of these thickenings be $K_{\pi\sigma}(\approx J_{\pi\sigma})$. Since the circles C'_i lie in $z = 0$, they also lie in the boundary of each $\Theta T \in K_{\pi\sigma}$. If $p : \Theta K(U) \to \Theta T$ is a thickened isomorphism, it is required to map $K(U)$ to T by conditions (i) and (iii) of Definition 5.3, and we call p a '$\mathcal{D}$-isomorphism' if $p(C_j) = C'_j$ for each j . From (iv) above, $\Theta K(U)$ and $\Theta K(\theta U)$ are $\mathcal{D}$-siomorphic to the same element ΘT of $K_{\pi\sigma}$.

Let $J(T)$ denote the family of Jordan chains of ΘT . Then for each $\mathcal{D}$-isomorphism $p : \Theta K(U) \to \Theta T$, $p(A(U))$ is a subset of $J(T)$ with m elements where m = card($A(U)$) . If, instead, we chose a $\mathcal{D}$-isomorphism $q : \Theta K(U) \to \Theta T$, then $q = \rho \circ p$ for some ρ in the group Q_T of (thickened) automorphisms of ΘT ; so $q(AU))$ is in the same orbit Ω of $\rho : J_m \to J_m$, where J_m is the set of all m-element subsets of $J(T)$. By (iv) above, if $V = \theta(U)$ then whatever $\mathcal{D}$-isomorphism $r : \Theta K(V) \to \Theta T$ we choose for V we find that $r(A(V))$ is in the same orbit Ω as before.

Hence, we have assigned to each orbit $\mathcal{U}.\underset{\sim}{H}$ of $\mathcal{Y}/\underset{\sim}{H}$ one member $\Theta T \in K_{\pi\sigma}$, and one orbit Ω in J_m/Q_T . The number of possible orbits Ω is clearly finite, and $K_{\pi\sigma}$ is finite since $K_{\pi\sigma} \approx J_{\pi\sigma}$. We therefore have a mapping κ , as in (iii) above, of $\mathcal{Y}/\underset{\sim}{H}$ into a finite set, where $\kappa(\mathcal{U}.\underset{\sim}{H})$ is the pair (K,Ω) .

It remains to prove that k is an injection. Therefore we must show that if U, W are $\mathcal{D}$-sets on M, with given $\mathcal{D}$-isomorphisms p, q respectively of $\Theta K(U)$, $\Theta K(W)$ to the same member $\Theta S \in \mathcal{K}_{\pi\sigma}$, and such that $p(A(U))$, $q(A(W))$ lie in the same orbit of Q_S, then U, W lie in the same orbit of $\underset{\sim}{H}$. But then there exists $\tau \in Q_S$ such that $\tau \circ p(A(U)) = q(A(W))$ so we have $\mathcal{D}$-isomorphism $\phi = q^{-1} \circ \tau \circ p : \Theta K(U) \approx \Theta K(W)$ such that $\phi(A(U)) = A(W)$. Now if R is any of the annuli of $M - \Theta K(U)$, as given by Prop. 5.4, one of its boundary components is $|c|$ for some $c \in A(U)$. Hence $\phi(c) \in A(W)$ so $|\phi(c)|$ is a boundary component of some annulus R' of $M - \Theta K(W)$; thus ϕ extends through R, to map R homeomorphically onto R'. In this way we extend ϕ throughout M to be a homeomorphism $M \to M$ that leaves the circles of $\mathcal{D}$ invariant. Hence $\phi(U) = W$ so U, W lie in the same orbit of $\underset{\sim}{H}$. This proves that κ in (i) is an injection, and completes the proof of the theorem.

8.4 <u>COROLLARY</u>. $\mathcal{Y}/\underset{\sim}{I}$ <u>is countable</u>

For, it is well-known that the homeotopy group $\underset{\sim}{D}$ is countable (see e.g. Birman [1]). Hence in (i) above, $D_{\mathcal{B}}$ is countable by (8.4), and $\mathcal{Y}/H_{\mathcal{B}}$ is finite.

9. THE CARDINALITY OF $\mathcal{Y}/H$.

It seems to be a difficult problem to list the sets $\mathcal{Y}/H_{\mathcal{D}}$, $\mathcal{Y}/H_{\mathcal{B}}$, or even to calculate their cardinals, except for small values of π and σ. In this Section, we look at the dependence of $\mathcal{Y}/H_{\mathcal{B}}$ upon $\mathcal{D}$. For simplicity we now assume that M is <u>orientable</u> and we write $U \sim \psi U$ when $U \in \mathcal{Y}$, $\psi \in H_{\mathcal{B}}$.

9.1 <u>LEMMA</u>. <u>If ∂M has just two components, then given $\mathcal{D}$-sets U, V on M, there exists a ∂-automorphism of M that carries U onto V.</u>

<u>Proof</u>. Let $U = \{\alpha_1, \ldots, \alpha_\sigma\}$, $V = \{\beta_1, \ldots, \beta_\sigma\}$ be two $\mathcal{D}$-sets on M. Since M is an orientable surface with two holes, σ is an even

number, by Corollary 5.6. Let the sole boundary component of ∂M in $\mathcal{D}$ be C .

Choose a thickening $A = \Theta K(U)$ in M ; it is a compact orientable surface with two holes, one of which is C . Also, by (5.1) $\chi(A) = \chi(M)$ so A has the same genus $g = \sigma/2$ as M . If $g = 0$, the lemma is obvious, and so we use induction on g . If $g \geq 1$ there are arcs $\alpha, \beta \in U$ such that the end-points p, q of α lie in distinct components of $C - \dot{\beta}$, otherwise A would be planar. Using the notation of Definition 5.2, remove from A the strip $S_\alpha - (D_p \cup D_q) = E_\alpha$ (say), and a similar strip E_β , to leave an orientable surface B of genus $g - 1$ that has $C_0 = \partial M - C$ as one of its two boundary components. The ends of E_α are two arcs that separate the curve $\partial B - C_0$ into two components, each of which contains an end of E_β . A similar construction, starting from a thickening $A' = \Theta K(V)$ leads to a surface B' , and strips $E_{\alpha'}$, $E_{\beta'}$ whose ends alternate on $(\partial B' - C_0)$ like those of E_α, E_β on $\partial B - C_0$. An inductive hypothesis yields a homeomorphism $\theta : B \approx B'$ which preserves C_0 and carries $U - \{\alpha,\beta\}$ to $V - \{\alpha',\beta'\}$. We extend θ over $E_\alpha \cup E_\beta$ in the obvious way to map them to $E_{\alpha'}$, $E_{\beta'}$, due care being taken over the alternation of their ends; and there results a ∂-automorphism $M \approx M$ that carries U to V as required.

9.2 COROLLARY. If ∂M has only 2 components (i.e. $h=1$) then $\mathcal{Y}/H_{\mathcal{B}}$ is a singleton.

Consider now the more general case when $\pi = \operatorname{card} \mathcal{D} \geq 2$. As in Lemma 8.3, let $\mathcal{D} = \alpha(A)$, $A = \{1,2,\ldots,\pi\}$. Recall from Cor. 5.6 that if $U \in \mathcal{Y}$ then $\operatorname{card} U = \sigma = -\chi(M)$. Also $\operatorname{card} \mathcal{B} = 1 + h > 2$.

9.3 THEOREM. The group Perm A acts on $\mathcal{Y}/H_{\mathcal{B}}$ in such a way that the isotropy subgroups are trivial, except when $\pi = 2$. If $\pi = 2$, and M is orientable, then $\sigma = h - 1$ and M is an annulus with $h - 1$ holes; and $\mathcal{Y}/H_{\mathcal{B}}$ is a singleton.

Proof. Given $\omega \in$ Perm A , we have a bijection $\beta = \alpha \circ \omega : A \to \mathcal{D}$, and hence, by Lemma 8.2, an automorphism $\phi_\omega \in H_{\mathcal{D}}$. If

$U = \{u_1,\ldots,u_\sigma\} \in \mathcal{Y}$, define $\omega.U = \{\phi_\omega(u_1),\ldots,\phi_\omega(u_\sigma)\}$. Then, if $U \sim V$ in $\mathcal{Y}$, there exists $\psi \in H_{\mathcal{B}}$ such that $U = \psi V$ so $\omega.U = \left(\phi_\omega \circ \psi \circ \phi_\omega^{-1}\right)(\omega.V)$, whence $\omega.U \sim \omega.V$. Hence we obtain an action of ω on $\mathcal{Y}/H_{\mathcal{B}}$.

To compute the isotropy subgroup J at $U.H_{\mathcal{B}}$, suppose $U \sim \omega U$ and $\omega \neq \mathrm{id}$. Then there exists $\psi \in H_{\mathcal{B}}$ such that $\psi(u_i) = \phi_\omega(u_i)$, $1 \leq i \leq \sigma$. Therefore, since $\psi \in H_{\mathcal{B}}$, then the end-points of $\phi_\omega(u_i)$ are on the same circles $C_{p(i)}$, $C_{q(i)}$ as those of u_i , and ω either fixes both $p(i)$, $q(i)$ or interchanges them. Since $\omega \neq \mathrm{id}$, then for some i , $p(i) \neq q(i)$ and ω interchanges them. Hence if $\sigma = 1$, then $\mathcal{D} = \{C_{p(1)}, C_{q(1)}\}$ so $\pi = \mathrm{card}\ \mathcal{D} = 2$.

Now suppose $\sigma > 1$. Since $K(U)$ is connected (by Cor. 5.5), then there exists $u_j \in U$ such that $q(i) = p(j)$, so $\omega(p(j)) = p(i)$. By the above 'interchange' property of ω , either $\omega(p(j)) = p(j)$ or $\omega(p(j)) = q(j)$. The first case conflicts with $p(i) \neq q(i)$; so $p(i) = q(j)$ and ω interchanges the ends of u_j as well as those of u_i . Hence all arcs of U that start on $C_{p(i)}$ must end on $C_{q(i)}$, and so $K(U)$ cannot be connected unless $\pi = 2$. Therefore if $\pi > 2$ then J is trivial.

If $\pi = 2$, then $A = \{1,2\}$ and J is Perm $A = \{\mathrm{id},\omega\} \approx Z_2$. Also, we have shown that $K(U)$ consists of the two circles $C_{p(i)}$, $C_{q(i)}$, joined by σ arcs like u_i , $(\sigma \geq 1)$. Now M is homeomorphic to a thickening $\Theta K(U)$; this is a union of two annuli and σ strips (like a Chinese lantern). Thus since M is orientable, then $\mathcal{B}$ has $\sigma + 2$ components, and $\sigma + 2 = 1 + h$, whence $\sigma = h - 1 = -\chi(M)$. Hence M is an annulus with $h - 1$ holes. If V is a $\mathcal{D}$-set which is also left fixed by ω , then each $\alpha \in V$ must run along one of the σ strips, so it is easy to see that $U = \psi(V)$ for some $\psi \in H$. Hence $\mathcal{Y}/H_{\mathcal{B}}$ is a singleton. This completes the proof.

REMARK. If M were not orientable, then $\Theta K(U)$ can have various forms since some of the strips may be twisted. (The resulting surface is of the type $S_{p,q,o}$ in the terminology of Griffiths [4], where a detailed analysis may be found.) For brevity we omit the appropriate extension to Theorem 9.3.

10. RECIPROCAL SETS

Next, we deal with the point that was raised after Definition 6.7, concerning the possible equivalence of any two reciprocals of a $\mathcal{D}$-set U on M. The proof of the following Proposition will indicate to what extent the relation '$\simeq$', of equivalence mod $\underset{\sim}{I}$, is stronger than that of equivalence mod $\underset{\sim}{H}_\partial$. The latter relation is denoted by '$\sim$'. Recall the notion of 'V-point' prior to 6.5.

10.1 PROPOSITION. If U is a $\mathcal{D}$-set on M, and V, W are two reciprocals of U, then $V \sim W$. Moreover, there exists $\eta \in H_\beta$ such that $\eta(W) = V$ and for each $\alpha \in U$, maps the W-point of α to the V-point of α, and leaves α invariant.

Proof. We use the reciprocal subdivisions of M, described in Prop. 6.3, that the reciprocal pairs (U,V) and (U,W) induce on M. Each subdivides every semi-annulus of $M - K(U)$ into discs and annuli, but possible in different ways.

Let R be one of the semi-annuli; then R has a boundary curve $C \notin \mathcal{D}$, while for its closure $\bar{R}$, $A = \bar{R} - R$ contains arcs of U and sub-arcs of circles in $\mathcal{D}$. For example, in Fig. 1 above if R is R_1, then $C = A_1$ and A is a Jordan curve; whereas if R is R_2, then $C = A_2$, and A contains the singular portion α' illustrated. Each arc α of U that forms part of A, is crossed just once at its V-point x by an arc $\delta \in V$; and δ ends on C at y. Thus $\delta \cap \bar{R}$ is a sub-arc γ of δ from x to y (see Fig. 3). Similarly, one arc $\delta' \in W$ meets A at its W-point u and ends at $v \in C$; let $\gamma' = \delta' \cap R$. There is a Jordan curve J on R that is homotopic to C and meets each of γ, γ' just once,

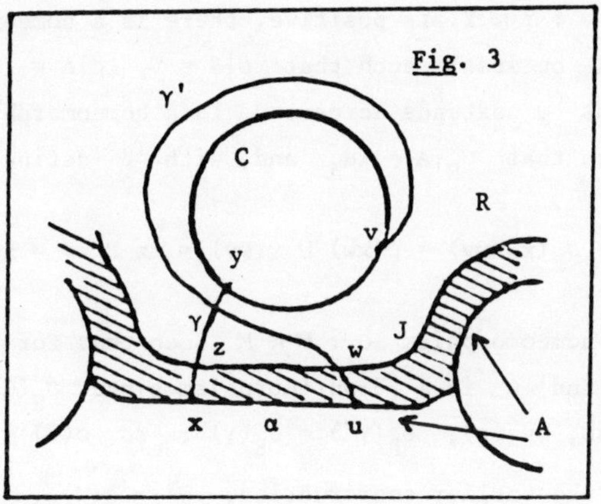

at z, w respectively. If J is sufficiently close to A, there is an isotopy θ of the annulus X between J and A, such that θ keeps each point of J fixed, leaves α invariant, and carries the sub-arc uw of γ' to the 'short' arc xw from x to w on R. Let β be the union of the arc xw with the sub-arc wv of γ'. Now δ' crosses α from R to another annulus R' on which δ' has a similar modification β' (like β) that passes through x. Let $\delta'' = \beta \cup \beta'$, and let W' be the set of all the modified curves δ'' as δ' runs through W. Thus $U \perp W'$ and by choosing the homotopies θ appropriately we obtain a ∂-automorphism $\phi : M \approx M$ such that $\phi(W) = W'$, φ is deformable to id_M, and $\phi | J_R = id$, for a curve J_R like J on each annulus R. Also φ leaves the arcs of U invariant.

Let Y be the closure of R − X, so Y is an annulus with boundary curves C, J. Any two of the p arcs $\gamma \cap Y (\gamma \varepsilon V)$ are disjoint and therefore isotopic on Y. Hence there is a positive homeomorphism λ that carries Y onto a standard oriented concentric annulus, and such that each arc $\gamma \cap Y$ is carried onto one of p equally spaced radii. Similarly a positive homeomorphism μ carries Y and the arcs $\gamma' \cap Y(\gamma' \varepsilon W)$ onto the same picture. Hence $\nu = \lambda^{-1} \circ \mu$ maps Y onto itself and carries each $\gamma \cap Y$ onto $\gamma' \cap Y$; in particular, in Fig. 3, vw is carried to yz.

Since $\nu : Y \approx Y$ is positive, there is a homeomorphism ρ of the annulus X on itself such that $\rho|J = \nu$, $\rho|A = id_A$ and $\rho(xw) = xz$. Thus ν extends across J to a homeomorphism $\sigma_R : R \approx R$, such that $\sigma_R|A = id_A$ and, with β defined as above,

$$\text{(i)} \qquad \sigma_R(\beta) = \sigma_R(xw+wv) = \rho(xw) \cup \nu(wv) = xz \cup zy = xy .$$

Hence there is a homeomorphism $\sigma : M \approx M$ such that for each annulus R, $\sigma|R = \sigma_R$; and σ is a ∂-automorphism since $\sigma_R(C) \subseteq C$ and $\sigma_R|A = id_A$. Also, by (i), $\sigma_R(\gamma'') = \sigma_R(\gamma)$, so $\sigma(V) = W'$.

But by the earlier construction, $W' = \phi(W)$, so if $\eta = \sigma^{-1}\circ\phi$ then $V = \eta(W)$ and $\eta(u) = \sigma^{-1}(\phi(u)) = \sigma^{-1}(x) = x$. Thus for each $\alpha \in U$, η maps its W-point to its V point. Also, by construction, η leaves invariant each arc of U, as required. This completes the proof.

10.2 REMARK. Observe that the homeomorphism σ_R in (i) is a 'Lickorish twist' and that σ is a composition of the σ_R's. This accords with Lickorish's theorem (see Birman [1]) that $D_{\mathcal{B}}$ is generated by the twists.

We are now in a position to show that, for a monic elliptic measure $f : M \to \mathbb{R}$, the set $\mathcal{S}(f)$ characterises the portrait $\Pi(f)$.

10.3 THEOREM. Let $f, g : M \to \mathbb{R}$ be two monic elliptic measures with the same sources, such that $\mathcal{S}(f) \sim \mathcal{S}(g)$. Then f, g have equivalent portraits.

Proof. There is a ∂-automorphism $\theta : M \to M$ such that $\theta(\mathcal{S}(f)) = \mathcal{S}(g)$. Therefore θ maps each stable manifold A of f to one of g. Thus, if c is the critical point of f on A, and B is the unstable manifold of f through c, then $f\circ\theta$ has $x = \theta(c)$ as a critical point through which $\theta(A)$, $B' = \theta(B)$ pass as stable and unstable manifolds; $\theta(A)$ has the same underlying point set as some stable manifold P of g, and P is crossed by exactly one unstable manifold of g and at some critical point y of g. Therefore the sets $V = \mathcal{U}(f\circ\theta)$, $W = \mathcal{U}(g)$ are each a reciprocal of $\mathcal{S}(g)$; and x, y are respectively the V-point and the W-point of A .

By Prop. 9.4 therefore, there exists $\eta \in H_\beta$ such that $\eta(V) = W$, and η carries V-points to W-points, while η leaves each arc of V invariant. This means that the functions g, $f\circ\theta\circ\eta$ have the same critical points, and unstable manifolds. But now Prop. 6.4 applies, so $g \sim f\circ\theta\circ\eta \sim f$; hence by Prop. 2.1, $\Pi(g) \sim \Pi(f)$; and the proof is complete.

11. THE REALISATION THEOREM

We shall now show that if M is orientable, every $\mathcal{D}$-set U on M can be 'realized' by a harmonic measure $f : M \to \mathbb{R}$ in the sense that for some $\theta \in I$ the arcs of U are mapped by θ to the stable manifolds of f , so $U \simeq \mathcal{S}(f)$. However, the word 'harmonic' now needs clarifying. We consider M with a Riemannian metric μ ; and a continuous function $g : M \to \mathbb{R}$ is harmonic provided $Lf = 0$ on $M - \partial M$, where L denotes the Laplace-Beltrami operator which is defined in terms of μ . To emphasise the role of μ , we write (M,μ) for M and then say that g is 'μ-harmonic'. Our previous rough assertion about f then obscures a weakness (which we do not know how to strengthen) that arises from our method of proof of Theorem 11.1 below; f turns out to be only ν-harmonic for <u>some</u> metric ν , not necessarily the given metric μ . For a planar region M , this amounts to changing the conformal type of M . The realisation theorem can be stated as follows.

11.1 THEOREM. <u>Let</u> U <u>be a</u> $\mathcal{D}$<u>-set on</u> (M,μ) , <u>where</u> M <u>is orientable</u>. <u>Then there is a metric</u> ν <u>on</u> M <u>and a</u> ν<u>-harmonic measure</u> $f_p : M \to \mathbb{R}$ <u>(levelled at</u> $p \in \mathbb{R}^{h+1}$) <u>such that</u> f_p <u>is</u> ND <u>and</u> $U \simeq \mathcal{S}(f_p)$.

<u>Proof</u>. Since the proof is lengthy we divide it into several steps.

<u>Step 1</u>. Suppose first that ∂M has just two components. Then by Cor. 5.7 we may assume that M is not an annulus, and card $\mathcal{D} = 1$, - so $\mathcal{D}$ consists of a single boundary curve C of M . We begin by

showing that there is a metric λ on M and an ND function $f_p : M \to \mathbb{R}$ which is λ-harmonic, and where $p = (0,1)$ while f_p is zero on C, and 1 on $\partial M - C$; automatically the normal derivative along the boundary never vanishes, by Courant-Hilbert [2] p.326.

As M is orientable, then $\chi(M) = -2g$ where g is the genus. Suppose first that $g = 1$. With any metric μ on M there is a μ-harmonic function f_p levelled at $p = (0,1)$, and we claim that it is ND. For, near a critical point, f_p is homeomorphic to some harmonic function $z \mapsto \text{Real}(z^{1+n})$ where $n \geq 1$ is the multiplicity. By Morse Theory, the sum of the multiplicities is $-\chi(M)$ where $\chi(M)$ here is -2; so either f_p has two critical points, each of multiplicity 1 - whence f_p is ND - or f_p has only one degenerate critical point X of multiplicity 2. But in the latter case, by the usual analysis of the growth of half-spaces, M would be the union of an annulus N with a disc D containing X, with a disc D containing X, where D is attached to one boundary of N along 4 mutually disjoint sub-arcs. Thus M would be planar which is impossible. Hence f_p is ND.

If $g > 1$, we represent M in $\mathbb{R}^3$ as a cylinder from which g discs have been removed and replaced by once-punctured toruses $T_1,\ldots,T_g$, the whole being invariant under a rotation R of the cylinder about its axis, through $2\pi/g$. The metric μ induced from $\mathbb{R}^3$ is invariant under R and hence so is the μ-harmonic function f_p on M, levelled at $p = (0,1)$. Therefore the critical points of f_p are also invariant under R, and each is of multiplicity 1 by an argument similar to the one above when $g = 1$. Hence f_p is ND. (A similar argument, using Moebius bands instead of punctured toruses, gives an ND function when M is non-orientable.)

As stated above, we may start then with a metric λ on M and an ND function f_p levelled at $p = (0,1)$; so, the set $\mathcal{S}(f_p)$ of its stable manifolds forms a $\mathcal{D}$-set. Given the $\mathcal{D}$-set U on M, then by Lemma 9.1 (since M is orientable) there is a ∂-automorphism $\theta : M \to M$ that maps each arc in U to a stable manifold of f_p. In case θ is not smooth, there exists a diffeomorphism $\phi : M \approx M$ so close to θ that, if $V = \phi^{-1}(\theta U)$ then $U \simeq V$, by Cor. 7.5. Now $\phi(V) = \theta(U) = S(f)$ so $U \simeq V = S(g)$ where $g = f_p \circ \phi$. But g is ν-harmonic, where ν is the metric induced on M by ϕ, for ϕ is then an isometry $(M,\nu) \to (M,\lambda)$. This proves the theorem when the set $\mathcal{B}$ of components of ∂M has two elements.

<u>Step 2</u>. Suppose therefore that card $\mathcal{B} = k > 2$. Then since M is orientable, every $\mathcal{D}$-set U on M has σ elements in U, where $\sigma = k - 2 + 2\gamma$ and γ is the genus of M. If card $\mathcal{D} = d > 1$, then by condition (ii) of Definition 5.1, $K(U)$ is connected, so there exist circles C_i, $C_j \in \mathcal{D}$, $(i \neq j)$ and $u \in U$ such that u has one end-point on each of C_i, C_j. We call u a <u>strap</u>, and so U has a strap if $d > 1$.

If $d = 1$, let $\mathcal{D}' = \mathcal{B} - \mathcal{D}$ so that card $\mathcal{D}' = k - d > 1$. Hence any reciprocal V of U has a strap. Suppose that we could prove the theorem for V; then there would exist a metric λ on M and a ND λ-harmonic measure $g : M \to \mathbb{R}$ such that $V \simeq S(g)$. Then there exists $\psi \in \underset{\sim}{I}$ such that $\psi(V) = S(g) \perp \psi(U) = W$. Also, $S(g) \perp U(g)$, so, by Proposition 10.1, there exists $\eta \in H_{\mathcal{B}}$ such that $\eta(W) = U(g)$. Therefore $U = U(g \circ \theta)$ where $\theta = \eta \circ \psi$; so $U = S(-g \circ \theta)$. As in Step 1, we conclude that $U \simeq S(h)$ where $h = -g \circ \phi$ and ϕ is a diffeomorphism close to θ. But then h is ND and ν-harmonic, where ν is induced from λ by ϕ. We may always suppose then, that $d > 1$ and U has a strap.

<u>Step 3</u>. In the simplest case now, M has genus zero and just three boundary components, (so M is planar). Then U is realised by one of the harmonic measures $f_{e(i)}$ in (3.1) (e.g. by $f_{e(0)}$ if u joins C_1 to C_2); it is ND by Griffiths [3].

<u>Step 4</u>. Suppose then that either $k > 3$ or $\gamma > 0$ (so $\sigma \gtrsim 2$, see Step 2). We must construct several subsets of M, as follows, the main ones being shown in Fig. 4. On M let U have a strap u joining C_i to C_j, and let $D \subseteq M$ be a planar domain with three boundary components C_i, C_j and (say) E such that $D - E$ is a

neighbourhood in M of $C_i \cup C_j \cup u$. Suppose too that D does not meet $\partial M - (C_i \cup C_j)$ and that if $\alpha_k \in U - \{u\}$ then $\alpha_k - D$ is connected. Let L be the closure of $M - D$, so that L is a compact sub-surface of M with $k - 1$ boundary curves of which $k - 2$ are those of $\partial M - (C_i \cup C_j)$ and the last is $E = L \cap D$. Let $\mathcal{D}_1$ consist of the components of $\mathcal{D}$ in ∂L, together with E, and let

$$V = \{\alpha_i \cap L | \alpha_i \in U - \{u\}\} .$$

Then V is a $\mathcal{D}_1$-set; to see this it suffices to show that some α in U is such that $\alpha \cap L \in V$, $\alpha \neq u$, and α meets E. But since $K(U)$ is connected and $\sigma \geqslant 2$ some $\alpha \neq u$ must have an end-point on $C_i \cup C_j$, so α meets E.

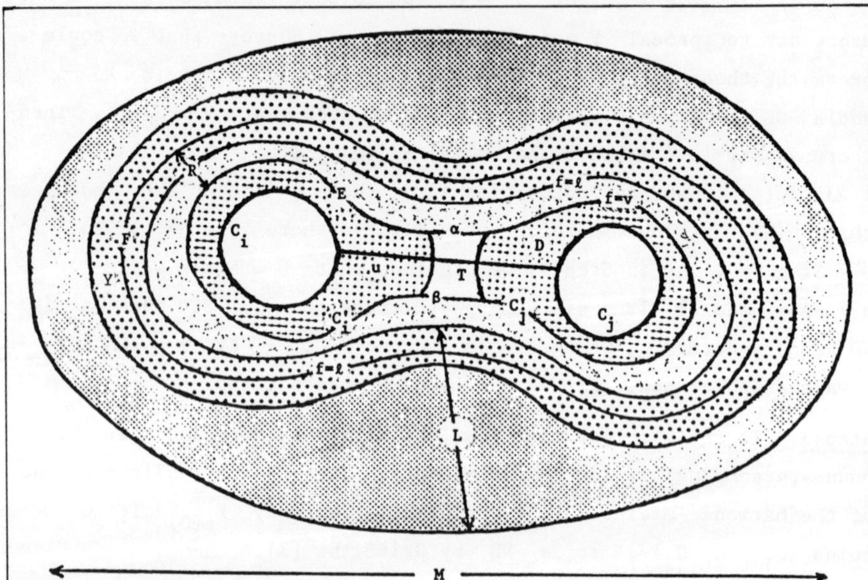

Fig. 4.

Observe that, if $\tau = \text{card}(V)$ then using the formula in Step 2 for $\sigma = \text{card}(U)$ we have

(i) $$\tau = \sigma - 1 = k - 3 + 2\gamma = -\chi(L) ,$$

and $\tau \geqslant 1$ since $\sigma \geqslant 2$. Also $\text{card}(\mathcal{D}_1) = d - 1 \geqslant 1$, so we can use induction on d .

<u>Step 5</u>. Now $(L,\mu) \subseteq (M,\mu)$ as Riemannian spaces. We therefore apply an obvious inductive hypothesis to the $\mathcal{D}_1$-set V on L . Thus, L has a metric $\bar{\mu}$, and there is a $\bar{\mu}$-harmonic ND measure $f : L \to \mathbb{R}$ levelled at some $p \in \mathbb{R}^{k+1}$, such that

(ii) $$V = S(f) .$$

Also L is a closed subspace of M , and $\bar{\mu}$ is a section over L of the tensor bundle over M of 3×3 symmetric positive-definite matrices. Hence by Steenrod [12] Theorems 12.2, 6.7, $\bar{\mu}$ has a smooth extension (still denoted by $\bar{\mu}$) to the whole of M . Thus, now $(L,\bar{\mu}) \subseteq (M,\bar{\mu})$.

By construction of E in Step 4, $E \in \mathcal{D}_1$ and is a source for f . Hence since f is ND , the normal derivative does not vanish on E . Let v denote the value of f, on E . Then there exists $\ell > v$ such that the set $\{x \in L | v \leqslant f(x) \leqslant \ell\}$ contains no critical points of f ; and one of its components, say R , is an annulus with two boundary components of which one is E , and the other is (say) F . For a sufficiently small $\varepsilon > 0$, the set

(iii) $$\{x \subset L | \ell - \varepsilon \leqslant f(x) \leqslant \ell + \varepsilon\}$$

has a component Y in L which is an annulus such that $F \subseteq \text{Int}\, Y$.

We now construct a surface N diffeomorphic to M such that $L \subseteq N \subseteq M$, and with certain other properties. To do this, let T be a narrow strip (whose width we adjust later) which crosses the strap u once, as in Figure 3 and whose ends lie on arcs α, β of the curve E . Then $N = L \cup T$ is the surface we require, that is (clearly) diffeomorphic to M . Also, N has the same boundary components as M , except that C_i and C_j are replaced by the 'key-hole' curves C_i' ,

C_j' in Figure 3, composed respectively of arcs of C_i, C_j and ∂T. Thus $\partial N = (\partial L - E) \cup C_i' \cup C_j'$.

On $(N,\bar{\mu})$ let $q : N \to \mathbb{R}$ denote the $\bar{\mu}$-harmonic function which takes the value v on $C_i' \cup C_j'$ and which agrees with f on all remaining components of ∂N. Therefore on $L \subseteq N$, the values of q lie between those of f, i.e. the oscillation of $q \lesssim \mathrm{osc}(f)$. Also the function $f - q : L \to \mathbb{R}$ is $\bar{\mu}$-harmonic, and it is zero at all boundary points except those of $\alpha \cup \beta$. Hence by a standard formula of Potential Theory (see Springer [11] p.218), $f - q$ is given at each $x \in L$ by

$$f(x) - q(x) = -\frac{1}{2\pi}\int_{\alpha\cup\beta} [q(t)-v]\partial_n g(x,t)ds_t$$

where ∂_n denotes the outward normal derivative, s denotes arc-length, and $g(x,t)$ is the value at $t \in \alpha \cup \beta$ of the Green's function of $(L,\bar{\mu})$ with pole x. In particular, if x lies in the annulus Y (see (iii)), and

$$G = \sup\{|\partial_n g(x,t)| \,\big|\, t \in \alpha \cup \beta,\ x \in Y\}$$

then

(iv) $$|f(x) - q(x)| < Gm.\mathrm{osc}(f)/2\pi$$

where m is the $\bar{\mu}$ - length of $\alpha \cup \beta$. Therefore if m is made sufficiently small, then the level $\{q=\ell\}$ of q on N has just one component J inside the annulus Y, and J cannot bound a disc in Y, by the maximum principle. Hence J separates N into two components P, Q of which P (say) contains the strap u. Then $g = q|P$ is a harmonic measure on P with values v, v, ℓ on the 3 boundary curves C_i', C_j', J of ∂P. Since $v < \ell$, then C_i', C_j' are sources for g.

Now P is a planar domain, and the values of g on ∂P are its absolute maxima and minima. Therefore, as explained in Griffiths [3], g has just one critical point c, which lies in Int P. There is then a stable manifold A of g through c, that joins C_i' to C_j'; and A is also a stable manifold of q.

Further, if m in (iv) is sufficiently small, then (see Kellogg [5] p.227), the derivatives of f and q on L - calculated in a fixed atlas of isothermal co-ordinate charts - can be made so close that, by stability, there is a ∂-automorphism ψ of L which carries the set $S(f)$ of stable manifolds of f to those of q. (The meaning of this statement should be clear, even though q is not constant on E.) We may also suppose that ψ preserves orientation.

Finally, we extend ψ across E to a diffeomorphism $\phi : M \to N$, as follows. Recall that $M = L \cup D$, where $D \cap L = E$; and D is a planar domain with $\partial D = C_i \cup C_j \cup E$. Since ψ preserves orientation and $\psi(E) = E$, then $\psi|E$ extends to a homeomorphism $D \to P$ that carries C_i, C_j, u respectively to C_i', C_j', A, and such that for each $\alpha_i \in U - \{e\}$, ϕ maps α_i to $\alpha_i \cap L \in V$. Thus, ψ is now extended to the required $\phi : M \to N$, and using (ii) we have - with an obvious notation -

$$U - \{u\} \overset{\phi}{\to} V = S(f) \overset{\psi}{\to} S(q|L); \quad u \overset{\psi}{\to} A \in S(q)$$

so ψ carries U to $S(q)$. As in Step 1 we now find a smooth approximation ψ' to ψ, to get $U \simeq S(q\circ\psi')$; and $q\circ\psi'$ is ν-harmonic where ν is the metric induced by ψ' from the metric $\bar{\mu}$.

This completes the proof, by induction on card $(\mathcal{D})$. Note that it does not guarantee that $q\circ\psi'$ is <u>monic</u>.

12. GENERIC FAMILIES OF FUNCTIONS

We can summarize our results in a 'generic' way as follows. Let $\mathcal{E}$ be the set of all monic elliptic measures on M that have $\mathcal{D}$ as their set of sources, and let $\underset{\sim}{R}$ denote the group of increasing homeomorphisms $\mathbb{R} \approx \mathbb{R}$. Our previous notion of equivalence of portraits suggests that we factor $\mathcal{E}$ by the action of $G = R \times H_{\mathcal{D}}$, where

$$f.(\rho,\theta) = \rho^{-1}\circ f\circ\theta \qquad (f \in \mathcal{E},\ \rho \in R,\ \theta \in H_{\mathcal{D}}) .$$

The orbit of f is then the set of homeomorphs of f and corresponds to the equivalence class of portraits that contains $\Pi(f)$. We then have a map into the finite set $\mathcal{Y}/H_{\mathcal{D}}$ (see Theorem 8.5):-

$$\eta : \mathcal{E}/\underset{\sim}{G} \to \mathcal{Y}/H_{\mathcal{D}}$$

induced by assigning to each $f \in \mathcal{E}$, its $\mathcal{D}$-set $\mathcal{S}(f)$.

12.1 <u>THEOREM</u>. *If M is orientable, then $\eta : \mathcal{E}/\underset{\sim}{G} \to \mathcal{Y}/H_{\mathcal{D}}$ is a bijection.*

Proof. By Theorem 10.4, η is an injection. To see that η is a surjection, let $U \in \mathcal{Y}$. By the Realisation Theorem 11.1, there is a ν-harmonic measure $f : M \to \mathbb{R}$ such that $U \simeq \mathcal{S}(f)$. If f is not already monic, then f can be C^2-approximated by a smooth $g : M \to \mathbb{R}$ such that (see 2.3) g is ND and monic, with critical points all of index 1, and such that f and g agree except near crit f. Therefore g is also a smooth elliptic measure (which may not be ν-harmonic); so $g \in \mathcal{E}$. If g is chosen sufficiently close to f, Cor. 7.5 applies, and then $\mathcal{S}(f) \simeq \mathcal{S}(g)$. Thus U and $\mathcal{S}(g)$ are in the same orbit $U.H_{\mathcal{D}}$ of $H_{\mathcal{D}}$ (indeed of $\underset{\sim}{I}$) so $U.H_{\mathcal{D}} = \eta(g.G)$. This completes the proof.

To show topological invariance, suppose M_1 is a surface homeomorphic to M, and $\mathcal{D}_1$ a set of its boundary components such that $\text{card}(\mathcal{D}_1) = \text{card}(\mathcal{D})$. Then an argument like that for (9.1) shows:

12.2 <u>LEMMA</u>. *There is a homeomorphism $\psi : M \to M_1$ which carries $\mathcal{E}/G$, $\mathcal{Y}(\mathcal{D})/H_{\mathcal{D}}$ bijectively to the corresponding items for $(M_1, \mathcal{D}_1)$.*

Now let $\mathcal{E}_s \subseteq \mathcal{E}$ denote the subset of <u>smooth</u> functions. Since $\mathcal{E}_s$ depends on $\mathcal{D}$, we write (temporarily) $\mathcal{E}_s = \mathcal{E}_s(\mathcal{D})$, and consider the set $\mathcal{F}$ of all smooth monic elliptic measures on M. By the argument used in the proof of Theorem 12.1, $\mathcal{F}$ is a dense open subset of the set of smooth elliptic measures in the C^2 topology. Also, $\mathcal{F}$ is a disjoint union of subspaces $\mathcal{E}_s(\mathcal{D})$ as $\mathcal{D}$ runs through the finite family of all possible sets of sources on M. A relation P, of 'having equivalent portraits', is defined on each such subspace; so passage to the quotient yields a set $\mathcal{F}/P$. By Theorem 12.1, we have:-

12.2 <u>THEOREM</u>. M <u>being orientable</u>, F/P <u>is finite</u>.

Hence, the natural map $F \to F/P$ splits F into mutually disjoint non-empty subsets:-

$$F = F_1 \dot{+} F_2 \dot{+} \dots \dot{+} F_c , \quad c = \mathrm{card}(F/P) ; \qquad (12.1)$$

and each F_i is <u>open</u> in F, by structural stability of the kind discussed after (3.3). <u>All functions in</u> F_i <u>have the same portrait</u> (in the sense of H_B-equivalence). In particular, if f, g in F_i lies in $E_s(D)$, $E_s(D_1)$ respectively, then $\Pi(f) \sim \Pi(g)$, and $\mathrm{card}(D) = \mathrm{card}(D')$.

Given $f \in F$, let f_i denote the (constant) value of f on the component C_i of ∂M . Thus f is an <u>extension</u> (over M) of the vector $(f_0, f_1, \dots, f_h) \in \mathbb{R}^{h+1}$. From (12.1), we conclude, therefore:-

12.3 <u>THEOREM</u>. <u>If</u> $v \in \mathbb{R}^{h+1}$, <u>and</u> M <u>is orientable, then among all possible extensions</u> $f \in F$ <u>of</u> v , <u>at most</u> c <u>distinct portraits</u> $\Pi(f)$ <u>can occur</u>.

The most familiar extensions of v are those arising from certain (not necessarily linear) elliptic operators on M of a kind for which the Dirichlet problem is always uniquely solvable. Hence Theorem 12.3 suggests questions of the sort: which operators yield the same decomposition (12.1)? We have no useful answers.

The Theorem also suggests a 'generic' form of the Dirichlet problem, which can be expressed roughly in the form: given a generic function $g : \partial M \to \mathbb{R}$, find all possible distinct types of portrait $\Pi(f)$ that can occur, when $f : M \to \mathbb{R}$ extends g , where f is smooth, ND , monic and such that any critical point in Int M is a saddle. This formulation is 'rough' for technical reasons, and a precise discussion will follow in Sections 14 below.

13. ELLIPTIFICATIONS

In this Section, we consider how to classify the portraits of ND functions on a <u>closed</u> surface. Thus let N be any smooth compact surface such that $\partial N = \emptyset$ and let $f : N \to \mathbb{R}$ be a smooth ND function. Then f has, say α, β, σ, maxima, minima and saddle points, respectively, where $\alpha + \beta - \sigma = \chi(N)$. Each maximum or minimum critical point of f lies interior to a disc, a component of a set of the form $\{f > h\}$ or $\{f < h\}$ respectively. For a fixed choice of such discs, all so small that none contains more than one critical point, let N_1 be their complement in N. Then N_1 is a compact surface whose boundary has $\alpha + \beta$ components and $f|N_1$ is an elliptic measure on N_1; its set $\mathcal{D}_1$ of sources consists of the boundaries of those discs that contain minima of f on N.

13.1 <u>Definition</u>. An <u>elliptification</u> of f is a triple of the form $(N_1, f|N_1, \mathcal{D}_1)$.

To define $\Pi(f)$ on N we use a Riemannian metric ν, and restrict ν to N_1, in order to define $\Pi(f|N_1)$, which is then the restriction of $\Pi(f)$ to N_1.

If $(N_2, f|N_2, \mathcal{D}_2)$ is a second elliptification of f, then $(N_1 \cap N_2\ f|N_1 \cap N_2, \mathcal{D}_3)$ is a third; and $\Pi(f|N_1 \cap N_2)$ is a common restriction of $\Pi(f|N_i)$ $(i = 1, 2)$. Since all three portraits contain the saddle points of f, we shall say that <u>all three are of the same type</u>. It is clear how to reconstruct $\Pi(f)$ from the portrait of an elliptification of f.

An argument like that used for G in the proof of Theorem 12.1 yields

13.2 <u>PROPOSITION</u>. <u>If</u> $f : N \to \mathbb{R}$ <u>is smooth and</u> ND <u>then any monic modification of</u> f <u>has the same portrait as any other</u>. <u>Moreover</u> $S(f) \simeq S(g)$.

It is not clear how to reconstruct f from a monic modification, so we now suppose that f itself is <u>monic</u>. The same then holds for any elliptification $g = (N_1\ f|N_1, \mathcal{D}_1)$ of f. Also let $N_{\alpha+\beta}$ denote a standard (compact) model of the result of making $\alpha + \beta$ holes in N, and let $\mathcal{D}$ consist of the first β of the boundary curves $C_1, \ldots, C_{\alpha+\beta}$ of $N_{\alpha+\beta}$. We can then form the set $Y = \mathcal{Y}(\mathcal{D})/H_{\mathcal{D}}$, relative to $N_{\alpha+\beta}$.

Then, if N is orientable, Theorems 12.1, 12.2 show that the equivalence class $\Pi(g)$ is determined uniquely within Y. Now Y has a finite set of representative $\mathcal{D}$-sets U, and so is equally well represented by the corresponding set

$$\Gamma(N, \alpha, \beta) \tag{13.1}$$

of linear graphs $K(U)$. Let $\sigma : Y \to \Gamma(N,\alpha,\beta)$ denote the corresponding mapping.

Now let $\mathrm{Port}_{\alpha,\beta}(N)$ denote the family of equivalence classes of portraits of monic, smooth, ND functions $f : N \to \mathbb{R}$ that have α maxima and β minima. The above description of the passage

$$\zeta : \Pi(f) \longrightarrow \Pi(g) \longrightarrow S(g) \overset{\sigma}{\longmapsto} \Gamma(N,\alpha,\beta) \tag{13.2}$$

therefore proves:-

13.3 THEOREM. Suppose N is an orientable closed connected surface. Then there is a bijection $\mathrm{Port}_{\alpha,\beta}(N) \approx \Gamma(N,\alpha,\beta)$ between portraits and graphs.

14. SYMMETRIC FUNCTIONS ON ∂M : THE GENERIC EXTENSION PROBLEM

The typical 'generic' function $q : \partial M \to \mathbb{R}$ is of the following form. Let $q_i = q|C_i$; then $q_i : C_i \to \mathbb{R}$ is a smooth non-degenerate function with an even number $(2m_i > 0)$ of critical points which alternate round C_i as maxima and minima. These divide C_i into $2m_i$ arcs which are sensed from the minimum end-point to the maximum. Then ∂M, divided into these sensed arcs, is called the portrait $\Pi(q)$ of q ; and we write

$$m(q) = (m_0, m_1, \ldots, m_h) \in \mathbb{N}^{1+h} \tag{14.1}$$

If $q' : \partial M \to \mathbb{R}$ is also generic, then q and q are homeomorphs iff $m(q) = m(q')$.

A smooth extension $f : M \to \mathbb{R}$ of q may behave awkwardly near ∂M (see Chapter I of Morse [7], so we avoid some technicalities by passing to the double, $2M$, of M (see (2.1)). Thus $2M$ is a compact surface without boundary, formed from two copies $M_<$, $M_>$ of M, united along their common boundary. The differentiable structure on M induces a unique one on $2M$, in which the involution ζ of $2M$ (see (2.1)) is smooth. We identify $M_<$ with M. Also, there is a Riemannian metric μ on $2M$ which is invariant under ζ (choose a product metric on M near ∂M, extend it smoothly over M, and use ζ to extend it over $M_>$). A continuous function $g : 2M \to \mathbb{R}$ will be called <u>symmetric</u> provided

$$g \circ \zeta = g \; . \qquad (14.1)$$

If g is also smooth then its portrait $\Pi(g)$ is invariant under ζ since ζ preserves critical points, level curves and - by the invariance of μ - the orthogonal trajectories. We write $g_\partial = g|\partial M$.

In particular if $f : M \to \mathbb{R}$ is a smooth ND function, then f has a symmetric extension $g : 2M \to \mathbb{R}$ defined by the equations

$$g|M = f \; , \quad g|M_> = f \circ \zeta \; ;$$

however, although g is continuous (since $\zeta|\partial M$ = identity) it may not be smooth on ∂M. This would not matter if the trajectories of f behave well enough near M, for then g is still portrayable. Alternatively, we could modify g arbitrarily closely to ∂M, to obtain a smooth ND symmetric $h : 2M \to \mathbb{R}$ such that h_∂ is a homeomorph of $f|\partial M = h_\partial$. All such alterations h are easily seen to be homeomorphs, and if $f|\partial M$ is generic then (see 14.1)

$$m(f|\partial M) = m(h_\partial) \; . \qquad (14.3)$$

We could then pose the 'Generic Dirichlet problem' in the form: <u>given a generic</u> $q : \partial M \to \mathbb{R}$, <u>classify the portraits of all smooth</u> ND <u>symmetric</u> $g : 2M \to \mathbb{R}$ <u>for which</u> $m(g_\partial) = m(q)$. However, this formulation allows functions g with 'topologically unnecessary' critical points in the following sense. Let μ_i denote the number of critical points of index i that g possesses on $2M$; then we have the well known Morse relation

$$\mu_0 - \mu_1 + \mu_2 = \chi(2M) = 2\chi(M) \ . \qquad (14.4)$$

We therefore impose the extra requirement on g that $\mu_0 + \mu_2$ be as small as possible - but to see how small we need the following analysis.

Thus let $g : 2M \to \mathbb{R}$ be symmetric as in (14.2) and also smooth. If g is ND and $g|\partial M$ is non-degenerate, then g has critical points of two kinds: a finite set K_∂ of critical points on ∂M, and a set of pairs

$$c_i \in \text{Int } M \ , \ \zeta(c_i) \in \text{Int } M_> \ , \ 1 \leq i \leq r \ . \qquad (14.5)$$

Then K_∂ is the disjoint union of sets $K_i = C_i \cap K_\partial$ $(i = 0,1,\ldots,h)$; we suppose as usual that $h > 0$ in (14.1) although we shall refer later to the rather special case when $h = 0$. Suppose also that g_∂ is generic, so $m(g_\partial) = (m_0,\ldots,m_h)$ as in (14.1). Firstly we see that $K_i \neq \emptyset$, by:-

14.1 LEMMA. $K_i = \text{crit}\ (g_\partial|C_i)$ so card $K_i = 2m_i > 0$.

Proof. On the arc A between any two critical points u, v of g_∂ on C_i, g_∂ is monotonic; so level curves of g on $2M$ cross A transversely. Hence A, being invariant under ζ in (14.2), must be a trajectory of g, so $u, v \in \text{crit}\ (g)$. The Lemma follows.

The critical points of g_∂ on C_i are alternately maxima and minima of g_∂ (but not perhaps of g). We describe them as follows.

From the proof of Lemma 14.1, we note that since g is ND, then u, v cannot both be saddles, since the trajectory A joins them. Hence if u (say) is a saddle, v has even index and the pair u, v contributes zero to the left hand side of (14.4). But it would be possible for u, v to be a maximum and a minimum, in which case the sum $\mu_0 + \mu_2$ is not 'as small as possible' in our sense following (14.4).

If this case is not allowed, we call g <u>alternating</u>. But then, as we move round C_i in the direction from u to v, the next critical point w of g after v must be a saddle like u; for if g_∂ rises (or falls) along A from u to v, it must fall (or rise) as we pass to w, since g_∂ is generic. After w we rise (or fall) to the next critical point z, which must therefore have the same (even) index as v; and so on as we traverse C_i. We have shown:-

14.2 <u>LEMMA</u>. <u>If</u> g <u>is alternating, then its critical points on</u> C_i <u>are alternately of index</u> 1 <u>and</u> $k = k(i) \neq 1$.

Letting $A = \{0,1,\ldots,h\}$, we can therefore define the set $Z(g) \subseteq A$ by

$$Z(g) = \{i \mid k(i) = 0\} \subseteq A .$$

Thus if $i \in Z(g)$ then the critical points of g on C_i are alternately saddles and <u>minima</u>. Since g is continuous and $2M$ is compact, $Z(g)$ is a <u>proper</u> subset of A.

By Lemma 14.2 also, the contribution from ∂M to the left side of (14.4) is zero, so we can now specify the allowable functions $g : 2M \to \mathbb{R}$ in our discussion of the Dirichlet problem.

14.5 <u>Definition</u>. We say that a smooth ND symmetric function $g : 2M \to \mathbb{R}$ is <u>minimal</u> iff it is (i) alternating and (ii) the points of crit (g) in Int M are saddles. (Thus, by (14.5), g has no maxima or minima off ∂M.)

14.6 <u>REMARKS</u>

(A) By the Morse relation (14.4), a minimal g has just $-\chi(M)$ saddles in Int M.

(B) If within the theory we wish to include surfaces M with only one boundary curve, we would need to have at least one maximum or minimum in Int M, so condition (ii) of Def. 14.5 would need to be modified accordingly. We leave the details of this case to the reader, as we shall not use it below.

Up to now, then, we have shown that the smooth ND symmetric minimal functions $g : 2M \to \mathbb{R}$ fall into classes

$$G(\underset{\sim}{m},Z) = \{g \mid m(g) = \underset{\sim}{m},\ Z(g) = Z\}$$

where $\underset{\sim}{m} \in \mathbb{N}^{1+h}$, $\emptyset \neq Z \subset A$.

We can therefore state the generic Dirichlet problem in the form: <u>given</u> $\underset{\sim}{m} \in N^{1+h}$ <u>and a proper subset</u> Z <u>of</u> A , <u>classify all portraits</u> $\Pi(g)$, <u>as</u> g <u>runs through</u> $G(\underset{\sim}{m},Z)$.

For this classification it is natural to use Theorem 13.3. Now, with $\underset{\sim}{m}$ and Z fixed, g must have α maxima and β minima, where

$$\beta = \sum_{j\in Z} m(j) \ , \quad \beta = \sum_{j\notin Z} m(j)$$

so that the equivalence class of $\Pi(g)$ lies in $\mathrm{Port}_{\alpha,\beta}$ (2M) - in the notation of Theorem 13.3 - and by (13.2) corresponds to a graph $\Delta = \xi(\Pi(g))$ in $\Gamma(2M,\alpha,\beta)$. However, we shall see in a moment that Δ is <u>involutary</u>, in that there is a structure-preserving map $\delta : \Delta \to \Delta$ such that $\delta^2 = \mathrm{id}$. (Thus δ is an involution.) Let

$$\Gamma_I(2M,\alpha,\beta) \subseteq \Gamma(2M,\alpha,\beta)$$

denote the subset of involutary graphs. We prove:-

14.7 <u>PROPOSITION</u>. <u>If</u> $g \in G(\underset{\sim}{m},Z)$, <u>then</u> $\xi(\Pi(g)) \in \Gamma_I(2M,\alpha,\beta)$.

<u>Proof</u>. By (13.2), $\xi(\Pi(g))$ is isomorphic to the graph $K(\bar{g})$, in an elliptification $(N_1,\bar{g},\mathcal{D})$ of g and $K(\bar{g})$ is the union of sources and stable manifolds of $\bar{g}$. Hence we describe $K(\bar{g})$ in more detail.

Recall from Def. 13.1 that N_1 is constructed by removing from 2M a disc D(p) for each $p \in \mathrm{crit}(2M)$ such that p is not a saddle; the boundary $\dot{D}(p)$ is a contour of g . Since g is minimal, $p \in \partial M$ so D(p) is invariant under the automorphism ζ of 2M in (14.2). Hence so is N_1 , and therefore

$$N_1 = 2N\ , \quad N = M \cap N_1\ .$$

Also $\bar{g} = g|N$, and this is symmetric relative to $\zeta|N$. The set $\mathcal{D}$ of sources of $\bar{g}$ consists of all contours $\dot{D}(p)$ where p lies on a C_j with $j \in Z$; and if p, q are successive minima of g on C_j , a stable manifold α of $\bar{g}$ joins $\dot{D}(p)$ to $\dot{D}(q)$ and $\alpha \subseteq C_j$. Hence $\bar{g}$ has m_j such stable manifolds α on C_j , each fixed under ζ ; and these together with the circles $\dot{D}(p)$ $(p \in C_j)$ form a connected subgraph of $K(\bar{g})$ that we call a <u>bracelet</u> B_j . Observe that each bracelet is invariant under ζ .

The remaining stable manifolds of $\bar{g}$ occur in pairs that are permuted by ζ . Hence $K(\bar{g})$ is an involutory graph with involution $\zeta|K(\bar{g})$. By the construction described earlier it is isomorphic to a graph in $\Gamma_I(2M,\alpha,\beta)$ and the proof is complete.

As a corollary we have the following theorem.

14.8 <u>THEOREM.</u> <u>If</u> M <u>is orientable, there is an injection</u>

$$\xi_* : G(m,Z) \to \Gamma_I(2M,\alpha,\beta)$$

<u>with</u> $$\xi_*(g) = \xi(\Pi(\bar{g})) .$$

(The injective property follows at once from Theorem 14.3, because the orientability of M implies that of 2M .)

14.9 <u>REMARK.</u> If $g \in G(\underset{\sim}{m},Z)$, the proof of Theorem 14.7 shows that the graph $\xi_*(g)$ is not merely involutary - it also has a 'bracelet' structure. It would be interesting to have a graph-theoretic characterisation of $\xi_*(g)$, and even more so to have an enumeration of these graphs. This then leaves us with a problem of Combinatorics. But there remains also a problem of Analysis: for given boundary data on ∂M , and two elliptic operators L_1, L_2 on M , the respective solutions to the classical Dirichlet Problem determine graphs Δ_1, Δ_2 in $\Gamma_I(2M,\alpha,\beta)$. How do we decide which Δ_i belongs to which L_j ? We have no answers to these problems.

References

1. Birman, Joan, in Discontinuous Groups and Riemann Surfaces. Edited by L. Greenberg, Annals Studies, 79, Princeton (1974).

2. Courant, R. and Hilbert, D. Methods of Mathematical Physics, Vol. 2, Academic Press, New York (1962).

3. Griffiths, H.B., 'The case of the jumping critical point', Math. Proc., Camb. Phil. Soc. (1976) 80, 511-520.

4. Griffiths, H.B., Surfaces, (Second Edition) Cambridge University Press, 1980.

5. Kellogg, O.D., Potential Theory. Springer (1929.

6. Morse, M., Calculus of Variations in the large. AMS Colloquium Publications, XVIII (1947).

7. Morse, M., Topological Methods in the Theory of Functionssof a Complex Variable. Annals Studies 15, Princeton, 1947.

8. Morse, M. and Van Schaak, G.B., 'The critical point theory under general boundary conditions', Ann. of Math 35 (1934) 545-571.

9. Munkres, J., Elementary Differential Topology. Annals Studies 54, Princeton (1963).

10. Peixoto, M.M., 'The Classification of flows on 2-manifolds' in Dynamical Systems. Academic Press (1973), 389-419.

11. Springer, G., Introduction to Riemann Surfaces. Addison Wesley, (1957).

12. Steenrod, N., The Topology of Fibre Bundles, Princeton (1951).

GEOMETRY AND TOPOLOGY (pp. 208–214)
edited by G. M. Rassias & G. M. Stratopoulos

DIFFERENTIATION OF INTEGRAL FORMULAS

H. GUGGENHEIMER

POLYTECHNIC UNIVERSITY

A procedure amounting to Gâteaux differentiation in the cone of support functions of convex bodies was introduced by Cavallin [1] in 1881 and systematically developed by G. Masotti-Biggigero [2,3] for the derivation of formulas of integral geometry depending on arbitrary functions (see also [4], p. 59; [6], p. 135). A similar procedure can be used on most integral formulas. As general examples, we treat the differentiation of Stokes's formula and study an invariance property of the integral of the product of the principal curvatures of compact, orientable hypersurfaces in Riemannian spaces.

1.

Let R^r be a Riemannian manifold with metric < , > and ∂M a compact, orientable hypersurface of R^r bounding a manifold M. All manifolds and functions introduced here and in the sequel are $C^{2+\alpha}$. At $x \in \partial M$, the tangent space $T_x(R)$ splits into $T_x(\partial M)$ and a normal line n. An ortho-normal basis $\{e_i\}$ of $T_x(R)$ is equivalent, by a rotation, to a basis $\{\bar{e}_i\}$ with $\bar{e}_j \in T_x(\partial M)$, $1 \le j \le r-1$, $\bar{e}_r = N \in n$. We shall always assume that N is chosen as an outer normal of M. The Riemannian metric induces in $T_x(R)$ a determinant function of r vectors $v_{(i)}=w^k{}_{(i)}e_k$ by $\det_x(v_{(1)},\dots,v_{(r)})=\det(w^k{}_{(i)})$ and, therefore, a vector product $v = [v_{(1)},\dots,v_{(r-1)}]$ by

$$\langle v,c\rangle = \det_x(v_{(1)},\dots,v_{(r-1)},c) \text{ for all } c \in T_x(R).$$

Using exterior differential forms in a neighborhood of x in R, from $dx = \sigma^i \bar{e}_i$ we get $\sigma^r = 0$ as local differential equation of ∂M and the volume elements $dS=|\sigma^1\wedge\dots\wedge\sigma^{r-1}|$ on ∂M, $dV = |\sigma^1\wedge\dots\wedge\sigma^r|$ on R.

Combining the skew symmetry of the determinant function and of exterior multiplication, we get the formulas of Hopf and Voss ([5], p. 213,341):

$$(r-1)!\, N\, dS = [dx, \dots, dx] \tag{1}$$

$$(r-1)!\, dS = \det_x(N, dx, \dots, dx). \tag{2}$$

In addition, we derive the k-th symmetric function H_k of the principal curvatures by ([5], p.338)

$$(r-1)!\, H_k\, dS = \det_x(N,\underbrace{dN,\dots, dN}_{k}, \underbrace{dx, \dots, dx}_{r-k-1}) \tag{3}$$

Let k_i denote the principal curvatures ([5], p.337). Then $H_o=1$, $H_1 = (r-1)^{-1}\Sigma k_i$, $H_{r-1} = \Pi\, k_i$.

In a neighborhood U(x) in R^r we use local coordinates (x^i). The corresponding basis of $T_x(R)$ is $\{\partial_i\}$, $\partial_i = \frac{\partial}{\partial x^i}$. (In [5], ∂_i is called i_i.) Let $\hat{\imath}$ be the multi-index $\hat{\imath} = (i+1, \dots, r, 1, \dots, i-1)$ and put $dx^{\hat{\imath}} = dx^{i-1}\wedge \dots \wedge dx^{i-1}$. let ϕ be an (r-1)-form on R^r. Then ϕ has a local representation $\phi = \phi_{\hat{\imath}} dx^{\hat{\imath}}$. The dual of ϕ ([5], p.329, #33) in the duality induced by the metric is

$$D\phi = \frac{1}{\sqrt{g}}\, \delta^{\hat{1}\ldots 1}_{1\ldots r}\, \phi_{\hat{i}}\, \delta_i \tag{4}$$

The divergence of ϕ is div ϕ = $*d\phi$.

Since $\phi = \langle N, D\phi \rangle\, dS$ by (1), (2), (3), and $d\phi = \text{div}\,\phi\, dV$, it follows that Stokes's theorem for an (r-1)-form,

$$\int_{\partial M} \phi = \int_M d\phi \tag{5}$$

can be written in the form of Gauss's theorem:

$$\int_{\partial M} \langle N, D\phi \rangle\, dS = \int_M \text{div}\,\phi \cdot dV \tag{6}$$

2.

Let f be a function f: $\partial M \to \mathbb{R}$. Let x_ϵ be the point at distance $\epsilon f(x)$ from x on the normal geodesic of initial vector N(x) starting at x. If ϵ is sufficiently small (e.g., if sup $\epsilon f(x)$ is less than the minimal focal distance of the normal field on the compact manifold ∂M) then $\partial M_\epsilon = \{x_\epsilon\}$ is differentiably homeomorphic to ∂M and bounds a manifold M_ϵ. For any (r-1)-form on R^r it follows from our hypotheses that

$$\lim_{\epsilon\to 0} \frac{1}{\epsilon}\left[\int_{\partial M_\epsilon} \phi - \int_{\partial M} \phi\right] = \lim_{\epsilon\to 0} \frac{1}{\epsilon}\left[\int_{M_\epsilon} d\phi - \int_M d\phi\right]. \tag{7}$$

Formula (7) is the Cavallin-Masotti Biggiogero derivative of Stokes's theorem (5).

It follows from (6) and $dV = |dS \wedge \bar{\sigma}^r|$ that

$$\int_{M_e} d\phi = \int_M d\phi + \int_{\partial M}\left[\int_o^{\epsilon f(x)} \text{div}\,\phi \cdot \bar{\sigma}^r\right] dS$$

$$= \int_M d\phi + \epsilon \int_{\partial M} \text{div}\,\phi \cdot f(x)\, dS + o(\epsilon) \tag{8}$$

since $\bar{\sigma}_r$ is the element of length on the normal geodesic. Also,

$$\int_{\partial M_\epsilon} \phi = \int_{\partial M} \phi_{\hat{i}} \,(x + \epsilon f(x)N + o(\epsilon))\,(dx + \epsilon d(f(x)N) + o(\epsilon))^{\hat{i}} \tag{9}$$

or

$$\int_{\partial M_\epsilon} \phi = \int_{\partial M} \phi + \epsilon \int_{\partial M} \left[f(x)N^j \frac{\partial \phi_{\hat{i}}}{\partial x^j}\, dx^i + \sum_j \phi_{\hat{i}}(x)\, dx^{i+1} \wedge \ldots \wedge d(fN)^j \wedge \ldots \wedge dx^{i-1} + o(\epsilon) \right]$$

$$= \int_{\partial M} \phi + \epsilon \int_{\partial M} \left\{ f(x)N^j \frac{\partial \phi_{\hat{i}}}{\partial x^j}\, dx^i + \left[\phi_{\hat{i}} N^j - \sum_j (-1)^{j-1} \phi_j N^i \right] \frac{\partial f}{\partial x^j} \right\} dx^{\hat{i}}$$

$$+ \epsilon \int_{\partial M} (r-1) H_1(x) f(x)\, \phi + o(\epsilon)\ .$$

With these computations, (7) yields the result:

<u>Proposition:</u> For all C^2 functions $f(x)$ on a compact, orientable C^2 hypersurface M of a Riemannian manifold which bounds a manifold M one has

$$\int_{\partial M} [\operatorname{div} \phi - (r-1)H_1(x) < N, D\phi >]\ f(x)\ dS$$

$$= \int_{\partial M} \left\{ f(x)N^j \frac{\partial \phi_{\hat{i}}}{\partial M} - \left[\phi_{\hat{i}} N^j - \sum_j (-1)^{j-1} \phi_j N^i \right] \frac{\partial f}{\partial x^i} \right\} dx^{\hat{i}}. \tag{10}$$

For f=1 we get the noteworthy special case

$$\int_{\partial M} \operatorname{div} \phi\ dS = \int_{\partial M} \left\{ N^j \frac{\partial \phi_{\hat{i}}}{\partial x^j} + (r-1)\ H_1(x) \phi_{\hat{i}} \right\} dx^{\hat{i}}\ . \tag{11}$$

3.

The analog of the Gauss curvature of M in R is H_{r-1}. the integral curvature can be defined as $K = \int_{\partial M} H_{r-1} dS$. For r=1(mod 2), this integral is a summand in the generalized Gauss-Bonnet formula for the Riemannian curvature induced on ∂M. If R is not an euclidean space, K does not have any direct relation with the Euler characteristic of ∂M.

The following definition is inspired by the notion of infintesimal deformation in elementary differential geometry ([5], definition 10-26). A function F(X) defined on smooth, compact hypersurfaces of a Riemannian manifold R is an <u>invariant of infinitesimal homotopies</u> if for every smooth homotopy X_t one has

$$\lim_{t\to 0} \frac{d}{dt} F(X_t) = 0.$$

All the definitions from Sec. 1 are taken over for the surfaces X that now do not necessarily bound a manifold. The homotopy is defined for all $t\in [0,1]$ and X_o=X. Let f(x,t) be the distance, measured on the normal at $x \in X$, from x to the point of intersection of the normal with X_t. The homotopy is <u>regular</u> if f(x,t) is defined for $0\le t<\epsilon$ and is <u>smooth</u> if $f(x,t) = tf_t(x,0) + o(t)$ in an interval $0\le t<\epsilon_1$. We compute $K(X_t)$=K(t). On X_t, $dx_t = d(x+tf_t(x,0)N + o(t))$. By (1), we get for the normal N_t and the volume element dS_t of X_t that

$$N_t dS_t = (r-1)!^{-1}[dx_t,\dots,dx_t] = [1 + (r-1)H_1(x)f_t(x,0)]\ N\ dS + t\sum_1^{r-1} f_i\bar{e}_i\ dS + o(t)$$

where we have put $df_t(x,0) = \sum_1^{r-1} f_i\bar{\sigma}^i$. Hence,

$$N_t = N + t\sum_1^{n-1} f_i\bar{e}_i + o(t)$$

and

$$(H_{r-1})_t dS_t = (r-1)!^{-1} \det_{x_t}(N_t,dN_t,\dots,dN_t)$$
$$= H_{r-1}dS + \frac{t}{(r-1)!} \det_x \Big(\sum_1^{n-1} f_i\bar{e}_i,\ dN,\dots,dN\Big)$$
$$+\frac{t}{(r-2)!} \det_x \Big(N,\ d\sum_1^{n-1} f_i\bar{e}_i,dN,\dots,dN\Big) + o(t).$$

Now the first determinant vanishes since all vectors are in the (r-1)-space $T_x(X)$. Therefore, also,

$$\det_x\Big(N,d\sum_1^{n-1} f_i\bar{e}_i,\ dN,\dots,\ dN\Big) = d\ \det_x\Big(N,\ \sum_1^{n-1} f_i\bar{e}_i,\ dN,\dots,\ dN\Big).$$

As a consequence,

$$K(t) = K(0) + o(t)$$

and $$\lim_{t\to 0} \frac{1}{t} [K(t) - K(0)] = 0.$$

Theorem: The integral curvature of a compact, smooth, orientable hypersurface of a Riemannian manifold is an invariant of infinitesimal, regular, smooth homotopies.

As an example, consider a curve c in the hyperbolic plane given in the Poincaré model $H=\{(x,y)/y>0\}$. Let k, s denote euclidean curvature and arclength, k_H, s_H the corresponding hyperbolic quantities, and ϕ the angle between positive x-axis and forward tangent. Then ([7], prop. 8.3),

$$K_H = \cos\phi + yk, \qquad d_{s_H} = \frac{ds}{y}$$

Hence,

$$K = \oint_c k_H \, ds_H = \oint \cos\phi \, ds_H + \oint k \, ds.$$

The second integral in the expression of the right hand side is the corresponding euclidean invariant, the winding number, which is an isotopy invariant. If c is, for example, a circle of euclidean center (x,y) and euclidean radius r_0 we get $K_H = \frac{y_0}{r_0}$

and $K = 2\pi \frac{\sinh r_0/y_0}{r_0/y_0}$

which is an invariant of infinitesimal homotopies but not of finite homotopies.

BIBLIOGRAPHY

[1] Cavallin, C.B.S., Question 6571, Educational Times 34(1881), 23.

[2] Masotti-Biggiogero, G.,Nuove formole di geometria integrale relative agli ovali. annali di mat. (4)58(1962), 85-108.

[3] Masotti-Biggiogero, G., Nuove formole di geometria integrale relative agli ovaloidi. Rend. Sc. Ist. Lombardo A96(1962), 666-685.

[4] Santaló, L.,Integral Geometry and Geometric Probability. Enc. of Math. and its Appl., vol. 1. Addison-Wesley, Reading, Mass. 1976.

[5] Guggenheimer, H., Differential Geometry. Dover, New York 1976.

[6] Guggenheimer, H., Applicable Geometry. Krieger, Huntington NY. 1977.

[7] Guggenheimer, H., Contact Elements, Contact Correspondences, and Contact Invariants. Annali di mat. (4)120(1979), 229-261.

GEOMETRY AND TOPOLOGY (pp. 215–218)
edited by G. M. Rassias & G. M. Stratopoulos

A GROUP THEORETIC PROPERTY OF HEINE-BOREL METRICS

LUDVIK JANOS

Department of Mathematics and Computer Science
California State University, Long Beach
Long Beach, California 90840

ABSTRACT. If X is a metrizable topological space, D(X) stands for the set of all compatible metrics on X. A metric $d \in D(X)$ is said to be Heine-Borel if every closed and d-bounded subset of X is compact. For $d \in D(X)$ we denote by I(d) the group of all d-isometries on X. We show that if X is connected and $d \in D(X)$ is such that the metric space (X,d) is uniformly locally compact then there exists $d^* \in D(X)$ such that d^* is Heine-Borel and $I(d) \subset I(d^*)$.

1. INTRODUCTION

If X is a metrizable topological space, we denote by D(X) the set of all compatible metrics on X. If $d \in D(X)$ and the metric space (X,d) has a certain property we say that d itself has this property. A metric $d \in D(X)$ is said to be Heine-Borel if every closed and d-bounded subset of X is compact. For $d \in D(X)$, we denote by I(D) the group of all d-isometries on X. For d_1, d_2 in D(X) we say d_2 is at least as symmetrical as d_1 if $I(d_1) \subset I(d_2)$. If $d \in D(X)$ and $a>0$, we denote by $d^{[a]}$ the a-bounded modification of d defined as $\min\{d,a\}$. We see that $d^{[a]} \in D(X)$ and it is trivial to observe that $I(d) \subset I(d^{[a]})$ for every $a>0$, so that the remetrization $d \to d^{[a]}$ is in general symmetry-increasing. In [4] it is shown that X has a Heine-Borel metric if and only if it is locally compact and σ-compact. If X is locally compact but not compact and $d \in D(X)$ is Heine-Borel, then $d^{[a]}$ cannot be Heine-Borel for any $a>0$, but it is evidently still uniformly locally compact (see [2]). One can easily construct examples where $I(d^{[a]})$ properly contains I(d). On the other hand the result in [3] implies that for the Euclidean metric $e \in D(\mathbb{R}^m)$ on $\mathbb{R}^m$ we have $I(e^{[a]}) = I(e)$ for every $a>0$, since e has the maximum symmetry property in the sense that if $d \in D(\mathbb{R}^m)$ and $I(e) \subset I(d)$ then $I(e) = I(d)$. The Euclidean metric is a special case of a Heine-Borel metric which in general does not possess this maximum symmetry property.

We observe that the remetrization $d \to d^{[a]}$ is in general symmetry increasing and results in a metric which is no longer Heine-Borel. The purpose of this note is to construct a symmetry increasing remetrization going in the opposite direction, i.e., resulting in a Heine-Borel metric.

THEOREM. *Let* X *be a connected locally compact metrizable space and let* $d \in D(X)$ *be a metric on* X *such that the metric space* (X,d) *is uniformly locally compact. Then there exists* $d^* \in D(X)$ *such that* d^* *is Heine-Borel and* $I(d) \subset I(d^*)$.

2. PROOF OF THE THEOREM.

A metric space (X,d) is said to be ϵ-chainable where $\epsilon>0$ if for any two points $x,y \in X$ there exists a finite sequence of points (an ϵ-chain) $x = x_0, x_1, \ldots, x_n = y$ so that $d(x_i, x_{i+1}) \le \epsilon$ for $i = 0,\ldots,n-1$. We denote by $d_\epsilon = d_\epsilon(x,y)$ the corresponding chain metric defined as the infimum of $\sum_{i=0}^{m} d(x_i, x_{i+1})$ taken over the set of all ϵ-chains between x and y. It is staightforward to verify that d_ϵ is a metric in D(X).

LEMMA (2.1) If the metric space (X,d) is ϵ-chainable then $I(d) \subset I(d_\epsilon)$.

Proof. Assume it is not so. Then there exists $T \in I(d)$ and $T \notin I(d_\epsilon)$. Hence there exist points $x,y \in X$ with $d_\epsilon(x,y) \neq d_\epsilon(Tx,Ty)$. We may assume $d_\epsilon(x,y) < d_\epsilon(Tx,Ty)$ since otherwise we may consider T^{-1}. Let $a>0$ be such that $d_\epsilon(Tx,Ty) - d_\epsilon(x,y) = 2a$. Then there is a chain $x = x_0, x_1, \ldots, x_n = y$ such that $\sum_{i=1}^{n-1} d(x_i, x_{i+1}) \le d_\epsilon(x,y) + a$. Since $T \in I(d)$, we have $\sum_{i=1}^{n-1} d(Tx_i, Tx_{i+1}) \le d_\epsilon(x,y) + a$. On the other hand, since Tx, Tx_1,…,Ty is an ϵ-chain, we have $\sum_{i=1}^{n-1} d(Tx_i, Tx_{i+1}) \ge d_\epsilon(Tx,Ty)$, so that $d(Tx,Ty) \le d_\epsilon(x,y) + a$ which is the desired contradiction with $d(Tx,Ty) - d_\epsilon(x,y) = 2a > 0$. □

We now introduce some notation. Let (X,d) be a metric space, $x \in X$, and $r>0$. We denote by $B(x,r)$ the open ball about x, i.e. $\{y\colon\ d(x,y) < r\}$; and by $CB(x,r)$ the closed ball about x, i.e. the set $\{y\colon\ d(x,y) \le r\}$.

If $S \subset X$ is a nonempty subset of X, we denote by $S(r)$ the set $\cup\{CB(x,r)\colon\ x \in S\}$. For $\epsilon>0$, and $\delta>0$, we say that a metric space (X,d) satisfies the axiom $A(\epsilon,\delta)$ if it is ϵ-chainable and if every closed δ-ball $CB(x,\delta)$ is compact.

LEMMA (2.2) *Let* $S\subset X$ *be a nonempty compact subset of a metric space* (X,d), *let* $0<\epsilon<r$, *and suppose that every closed r-ball in* (X,d) *is compact. Then* $S(\epsilon)$ *is compact.*

Proof. Let $\{y_n\}$ be a sequence in $S(\epsilon)$. Since every y_n belongs to some closed ball, say $CB(s_n,\epsilon)$, we obtain a sequence s_n in S. Now, since S is compact there is a subsequence of it, that we denote again by $\{s_n\}$ which converges to some s in S. Consider the open ball $B(s,\lambda)$ with $\lambda=\frac{1}{2}(\epsilon+r)$, and observe that the corresponding subsequence $\{y_n\}$ is eventually in this ball. Since $B(s,\lambda)$ has compact closure there exists a subsequence of $\{y_n\}$ which converges to some point y. But since $S(\epsilon)$ is closed which can be seen easily from the fact that $S(\epsilon)=\{y: d(y,S)\leq\epsilon\}$ the point y belongs to $S(\epsilon)$. □

If (X,d) is a metric space, $x\in X$, $\epsilon>0$, and $n\in\mathbb{N}$, we denote by $Y(x,\epsilon,n)$ the set of points of X which can be joined with x by an ϵ-chain of lenght $\leq n$.

COROLLARY (2.3) *Suppose that the metric space* (X,d) *satisfies the axiom* $A(\epsilon,\delta)$ *with* $0<\epsilon<\delta$. *Then, for every* $x\in X$ *and* $n\in\mathbb{N}$, *we have that* $Y(x,\epsilon,n)$ *is compact.*

Proof. We see that $Y(x,\epsilon,1)=CB(x,\epsilon)$ which is compact since $\epsilon<\delta$. By Lemma (2.2), it follows that $Y(x,\epsilon,2)=(CB(x,\epsilon))(\epsilon)$ is compact. The proof is completed by induction on n. □

From the above corollary it follows that X is σ-compact since $X=\cup\{Y(x,\epsilon,n): n\in\mathbb{N}\}$.

LEMMA (2.4) *If a metric space* (X,d) *satisfies the axiom* $A(\epsilon,\delta)$ *with* $0<\epsilon<\delta$, *then the chain metric* d_ϵ *is Heine-Borel.*

Proof. Let $x_0\in X$ and $r>0$, and let $B^*(x_0,r)$ be the r-ball about x_0 relative to the metric d_ϵ. If $y\in B^*(x_0,r)$, then there exists an ϵ-chain $x=x_0,x_1,\dots,x_n=y$ such that $\sum_{i=0}^{n-1} d(x_i,x_{i+1})<r$. We may consider only those chains where $d(x_i,x_{i+1})+d(x_{i+1},x_{i+2})>\epsilon$, since otherwise the point

x_{i+1} can be left out due to the triangle inequality $d(x_i,x_{i+2})\leq d(x_i,x_{i+1})+d(x_{i+1},x_{i+2})\leq\epsilon$. This shows that $\frac{n}{2}\epsilon<r$ if n is even, and $\frac{n-1}{2}\epsilon<r$ if n is odd. This implies that $n\leq\frac{2r}{\epsilon}+1$

independently of $y \in B^*(x_0,r)$ so that $B^*(x_0,r) \subset Y(x_0,\epsilon,m)$ for a suitable m. Corollary (2.3) implies that the closure of $B^*(x_0,r)$ is compact. Hence the proof is complete since x_0 and r are arbitrary. □

The following lemma is an immediate consequence of (2.1).

LEMMA (2.5) If a metric space (X,d) satisfies the axiom $A(\epsilon,\delta)$ with $0<\epsilon<\delta$, then the metric d_ϵ is Heine-Borel and $I(d) \subset I(d_\epsilon)$.

This is all we need to prove our theorem. Since if X is connected it is ϵ-chainable for every $\epsilon>0$. If $d \in D(X)$ is uniformly locally compact then there exists a $\delta>0$ such that all the closed δ-balls are compact. Choosing, for example, $\epsilon = \frac{\delta}{2}$, we see that the metric space (X,d) satisfies the axiom $A(\epsilon,\delta)$ and our theorem follows from Lemma (2.5).

The author wishes to thank his friends Nakhlé Asmar and Saleem Watson for typesetting the paper, and for helping in its final form.

References

1. Leonard Blumenthal, Theory and application of distance geometry. Chelsea P.C., Bronx, New York, 1970

2. John L. Kelly, General topology. D. Van Norstrand New York, 1955

3. Robert Williamson and Ludvik Janos, A group-theoretic property of the Euclidean metric. Proc. Amer. Mat. Soc. 98 (1986), 150-152.

4. Robert Williamson and Ludvik Janos, Constructing metrics with Heine-Borel property. Proc. Amer. Mat. Soc. 100 (1987), 567-573.

GEOMETRY AND TOPOLOGY (pp. 219–234)
edited by G. M. Rassias & G. M. Stratopoulos

DEGENERATE CRITICAL POINTS IN FLUID MECHANICS

George H. Knightly*
University of Massachusetts
and
D. Sather
University of Colorado

ABSTRACT

Several nonlinear problems from fluid mechanics give rise to operator equations that are special cases of the abstract problem considered here. Nontrivial states branching from the trivial solution of this problem are sought and splitting methods are used to obtain the bifurcation equations. Methods utilizing topological degree are given to solve this system when the reduced bifurcation equations have a gradient structure. The technique is valid even at degenerate critical points of the functional involved and leads to the existence of nontrivial states of the abstract problem. Some additional properties of the solutions are discussed and applications of the theory to fluid mechanics are described.

1. INTRODUCTION

In an earlier paper[5] we presented a new method for obtaining solutions of Bénard-type convection problems using properties of the topological degree of the mapping defined by the fluid equations, thereby extending our theory[6] for Bénard problems to admit certain types of (possibly) degenerate critical points. In the present paper we show that our approach[5] applies also to a class of nonlinear operator equations of the form

*This research was supported in part by ONR grant N00014-88-K-0098

$$0 = u - \lambda(L - \gamma M)u - F(u) - \gamma G(u), \quad u \in H, \quad \lambda \in \mathbf{R}^1, \quad \gamma \in \mathbf{R}^1 \qquad (*)$$

where H is a Hilbert space, L and M are linear operators and F and G are generated by bilinear operators. Several Bénard-type problems[5;6;7] lead to equations of the form (*) as do some rather different fluid problems, such as the Taylor problem[12] and the problems of spiral flows[8;9].

One can give a setting[8] in which it is possible to determine a complete set of bifurcation equations for an operator equation of the form (*) by using the structure parameter γ as an "amplitude" parameter rather than regarding γ as merely a constant to be incorporated into the operators M and G in (*). Complete bifurcation diagrams are obtained by such an approach because, e.g., the "reduced" bifurcation equations contain linear as well as <u>both</u> quadratic and cubic terms. This is in contrast to many other approaches[3;4;11;13] in bifurcation theory based upon splitting methods in which either linear and quadratic or linear and cubic terms appear in the reduced equations. If these reduced bifurcation equations (or reduced selection equations[5]) for (*) are a gradient system, i.e., have the form

$$0 = \nabla h \qquad (1.1)$$

for a smooth $h:\mathbf{R}^n \to \mathbf{R}^1$, then the earlier approach[5] becomes feasible. That approach was based upon the following well-known result.

LEMMA 1.1. If h has an isolated relative extremum at $x = 0$, then

$$\text{ind}(\nabla h, 0) = \begin{cases} 1, & \text{if the extremum is a minimum,} \\ (-1)^n, & \text{if the extremum is a maximum.} \end{cases} \qquad (1.2)$$

Here $\text{ind}(\nabla h, 0)$ denotes the topological index of ∇h at $x=0$, i.e., the local topological degree of ∇h at $x = 0$ relative to a ball centered at $x = 0$ containing no zero of ∇h other than $x = 0$.

In our approach to determining nontrivial solutions of (*) we first construct a suitable functional h in order to use Lemma 1.1 to solve the reduced bifurcation equations, even at a (possibly) degenerate critical point of h. Then a homotopy argument is given to generate a solution of the full bifurcation equations and hence, because of the splitting technique, a solution of (*).

The outline of the paper is as follows. In Section 2 we obtain the bifurcation equations for (*). The reduced bifurcation system is studied in Section 3, where the functional h is introduced and the main results on the existence of nontrivial solutions of (*) are developed. Also in Section 3 we present some ideas for obtaining solutions in special, nondegenerate cases (see Remark 3.3). Applications of the work in Section 3 to some problems of fluid mechanics are discussed in Section 4.

2. THE BIFURCATION EQUATIONS

In this section we derive the bifurcation equations for (*), based on a set of hypotheses that are verified for several applications (see the discussion in Section 4). Standard splitting methods are used, except that the structure parameter γ plays the role of an amplitude parameter.

In the case of a one dimensional null space N the following assumptions reduce to those made in our previous work[8]. Here H is a real Hilbert space with norm $\|\cdot\|$ generated by an inner product $(\, , \,)$ and δ_{ij} denotes the usual Kronecker delta symbol.

(H1) The linear operator $L:H \to H$ is self-adjoint and compact, and, if μ_0 denotes the smallest positive characteristic value of L, then $I - \mu_0 L$ has nullspace $N \equiv N(I - \mu_0 L)$ of dimension n and N is spanned by $\psi_1, \ldots, \psi_n$ with $(\psi_i, \psi_j) = \delta_{ij}$.

(H2) The linear operator $M:H \to H$ is compact and satisfies $M:N \to N^{\perp}$, where $N^{\perp}$ denotes the orthogonal complement of N in H.

(H3) The critical characteristic value, $\lambda_c = \lambda_c(\gamma)$, of

$L_\gamma \equiv L - \gamma M$, i.e., the positive characteristic value of L_γ of least magnitude, also corresponds to an n-dimensional nullspace of $I - \lambda_c L_\gamma$ and is of the form

$$\lambda_c(\gamma) = \mu_0 - \mu_0^3 b\gamma^2 + \Lambda(\gamma), \ |\gamma| < \gamma_1, \qquad (2.1)$$

where Λ is real and analytic and satisfies $\Lambda(\gamma) = O(\gamma^3)$ as $\gamma \to 0$. Here we assume that $(MKM\psi_i, \psi_j) = b\delta_{ij}$, where b is independent of i,j and $K: N^\perp \to N^\perp$ denotes the inverse of the restriction of $(I - \mu_0 L)$ to $N^\perp$.

(H4) The nonlinear operators $F(w) = \Phi(w,w)$ and $G(w) = \Gamma(w,w)$ are generated by bounded bilinear operators $\Phi: H \times H \to H$ and $\Gamma: H \times H \to H$, where, in addition, $\Phi: N \times N \to N^\perp$.

The assumption (2.1) in (H3) is the "natural" expansion for λ_c and also is made for the sake of convenience. It, or a suitable modification, holds in many linear problems in fluid mechanics even if M is not symmetric (e.g., see the references [6;7;8;9;12]). The modifications required when H is a complex Hilbert space are somewhat more involved but may be formulated as in an earlier paper[6] in terms of real operators and group representations (see also the other references[5;13]).

Let $P: H \to N^\perp$ denote the orthogonal projection of H onto $N^\perp$ and let $S = I - P: H \to N$ denote the orthogonal projection of H onto N. For γ sufficiently small, we shall seek solutions of (*) of the form

$$u = \gamma(\psi + \gamma\Psi), \quad \lambda = \mu_0 - \mu_0\gamma^2(\mu_0^2 b - t), \qquad (2.2)$$

where $\psi \in N$, $\Psi \in N^\perp$ and $t \in \mathbb{R}^1$ are to be determined. Note that if $t = t_0 + O(\gamma)$ and if $\lambda_c = \lambda_c(\gamma)$ is defined as in (H3), then $\lambda = \lambda_c + \mu_0 t_0 \gamma^2 + O(\gamma^3)$. Thus, for γ sufficiently small, a solution of (*) of the form (2.1) is subcritical if $t_0 < 0$ and supercritical if $t_0 > 0$.

Substituting (2.2) into equation (*), using the projection P onto $N^\perp$ and S onto N, and making use of (H2) and (H4), one obtains the following equations in $N^\perp$ and N:

$$\begin{aligned} 0 = {} & (I - \mu_0 L)\Psi + \mu_0 M\psi - F(\psi) \\ & + \gamma P[\mu_0 M\Psi - \Phi(\psi,\Psi) - \Phi(\Psi,\psi) - G(\psi)] \\ & + \gamma^2 P[\mu_0(t - \mu_0^2 b)M\psi + \mu_0(\mu_0^2 b - t)L\Psi \\ & - F(\Psi) - \Gamma(\psi,\Psi) - \Gamma(\Psi,\psi)] \\ & + \gamma^3 P[\mu_0(t - \mu_0^2 b)M\Psi - G(\Psi)], \end{aligned} \tag{2.3a}$$

$$\begin{aligned} 0 = {} & (\mu_0^2 b - t)\psi + S[\mu_0 M\Psi - \Phi(\psi,\Psi) - \Phi(\Psi,\psi) - G(\psi)] \\ & - \gamma S[F(\Psi) + \Gamma(\psi,\Psi) + \Gamma(\Psi,\psi)] + \gamma^2 S[\mu_0(t - \mu_0^2 b)M\Psi - G(\Psi)]. \end{aligned} \tag{2.3b}$$

Since $K \equiv [(I - \mu_0 L)|_{N^\perp}]^{-1}$ is bounded on $N^\perp$, given $\rho > 0$ there exists $\gamma_0 > 0$ such that, if $(\psi,t) \in N \times \mathbf{R}^1$ with $|t| < \rho$ and $\|\psi\| < \rho$, then by the implicit function,theorem (2.3a) can be solved uniquely for $\Psi = \Psi(\psi,t,\gamma)$ in $N^\perp$ provided that $|\gamma| < \gamma_0$. In fact, Ψ is analytic and of the form

$$\Psi = -\mu_0 KM\psi + KF(\psi) + \gamma\Psi_1, \tag{2.4}$$

where $\Psi_1 = \Psi_1(\psi,t,\gamma) \in N^\perp$ is bounded with bound depending only on ρ. It is important here that ρ can be arbitrarily large provided that γ_0 is chosen sufficiently small. Subbing Ψ into (2.3b), taking the inner product with ψ_j, and making use of the definition of b in (H3), one obtains, for ψ of the form

$$\psi = \sum_{j=1}^{n} \beta_j \psi_j \tag{2.5}$$

with $\beta = (\beta_1,\ldots,\beta_n) \in \mathbf{R}^n$, the <u>bifurcation</u> <u>equations</u> associated with (*):

$$0 = -t\beta + Q(\beta) + C(\beta) + r(\beta,t,\gamma) \equiv D(\beta,t,\gamma). \tag{2.6}$$

Here $Q = (Q_1,\dots,Q_n)$, $C = (C_1,\dots,C_n)$, where, for $p,q,m,j = 1,2,\dots,n$,

$$Q_j(\beta) = \sum_{p,q=1}^{n} b_{pqj}\beta_p\beta_q , \tag{2.7}$$

$$C_j(\beta) = \sum_{p,q,m=1}^{n} b_{pqmj}\beta_p\beta_q\beta_m , \tag{2.8}$$

with

$$b_{pqj} = \mu_0(MK\Phi(\psi_p,\psi_q) + \Phi(\psi_p,KM\psi_q) + \Phi(KM\psi_p,\psi_q),\psi_j) - (\Gamma(\psi_p,\psi_q),\psi_j),$$

$$b_{pqmj} = -(\Phi(\psi_p,K\Phi(\psi_q,\psi_m)) + \Phi(K\Phi(\psi_p,\psi_q),\psi_m),\psi_j),$$

and, for $|t| < \rho$, $|\beta| < \rho$, $|\gamma| < \gamma_0$, the remainder terms r_j are defined by

$$\begin{aligned} r_j(\beta,t,\gamma) = \gamma(\mu_0 M\Psi_1 &- \Phi(\psi,\Psi_1) - \Phi(\Psi_1,\psi) \\ &- F(\Psi) - \Gamma(\psi,\Psi) - \Gamma(\Psi,\psi),\psi_j) \\ &+ \gamma^2(\mu_0(t - \mu_0^2 b)M\Psi - G(\Psi),\psi_j), \end{aligned} \tag{2.9}$$

where $\Psi_1 = \Psi_1(\psi,t,\gamma)$ and ψ is given by (2.5). Moreover, $r = (r_1,\dots,r_n)$ is analytic in (β,t,γ) and, for some r_0 depending only on ρ, satisfies

$$|r(\beta,t,\gamma)| \le r_0|\gamma|, \quad |\beta| < \rho, \quad |t| < \rho, \quad |\gamma| < \gamma_0. \tag{2.10}$$

Note that if, for *fixed* γ satisfying $|\gamma| < \gamma_0$, (β^*,t^*,γ) is a solution of (2.6) with $|t^*| < \rho$ and $|\beta^*| < \rho$, then $(u^*,\lambda^*) \in H \times \mathbf{R}^1$ given by

$$u^* = \gamma(\psi^* + \gamma\Psi(\psi^*,t^*,\gamma)), \quad \lambda^* = \mu_0 - \mu_0\gamma^2(\mu_0^2 b - t^*), \tag{2.11}$$

with $\psi^* = \sum_{j=1}^{n} \beta_j^*\psi_j$, is a solution of equation (*).

In the next section we solve the bifurcation equations (2.6) under some additional assumptions on the terms.

3. SOLUTIONS OF EQUATION (*).

In this section we describe the use of topological degree in solving the bifurcation equations (2.6), thereby solving (*). Briefly, the idea is to solve the reduced bifurcation equations

$$0 = D(\beta,t,0) = -\beta t + Q(\beta) + C(\beta), \tag{3.1}$$

obtained from (2.6) by letting $\gamma \to 0$, then to obtain a corresponding solution of (2.6) for fixed t by a homotopy argument in γ. The solution obtained for (2.6) yields a solution of (*) by means of equations (2.11). In this way one easily establishes the following theorem, in which $B(\beta,\rho)$ denotes the ball in $\mathbf{R}^n$ having center at β and radius ρ.

THEOREM 3.1. Let $\beta = \beta^*$, $t = t^*$ satisfy (3.1) and suppose, for $t = t^*$ fixed, that β^* is an isolated solution of $D(\beta,t^*,0) = 0$ with topological index $\operatorname{ind}(D(\beta,t^*,0),\beta^*) \neq 0$. Then there exists $\gamma_0 > 0$ such that, for $|\gamma| < \gamma_0$, equations (2.6) have a real solution $\beta = \beta(\gamma)$, $t = t^*$ satisfying $\beta(\gamma) \to \beta^*$ as $\gamma \to 0$. Moreover, equation (*) has a solution (u^*,λ^*) of the form (2.11). In fact (u^*,λ^*) satisfy

$$u^* = \gamma \sum_{j=1}^{n} \beta_j^* \psi_j + U^*(\gamma), \tag{3.2}$$

$$\lambda^* = \lambda_c + \gamma^2 \mu_0 t^* - \Lambda(\gamma), \tag{3.3}$$

where $U^*(\gamma) = O(\gamma^2)$ as $\gamma \to 0$ and, as in (H3), $\Lambda(\gamma) = O(\gamma^3)$ as $\gamma \to 0$.

Proof. Since β^* is an isolated zero of $D(\beta,t^*,0)$, there is a ball $B = B(\beta^*,\rho)$ in $\mathbf{R}^n$ whose closure contains no zero of $D(\beta,t^*,0)$

other than β^*. Since the vector field $D(\beta,t,\gamma)$ is continuous for (β,t,γ) near $(\beta^*,t^*,0)$, there exists $\gamma_1 > 0$ such that $D(\beta,t^*,\gamma)$ does not vanish on the boundary of B for $|\gamma| < \gamma_1$. It follows that $D(\beta,t^*,\gamma)$ is homotopic to $D(\beta^*,t^*,0)$ for $|\gamma| < \gamma_1$ so that

$$d(D(\beta,t^*,\gamma),B,0) = d(D(\beta,t^*,0),B,0) = \text{ind}(D(\beta,t^*,0),\beta^*) \neq 0. \quad (3.4)$$

According to the basic existence theorem of topological degree theory, for each γ in $|\gamma| < \gamma_1$ there exists a point $\beta(\gamma)$ in B satisfying $0 = D(\beta(\gamma),t^*,\gamma)$. Since the radius of B may be taken small, it follows that $\beta(\gamma) \to \beta^*$ as $\gamma \to 0$ and that $\beta(\gamma) \neq 0$ for small γ.

In order to obtain a nontrivial solution of (*) via Theorem 3.1 one needs to find a nontrivial isolated solution of the reduced bifurcation equations (3.1) with nonzero index. This is possible when (3.1) is a gradient system, as specified in the following hypothesis.

(H5) There exist real homogeneous polynomials $q(\beta)$ and $c(\beta)$, from $\mathbf{R}^n$ into $\mathbf{R}^1$, with the properties

$$\nabla q(\beta) = Q(\beta); \quad q(k\beta) = k^3 q(\beta) \quad \text{for } k \in \mathbb{R}^1; \ q(\beta) \not\equiv 0, \quad (3.5)$$

$$\nabla c(\beta) = C(\beta); \quad c(k\beta) = k^4 c(\beta) \quad \text{for } k \in \mathbf{R}^1; \ c(\beta) > 0 \text{ if } \beta \neq 0. \quad (3.6)$$

If (H5) holds, then (3.1) becomes

$$0 = D(\beta,t,0) = -t\beta + \nabla q(\beta) + \nabla c(\beta). \quad (3.7)$$

Let $h:\mathbf{R}^n \to \mathbf{R}^1$ be defined by

$$h(\beta) = \begin{cases} \dfrac{g(\beta) + c(\beta)}{|\beta|^2}, & \text{if } \beta \neq 0 \\ 0, & \text{if } \beta = 0 \end{cases} \quad (3.8)$$

Note that if the extremum problem

$$\inf_{\beta \in \mathbf{R}^n} \{h(\beta)\} \tag{3.9}$$

has a (finite) minimum at $\beta^* \neq 0$, then

$$0 = \nabla h(\beta^*) = |\beta^*|^{-2}(-2h(\beta^*)\beta^* + \nabla q(\beta^*) + \nabla c(\beta^*)). \tag{3.10}$$

Hence, if we set $t^* = 2h(\beta^*)$ then (β^*,t^*) is a nontrivial solution of (3.7).

The hypotheses $q(\beta) \not\equiv 0$, $c(\beta) > 0$ in (H5) imply that the extremum problem

$$\max_{\substack{|\xi|=1 \\ \xi \in \mathbf{R}^n}} \left\{\frac{q^2(\xi)}{c(\xi)}\right\} \tag{3.11}$$

always has at least one nontrivial solution. Thus, there exists an element $\xi_0 \in \mathbf{R}^n$, $|\xi_0| = 1$, such that $q_0^2/c_0 = q_0^2(\xi_0)/c(\xi_0)$ is the absolute maximum of q^2/c on $|\xi| = 1$, and $0 < q_0^2/c_0 < \infty$. For $\beta \in \mathbf{R}^n$, $\beta \neq 0$ we set $\beta = \rho\xi$ where $|\xi| = 1$. Then

$$h(\beta) = c(\xi)[\rho + q(\xi)/2c(\xi)]^2 - q^2(\xi)/4c(\xi) \geq -q_0^2/4c_0 \tag{3.12}$$

for all $\beta \in \mathbf{R}^n$, $\beta \neq 0$. Since $h(\beta^*) = -q_0^2/4c_0$ when $\beta^* = -(q_0/2c_0)\xi_0$, it follows that h has a (finite) <u>negative</u> minimum at $\beta^* \neq 0$. In view of (3.10), if $t^* = 2h(\beta^*)$ then (β^*,t^*) is a nontrivial solution of the system (3.7) <u>with</u> <u>$t^* < 0$</u>. When β^* is an isolated solution of $\nabla h(\beta) = 0$ we have, according to Lemma 1.1,

$$\operatorname{ind}(\nabla h,\beta^*) = 1. \tag{3.13}$$

If we can show that

$$\operatorname{ind}(D(\beta,t^*,0),\beta^*) = \operatorname{ind}(\nabla h,\beta^*), \tag{3.14}$$

then Theorem 3.1 applies and yields a nontrivial solution of (*) for each γ in $|\gamma| < \gamma_0$.

To establish (3.14), it suffices to consider the case in which none of the relevant Jacobians vanish; a standard limiting argument using Sard's lemma then yields (3.14) in the general case. If $J(\beta)$ denotes the Jacobian of $D(\beta,t^*,0)$ at β, then from (3.7)

$$J(\beta) = \det[-t^*I + \nabla\nabla q(\beta) + \nabla\nabla c(\beta)], \tag{3.15}$$

where $\nabla\nabla$ denotes the Hessian of a functional and I is the $n \times n$ identity matrix. At a critical point $\beta = \beta^*$ of $h(\beta)$ one easily uses (3.10) to calculate

$$\nabla\nabla h(\beta^*) = |\beta^*|^{-2}\{-t^*I + \nabla\nabla q(\beta^*) + \nabla\nabla c(\beta^*)\}. \tag{3.16}$$

Then

$$\operatorname{ind}(D(\beta,t^*,0),\beta^*) = \frac{J(\beta^*)}{|J(\beta^*)|} = \frac{\det(\nabla\nabla h(\beta^*))}{|\det(\nabla\nabla h(\beta^*))|} = \operatorname{ind}(\nabla h,\beta^*),$$

which establishes (3.14) and completes the proof of the next theorem.

THEOREM 3.2. Let the operators in (*) satisfy (H1),...,(H4) and suppose there are functionals q and c for which the associated reduced bifurcation equations (3.1) satisfy (H5). If the critical point β^* giving the global minimum of h is isolated, then there exists $\gamma_0 > 0$ such that, for $|\gamma| < \gamma_0$, equation (*) has a nontrivial solution (u^*,λ^*) of the form (2.11) and satisfying (3.2), (3.3) with $t^* = 2h(\beta^*)$.

REMARK 3.1. A nontrivial state (u,λ) of (*) is <u>subcritical</u> if $\lambda < \lambda_c$. Since, in Theorem 3.2, $t^* = 2h(\beta^*) = -q_0^2/4c_0 < 0$, it follows from (3.3) that the solutions of (*) obtained in Theorem 3.2 are subcritical for small γ. In addition, from the continuity of $D(\beta,t,\gamma)$ in t for fixed γ, $|\gamma| < \gamma_0$, it is clear that $D(\beta,t,\gamma)$ is homotopic to $D(\beta,t^*,\gamma)$ for $|t - t^*| < t_0$. Consequently, for fixed γ, $|\gamma| < \gamma_0$, the subcritical state of Theorem 3.2 is a point on a set of solutions $(\beta(t,\gamma),t,\gamma)$ for $|t - t^*| < t_0$.

REMARK 3.2. The functional

$$\delta(\beta,t) = -\frac{1}{2}t|\beta|^2 + q(\beta) + c(\beta) \tag{3.17}$$

may perhaps be regarded as a <u>generalized</u> <u>dissipation</u> for problem (*) in the context of Theorem 3.2. $D(\beta,t,0)$ is its gradient and, when the set of solutions of (*) obtained in Remark 3.1 is actually a curve depending smoothly on t, one can show that

$$\frac{d\delta}{dt} = -\frac{1}{2}|\beta|^2 + \nabla\delta\cdot\frac{d\beta}{dt} = -\frac{1}{2}|\beta^*|^2 + O(\gamma) \tag{3.18}$$

is negative for small γ. Thus, δ decreases as t increases through t^*. Since $\delta(\beta^*,t^*) = 0 = \delta(0,t)$, we see that the generalized dissipation decreases along the solution curve, from positive values for $t < t^*$ to negative values for $t > t^*$. The states found in Theorem 3.2 mark an intermediate value λ^* of the "load" parameter λ, between the minimum load at which nontrivial states exist and the critical value λ_c of the linear problem. If δ is a determining functional for selection of a state of the system, then for $\lambda^* < \lambda < \lambda_c$ one expects the system to prefer the nontrivial states of Remark 3.1 to the trivial state since the former states have smaller generalized dissipation, $\delta < 0$.

REMARK 3.3. If the reduced system (3.1) has additional structure (e.g., because of symmetry considerations), it may be possible to deduce considerably more about the solutions of (*) than is given in the above theorems and remarks. For instance, suppose that if $\beta_j = \xi$, $j = 1,\dots,n$, then all the component equations of (3.1) become identical of the form

$$0 = -t\xi + q\xi^2 + c\xi^3 + \xi s(\xi,\gamma) \tag{3.19}$$

for some constants q and c. Cancelling a factor ξ and setting $\gamma = 0$ we obtain the reduced bifurcation equation

$$0 = -t + q\xi + c\xi^2. \tag{3.20}$$

In this case the full hypothesis (H5) is unnecessary. Instead, we merely assume $c \neq 0$, say $c > 0$. Then (3.20) has solutions for all t. (Here it is even natural to use ξ as parameter instead of t). Since the remainder $s(\xi,\gamma)$ in (3.19) can be made small for $|\xi| < \rho$, $|\gamma| < \gamma_0$, where ρ is arbitrary but fixed and γ_0 is chosen sufficiently small, a direct application of the implicit function theorem leads to the existence for small γ of a branch of solutions $(u^*(\xi),\lambda^*(\xi))$ of (*) of the form (3.2), (3.3) with $\beta^* = (\xi,\ldots,\xi)$, for $|-t^* + q\xi + c\xi^2| < k|\gamma|$, $|\xi| < \rho$, $|\gamma| < \gamma_0$. Note, in particular, that this branch retains approximately the parabolic form of (3.20). It is, in fact, analytic in ξ so that one can follow the generalized dissipation on the branch, as discussed in Remark 3.2.

4. APPLICATIONS TO FLUID MECHANICS

The theory developed in the preceding section helps to unify several application areas. We discuss briefly four such problems from fluid mechanics (further details are contained in the references[1;2; 5-10;12]).

CELLULAR CONVECTION. These generalized Bénard problems concern the stationary flow of an infinite layer of viscous fluid between two rigid horizontal walls and heated uniformly from below. The fluid density is assumed constant except in the buoyancy term where it is taken to be quadratic in the temperature. A conduction state having zero velocity is easily found and secondary states branching from this trivial state are sought. The governing Boussinesq-type equations lead[6] to an equation of the form (*) when the modifications of the theory necessary when H is a complex Hilbert space are included (see also the other references[5;7]).

Properties corresponding to (H1)-(H5) can be established[6] (motivated in the case of $c(\beta) > 0$) and arguments involving the usual implicit function theorem given in the nondegenerate case to determine stable, subcritical, hexagonal-cellular solutions of (*). In the degenerate case such hexagonal states of (*) are obtained[5] by methods similar to those leading to Theorem 3.2. In fact, the present techniques apply to the Bénard problem[6] to show <u>there exist subcritical hexagonal states</u> having the properties stated in Theorem 3.2, Remark 3.1 and Remark 3.2 whenever the critical point β^* giving the global minimum of the functional h is isolated.

For these Bénard-type problems the ideas of Remark 3.3 were pursued extensively[7] to generate a bifurcation diagram for small γ that contains three types of nontrivial branches: (i) transcritical branches of hexagonal cellular states. On the subcritical parts of these branches, λ first decreases then increases and extends to $\lambda = \lambda_c$. (ii) Supercritical ("pitchfork" bifurcations) branches of roll solutions also bifurcating from the conduction state at $\lambda = \lambda_c$. (iii) Transition states that join hexagonal and roll states in the supercritical region. In particular, the hexagonal states are obtained from equations of the form (3.19), (3.20) as described in Remark 3.3.

ROTATING COUETTE-POISEUILLE CHANNEL FLOW. This problem deals with an infinite layer of viscous, incompressible fluid between rigid walls normal to the y-axis. One wall translates uniformly in the x-direction, a pressure gradient is maintained in the x-direction and the entire system rotates at constant angular speed about the z-axis. The Navier-Stokes equations, written in a rectangular coordinate system that rotates with the physical system about the z-axis, are shown[10;8] to have an explicit, primary, Couette-Poiseuille state depending only on y. Secondary states are sought as roll-type perturbations, periodic in z and independent of x, of this trivial state and the governing system is shown[8] to correspond to an operator equation of the form (*), with G = 0, on a real Hilbert space. Here λ is essentially the Reynolds number and γ measures the ratio of Poiseuille to Couette components in the primary flow. Since γ is small in our theory, the problem may be regarded as that of the perturbation of Couette flow by a small amount of Poiseuille flow.

The hypotheses (H1)-(H4) are verified directly[8] for this problem. Here the nullspace N in (H1) is one-dimensional and it is shown that the system (3.1) is a single bifurcation equation

$$0 = -\beta t + c\beta^3 + \beta s(\beta,t,\gamma) \tag{4.1}$$

of the form (3.19) with $q = 0$; the property $c > 0$ can be motivated[8]. Consequently, the approach of Remark 3.3 yields the <u>existence of a smooth branch of nontrivial states</u> having the properties stated in Remark 3.3. Thus, for fixed, small γ one has a supercritical branch of roll-type states bifurcating from the basic Couette-Poiseuille rotating flow at $\lambda = \lambda_c(\gamma)$ and having, approximately, the parabolic form of (3.20) $(q = 0)$.

ROTATING PLANE COUETTE FLOW. This problem also concerns the flow of an infinite layer of a viscous, incompressible fluid between rigid parallel walls normal to the y-direction. The entire system rotates about the x-axis and one wall translates uniformly, in its own plane, at an angle χ with the z-direction. One can develop[1;2] this plane problem as a narrow gap limit of flow between concentric rotating cylinders when the inner cylinder also translates axially. An explicit primary solution of the Navier-Stokes equations satisfying the stated conditions can be given[1;2] and is independent of the coordinate in the direction of the basic spiral flow angle χ. Secondary states are sought as perturbations of this basic flow and possessing a spiral flow angle ψ (i.e., independent of the coordinate in the ψ-direction). For this problem λ is essentially the Reynolds number and the structure parameter γ is a measure of $\psi - \chi$. The governing system for the stationary problem can be shown[9] to correspond to an operator equation of the form (*) with $G = 0$.

In this case, however, the (double) real characteristic value of

L_γ at $\gamma = 0$ splits for $0 < |\gamma| < \gamma_1$ into a conjugate pair of complex characteristic values so that we must take H to be a complex Hilbert space, the techniques of Hopf bifurcation are then implemented to yield the existence of a smooth branch of nontrivial, time-periodic, supercritical, spiral states of the time-dependent problem for each γ in some interval $|\gamma| < \gamma_0$. For $\gamma = 0$ the branch is stationary.

The theory developed here has, of course, some limitations, even for equations of the form (*) with operators having many of the properties in (H1)-(H4). For example, an equation (*) can be derived[12] for secondary states of the Taylor problem, concerning the flow of a viscous, incompressible fluid between concentric rotating cylinders, in the narrow gap limit. The operators in (*) for this problem do satisfy much of (H1)-(H4) with $n = 4$, but (H3) is violated because $\lambda_c(0)$ splits into two characteristic values of L_γ each of which corresponds to a two-dimensional nullspace when $\gamma \neq 0$. Nevertheless, a bifurcation analysis for the problem can be developed[12]. Thus, a more general abstract theory for degenerate critical points of (*) may be possible and would be of interest from the point of view of the applications to fluid mechanics.

REFERENCES

[1] Hung, W. L., Joseph, D. D. and Munson, B. R., "Global stability of spiral flow, Part II," J. Fluid Mech. 51, 593-612 (1972).

[2] Joseph, D. D., Stability of Fluid Motions I, Springer Tracts in Natural Philosophy 28, Springer Verlag, New York, 1976.

[3] Kirchgässner, K., "Bifurcation in nonlinear hydrodynamic stability," SIAM Review 17, 652-683 (1975).

[4] Kirchgässner, K. and Kielhöfer, H., "Stability and bifurcation in fluid dynamics," Rocky Mountain J. Math. 3, 275-318 (1973).

[5] Knightly, G. H. and Sather, D., "Degenerate critical points in convection problems," Algebraic and Differential Topology-Global Differential Geometry, G. M. Rassias, Ed., Teubner, Leipzig, 170-183.

[6] Knightly, G. H. and Sather, D., "A selection principle for Bénard-type convection", Arch. Rational Mech. Anal. 88, 163-193 (1985).

[7] Knightly, G. H. and Sather, D., "Stability of cellular convection," Arch. Rational Mech. Anal. 97, 271-297 (1987).

[8] Knightly, G. H. and Sather, D., "Structure parameters in rotating Couette-Poiseuille channel flow," Rocky Mountain J. Math. to appear.

[9] Knightly, G. H. and Sather, D., "Structure parameters and spiral states in rotating plane Couette flow," to appear.

[10] Potter, M. C., "Stability of plane Couette-Poiseuille flow," J. Fluid Mech. 24, 609-619 (1966).

[11] Sather, D., "Branching of solutions of nonlinear equations," Rocky Mountain J. Math. 3, 203-250 (1973).

[12] Sather, D., "Primary and secondary steady flows of the Taylor problem," J. Differential Equations 71, 145-184 (1988).

[13] Sattinger, D. H., "Bifurcation and symmetry breaking in applied mathematics," Bull. (N.S.) Amer. Math. Soc. 3, 779-819 (1980).

GEOMETRY AND TOPOLOGY (pp. 235–250)
edited by G. M. Rassias & G. M. Stratopoulos

NONMONOTONEITY OF PICARD DIMENSIONS

Mitsuru Nakai* and Toshimasa Tada**

* Department of Mathematics, Nagoya Institute of Technology, Gokiso, Showa, Nagoya 466, Japan
** Department of Electrical and Computer Engineering, Nagoya Institute of Technology, Gokiso, Showa, Nagoya 466, Japan and Department of Mathematics, Daido Institute of Technology, Daido, Minami, Nagoya 457, Japan

ABSTRACT. The purpose of this paper is to show the existence of a density on the punctured unit disk which in a sense behaves more wildly at the origin than one chosen in advance and yet the Picard dimension of which at the origin is an arbitrarily assigned positive integer in advance.

0. INTRODUCTION

We will take the punctured sphere $R: 0 < |z| \leqq \infty$ as our base surface so that every topological notion will be considered relative to $R = \{0 < |z| \leqq \infty\}$. Then the unit circle $|z| = 1$ is the relative boundary $\partial\Omega$ of the punctured unit disk $\Omega: 0 < |z| < 1$ and the origin $z = 0$ is the ideal boundary $\delta\Omega$ of Ω.

By a *density* P on Ω we mean a nonnegative locally Hölder continuous function $P(z)$ on the closure $\overline{\Omega} = \Omega \cup \partial\Omega: 0 < |z| \leqq 1$ so that P may or may not have a singularity at $z = 0$. We say that P is a C^p density on Ω if P is a nonnegative and of class C^p on $\overline{\Omega}$ $(p = 1, 2, \cdots, \infty)$.

1980 *Mathematics Subject Classification.* Primary 31C35; Secondary 30F25, 31A10, 31A35.

To complete the present work the first (second, resp.) named author was supported in part by Grant-in-Aid for Scientific Research, No. 61540098 (No.62302003, resp.) Japanese Ministry of Education, Science and Culture.

With a density P on Ω we associate the class $PP(\Omega;\partial\Omega)$ of nonnegative real valued continuous functions u on $\overline{\Omega}$ vanishing on $\partial\Omega$ such that u satisfies the following selfadjoint elliptic equation of the second order

$$\Delta u(z) = P(z)u(z) \tag{1}$$

on Ω where Δ is the Laplacian $4\partial^2/\partial z\partial\bar{z}$. In the notation $PP(\Omega;\partial\Omega)$ the first P stands for the density P and the second for "positive" so that if the density P is replaced by another density Q then the class is denoted by $QP(\Omega;\partial\Omega)$. We denote by $PP_1(\Omega;\partial\Omega)$ the subclass of $PP(\Omega;\partial\Omega)$ consisting of functions u with the normalization

$$-\int_{\partial\Omega} \frac{\partial}{\partial|z|} u(z)|dz| = 2\pi. \tag{2}$$

An application of the Choquet theorem (cf. e.g. Phelps[7]) we see that (cf. e.g. Nakai[4]) there exists a bijective correspondence $u \leftrightarrow \mu$ between $PP_1(\Omega;\partial\Omega)$ and the set of probability measure μ on the ex.$PP_1(\Omega;\partial\Omega)$ of extreme points of the convex set $PP_1(\Omega;\partial\Omega)$ such that

$$u = \int_{\mathrm{ex}.PP_1(\Omega;\partial\Omega)} v\,d\mu(v)$$

where ex.$PP_1(\Omega;\partial\Omega)$ is topologized by the uniform convergence on every compact subset of Ω. Thus the set ex.$PP_1(\Omega;\partial\Omega)$ is essential for the class $PP_1(\Omega;\partial\Omega)$ or $PP(\Omega;\partial\Omega)$, and the cardinal number $\#(\mathrm{ex}.PP_1(\Omega;\partial\Omega))$ of the set ex.$PP_1(\Omega;\partial\Omega)$ is referred to as the *Picard dimension* of a density P at the ideal boundary $\delta\Omega$ of Ω, $\dim P$ in notation, i.e.

$$\dim P = \#(\mathrm{ex}.PP_1(\Omega;\partial\Omega)). \tag{3}$$

It can be seen that $\dim P$ is at least one and at most cardinal number of the continuum. In the case $\dim P = 1$ we say that the *Picard principle* is valid for P (cf. e.g. Brelot[2],[3]).

The purpose of this paper is to prove the following result:

THE MAIN THEOREM. *There exists a* C^∞ *density* $Q_{P,m}$ *on* Ω *for any density* P *on* Ω *and any positive integer* m *such that* $Q_{P,m} \geq P$ *on* Ω *and yet the Picard dimension* $\dim Q_{P,m}$ *of* $Q_{P,m}$ *is* m.

1. RELATIVE PICARD DIMENSIONS FOR ADMISSIBLE SUBREGIONS

1. Let W be a subregion of Ω which is relatively noncompact in the punctured sphere $R = \{0 < |z| \leqq \infty\}$. We denote by ∂W the relative boundary of W and assume that any component of ∂W is a continuum and that $\partial\Omega$ is an isolated component of ∂W. Such a W will be referred to as an *admissible* subregion of Ω. For a density P on Ω we consider a class $PP(W;\partial W)$ of nonnegative solutions u of the equation (1): $\Delta u = Pu$ on W with vanishing boundary values on ∂W and a subclass $PP_1(W;\partial W)$ of $PP(W;\partial W)$ consisting of those u with the normalization (2):

$$-\int_{\partial\Omega} \frac{\partial}{\partial |z|} u(z)\,|dz| = 2\pi.$$

In the notation $PP(W;\partial W)$ or $PP_1(W;\partial W)$, like $PP(\Omega;\partial\Omega)$ or $PP_1(\Omega;\partial\Omega)$, the first P stands for the density P and the second for "positive". Similar to the Picard dimension we define the *relative Picard dimension*, $\dim(P,W)$ in notation, of a density P at $\delta\Omega$: $z = 0$ relative to W by

(4) $$\dim(P,W) = \#(\text{ex}.PP_1(W;\partial W)).$$

We now prove the following which is essential for the proof of our main theorem:

THEOREM 1. *For any density P on Ω and any positive integer m there exists an admissible subregion $\Omega_{P,m}$ of Ω with smooth boundary $\partial\Omega_{P,m}$ of class C^∞ such that the relative Picard dimension $\dim(P,\Omega_{P,m})$ of P at $\delta\Omega$ relative to $\Omega_{P,m}$ is m.*

2. The required subregion $\Omega_{P,m}$ of Ω is constructed by five sequences $\{a_n\}$, $\{b_n\}$, $\{p_n\}$, $\{q_n\}$, and $\{\theta_n\}$ as described in the sequel.

First we choose $\{a_n\}_1^\infty$, $\{b_n\}_1^\infty$, $\{p_n\}_1^\infty$, and $\{q_n\}_1^\infty$ as arbitrary sequences in the interval $(0,1)$ convergent to zero such that

$$a_{n+1} < q_n < p_n < b_n < a_n \quad (n = 1,2,\cdots).$$

Although these are entirely arbitrary, we will fix them throughout our construction once they are chosen.

The sequence $\{\theta_n\}_1^\infty$ is to be chosen in $(0,\pi/4m)$. We do not require the sequence anything more than lying in the above interval for a while but it will be specified later to converge to zero rapidly enough in a manner depending on $\{a_n\}$, $\{b_n\}$, the density P on Ω, and the positive integer m given in advance.

3. Consider circular arcs $\alpha_n^\pm$, $\beta_n^\pm$ and line segment $\gamma_n^\pm$ $(n = 1,2,\cdots)$ given by

$$\begin{cases} \alpha_n^+ = \{|z| = a_n,\ \theta_n \leqq \arg z \leqq \pi/4m\}, \\ \alpha_n^- = \{|z| = a_n,\ -\pi/4m \leqq \arg z \leqq -\theta_n\}, \end{cases}$$

$$\begin{cases} \beta_n^+ = \{|z| = b_n,\ \theta_n \leqq \arg z \leqq \pi/4m\}, \\ \beta_n^- = \{|z| = b_n,\ -\pi/4m \leqq \arg z \leqq -\theta_n\}, \end{cases}$$

$$\begin{cases} \gamma_n^+ = \{q_n \leqq |z| \leqq p_n,\ \arg z = \pi/2m\}, \\ \gamma_n^- = \{q_n \leqq |z| \leqq p_n,\ \arg z = -\pi/2m\}. \end{cases}$$

Take simple arcs $\sigma_n^+, \tau_n^+, \upsilon_n^+$ $(n = 1,2,\cdots)$ in

$$\{b_n < |z| < a_n,\ 0 < \arg z < \theta_n\},$$

$$\{p_n < |z| < b_n,\ \pi/4m < \arg z < \pi/2m\},$$

$$\{a_{n+1} < |z| < q_n,\ \pi/4m < \arg z < \pi/2m\},$$

respectively which joins two points $a_n e^{i\theta_n}$ and $b_n e^{i\theta_n}$, $b_n e^{\pi i/4m}$ and $p_n e^{\pi i/2m}$, $q_n e^{\pi i/2m}$ and $a_{n+1} e^{\pi i/4m}$, respectively (see Fig. 1). Similarly take simple arcs $\sigma_n^-, \tau_n^-, \upsilon_n^-$ $(n = 1,2,\cdots)$ in

$$\{b_n < |z| < a_n,\ -\theta_n < \arg z < 0\},$$

$$\{p_n < |z| < b_n,\ -\pi/2m < \arg z < -\pi/4m\},$$

$$\{a_{n+1} < |z| < q_n,\ -\pi/2m < \arg z < -\pi/4m\},$$

respectively joining two points $a_n e^{-i\theta_n}$ and $b_n e^{-i\theta_n}$, $b_n e^{-\pi i/4m}$ and $p_n e^{-\pi i/2m}$, $q_n e^{-\pi i/2m}$ and $a_{n+1} e^{-\pi i/4m}$, respectively (see Fig. 1). We denote by Γ_1^+ the Jordan arc formed by infinite number of arcs $\{|z| = a_1,\ \pi/4m \leqq \arg z \leqq \pi/m\}$, $\alpha_n^+, \sigma_n^+, \beta_n^+, \tau_n^+, \gamma_n^+, \upsilon_n^+$ $(n = 1,2,\cdots)$. Similarly we denote by Γ_1^- the Jordan arc formed by infinite number of arcs $\{|z| = a_1,\ -\pi/m \leqq \arg z \leqq -\pi/4m\}$, $\alpha_n^-, \sigma_n^-, \beta_n^-, \tau_n^-, \gamma_n^-, \upsilon_n^-$ (n =

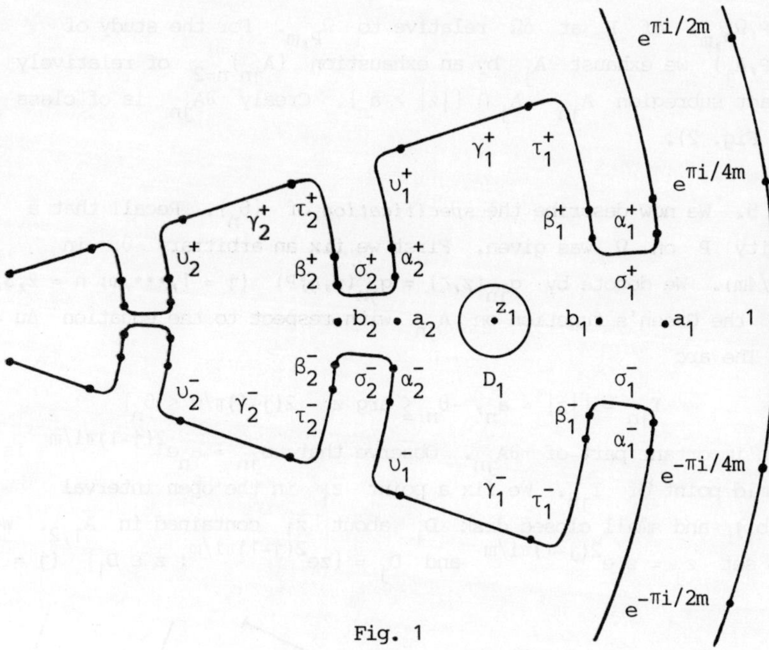

Fig. 1

$1,2,\cdots$). We can and hence will choose $\sigma_n^\pm$, $\tau_n^\pm$, $\upsilon_n^\pm$ suitablly so as to make $\Gamma_1^\pm$ smooth curves of class C^∞.

4. Let $\Gamma_j^\pm$ $(j = 2,\cdots,m)$ be the jordan arcs in Ω obtained by rotating $\Gamma_1^\pm$ around $\delta\Omega$:

$$\Gamma_j^\pm = \{ze^{2(j-1)\pi i/m} : z \in \Gamma_1^\pm\}.$$

Consider subregions S_j $(j = 1,\cdots,m)$ of Ω bounded by Jordan arcs $\Gamma_j^+ \cup \Gamma_{j+1}^-$ where $\Gamma_{m+1}^- = \Gamma_1^-$. We thus obtain an admissible subregion $\Omega_{P,m}$ of Ω by subtracting closed subregions $\overline{S}_j$ $(j = 1,\cdots,m)$ from Ω:

$$\Omega_{P,m} = \Omega - \bigcup_{j=1}^{m} \overline{S}_j.$$

The relative Picard dimensions $\dim(P,A_j)$ $(j = 1,\cdots,m)$ of P at $\delta\Omega$ relative to the subregions

$$A_j = \Omega_{P,m} \cap \{|z| < b_1,\ (2j-3)\pi/m < \arg z < (2j-1)\pi/m\}$$

of Ω are essential for determining the relative Picard dimension $\dim(P,\Omega_{P,m})$ of P at $\delta\Omega$ relative to $\Omega_{P,m}$. For the study of $\dim(P,A_j)$ we exhaust A_j by an exhaustion $\{A_{jn}\}_{n=2}^{\infty}$ of relatively compact subregion $A_{jn} = A_j \cap \{|z| > a_n\}$. Crealy ∂A_{jn} is of class C^{∞} (see Fig. 2).

5. We now describe the *specification* of $\{\theta_n\}$. Recall that a density P on Ω was given. First we fix an arbitrary θ_1 in $(0,\pi/4m)$. We denote by $g_{jn}(z,\zeta) = g_{jn}(z,\zeta;P)$ $(j = 1,\cdots,m;\ n = 2,3,\cdots)$ the Green's function on A_{jn} with respect to the equation $\Delta u = Pu$. The arc

$$I_{jn} = \{|z| = a_n,\ -\theta_n \leqq \arg z - 2(j-1)\pi/m \leqq \theta_n\}$$

is an important part of ∂A_{jn}. Observe that $a_{jn} = a_n e^{2(j-1)\pi i/m}$ is the mid point of I_{jn}. We fix a point z_1 in the open interval (a_2,b_1) and small closed disk D_1 about z_1 contained in $A_{1,2}$. We also set $z_j = z_1 e^{2(j-1)\pi i/m}$ and $D_j = \{ze^{2(j-1)\pi i/m} : z \in D_1\}$ $(j = 1,$

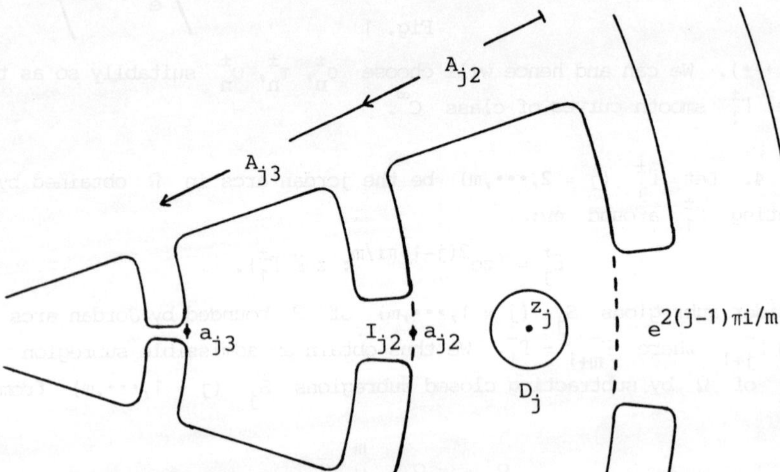

Fig. 2

$\cdots,m)$. Consider the kernel function $h_{jn}(\zeta,z)$ given by

$$h_{jn}(\zeta,z) = \left(\frac{\partial}{\partial n_\zeta} g_{jn}(\zeta,z)\right) / \left(\frac{\partial}{\partial n_\zeta} g_{jn}(\zeta,z_j)\right)$$

for (ζ,z) in $\partial A_{jn} \times A_{jn}$, where $\partial/\partial n_\zeta$ denotes the outer normal differentiation with respect to the region A_{jn}. It is well defined since $\partial g_{jn}(\zeta,z)/\partial n_\zeta$ is strictly positive for (ζ,z) in $\partial A_{jn} \times A_{jn}$. As a consequence of the Harnack inequality the function $(\zeta,z) \mapsto \partial g_{jn}(\zeta,z)/\partial n_\zeta$ is seen to be continuous on $\partial A_{jn} \times A_{jn}$ and the same is true for the function $(\zeta,z) \mapsto h_{jn}(\zeta,z)$. Therefore $\zeta \mapsto h_{jn}(\zeta,z)$ is uniformly continuous on ∂A_{jn} for each z in $D_j \subset A_{jn}$. Hence the function

$$\psi_{jn}(\zeta) = \sup_{z \in D_j} |h_{jn}(\zeta,z) - h_{jn}(a_{jn},z)|$$

is nonnegative and continuous on $I_{jn} \subset \partial A_{jn}$, with $\psi_{jn}(a_{jn}) = 0$. Here ψ_{jn} depend on $\theta_2,\cdots,\theta_{n-1}$ together with choices of $\sigma_\ell^\pm$ $(\ell = 2,\cdots, n-1)$, $\tau_\ell^\pm$, $\upsilon_\ell^\pm$ $(\ell = 1,\cdots,n-1)$ (see Fig. 1) but do not depend on θ_n. If we take θ_n sufficiently small in $(0,\pi/4m)$, then

$$\max_{1 \leq j \leq m} \sup_{\zeta \in I_{jn}} \psi_{jn}(\zeta) < 2^{-n}.$$

Therfore we can and thus will choose θ_n $(n = 2,3,\cdots)$ successively in $(0,\pi/4m)$ so small that

(5) $$\max_{1 \leq j \leq m} \sup_{\zeta \in I_{jn}} \left(\sup_{z \in D_j} |h_{jn}(\zeta,z) - h_{jn}(a_{jn},z)|\right) < 2^{-n} \quad (n = 2,3,\cdots).$$

This is the specification of $\{\theta_n\}_1^\infty$ we announced in no.2.

6. We will proceed to the analysis of the space $PP(A_j;\partial A_j)$ $(j = 1,2,\cdots)$. As a consequence of the foregoing construction we now see that $PP(A_j;\partial A_j)$ has a single generator k_j, say, i.e. $PP(A_j;\partial A_j)$ is the one dimensional space $\{ck_j : c \in [0,+\infty)\}$. To determine k_j, consider functions k_{jn} on A_{jn} given by

$$k_{jn}(z) = h_{jn}(a_{jn},z) \quad (n = 2,3,\cdots).$$

The function k_{jn} belongs to $PP(A_{jn};\partial A_{jn}-\{a_{jn}\})$, the class of nonnegative solutions of $\Delta u = Pu$ on A_{jn} with vanishing boundary values on $\partial A_{jn} - \{a_{jn}\}$. Observe that $k_{jn}(z_j) = 1$ and therefore, by the Harnack inequality, $\{k_{jn}\}_{n=2}^{\infty}$ forms a normal family, i.e. there exists a subsequence $\{k_{j\nu(n)}\}_{n=1}^{\infty}$ of $\{k_{jn}\}_{n=2}^{\infty}$ such that

$$(6) \qquad k_j(z) = \lim_{n \to \infty} k_{j\nu(n)}(z)$$

exists on A_j and the convergence is uniform on each compact subset of A_j and hence of $\overline{A}_j$ by an obvious application of the maximum principle. Thus clearly $k_j \in PP(A_j;\partial A_j)$ and $k_j(z_j) = 1$. This is a required generator k_j.

Choose an arbitrary element v in $PP(A_j;\partial A_j)$ with $v(z_j) = 1$. The Green's formula yields

$$v(z) = -\frac{1}{2\pi}\int_{\partial A_{jn}} v^* dg_{jn}(\cdot,z) \quad (z \in A_{jn}).$$

Here the boundary function v on ∂A_{jn} is nonvanishing only on I_{jn} and thus we can rewrite the above formula as

$$v(z) = \int_{I_{jn}} \left(-\frac{1}{2\pi}\frac{\partial}{\partial n_\zeta} g_{jn}(\zeta,z)v(\zeta)\right)|d\zeta| \quad (z \in A_{jn}).$$

Using a positive measure μ_{jn} on I_{jn} defined by

$$(7) \qquad d\mu_{jn}(\zeta) = -\frac{1}{2\pi}\frac{\partial}{\partial n_\zeta} g_{jn}(\zeta,z_j)v(\zeta)|d\zeta|$$

and recalling the definition of $h_{jn}(\zeta,z)$ we obtain

$$(8) \qquad v(z) = \int_{I_{jn}} h_{jn}(\zeta,z)d\mu_{jn}(\zeta) \quad (z \in A_{jn})$$

for every $n = 2,3,\cdots$. For $z = z_j$ this implies that $\mu_{jn}(I_{jn}) = 1$ which is crucial in the estimation below. Take an arbitrary z in D_j. Then (5), (7), and (8) give

$$\begin{aligned}
&|v(z) - k_{j\nu(n)}(z)| \\
&= \left|\int_{I_{j\nu(n)}} h_{j\nu(n)}(\zeta,z)d\mu_{j\nu(n)}(\zeta) - h_{j\nu(n)}(a_{j\nu(n)},z)\right| \\
&= \left|\int_{I_{j\nu(n)}} \left(h_{j\nu(n)}(\zeta,z) - h_{j\nu(n)}(a_{j\nu(n)},z)\right)d\mu_{j\nu(n)}(\zeta)\right| \\
&\leq \int_{I_{j\nu(n)}} |h_{j\nu(n)}(\zeta,z) - h_{j\nu(n)}(a_{j\nu(n)},z)|d\mu_{j\nu(n)}(\zeta)
\end{aligned}$$

$$< 2^{-\nu(n)}.$$

We have shown that $\sup_{z\in D} |v(z) - k_{j\nu(n)}(z)| \leqq 2^{-\nu(n)}$. On letting $n \to \infty$ and using (6) we see that $v = k_j$ on D and therefore on A_j. Thus we have proved $\dim(P, A_j) = 1$.

7. For the analysis of the space $PP(\Omega_{P,m}; \partial\Omega_{P,m})$ we set $A = A_1$,

$$B = B_1 = \Omega_{P,m} - (A_1 \cap \{|z| \leqq a_2\}),$$

$$J_1 = \{|z| = b_1, |\arg z| \leqq \theta_1\},$$

$$I_2 = I_{1,2} = \{|z| = a_2, |\arg z| \leqq \theta_2\}$$

and consider a bounded operator K of $C(I_2)$ to $C(J_1)$ defined by

$$(K\phi)(z) = P_\phi^B(z) \quad (\phi \in C(I_2),\ z \in J_1).$$

Here P_ϕ^B is the unique bounded solution (cf. e.g. Nakai-Tada[5]) of $\Delta u = Pu$ on B with boundary values $\tilde{\phi}$ on ∂B which is ϕ on I_2 and zero on $\partial B - I_2$. In addition to the operator K we also consider a bounded operator L of $C(J_1)$ to $C(I_2)$ defined by

$$(L\phi)(z) = P_\phi^A(z) \quad (\phi \in C(J_1),\ z \in I_2)$$

where P_ϕ^A is the unique bounded solution of $\Delta u = Pu$ on A with boundary values $\tilde{\phi}$ on ∂A which is ϕ on J_1 and zero on $\partial A - J_1$. Let u be any function in $PP(A; \partial A)$. We will construct a function v in $PP(\Omega_{P,m}; \partial\Omega_{P,m})$ such that v and $v - u$ are bounded on B and A, respectively. This will be done by using the bounded operator

$$M = L \circ K$$

of $C(I_2)$ to $C(I_2)$.

We denote by $\|\phi\|_1$ ($\|\phi\|_2$, resp.) the supremum norm of a function ϕ in $C(J_1)$ ($C(I_2)$, resp.): $\|\phi\|_1 = \sup_{z\in J_1} |\phi(z)|$ ($\|\phi\|_2 = \sup_{z\in I_2} |\phi(z)|$, resp.). Then we have

$$\|M^n u\|_2 \leqq \rho^n \|u\|_2 \quad (n = 1, 2, \cdots)$$

where $\rho = \|L1\|_2$. Since $\rho < 1$ by the maximum principle, we can define, using C. Neumann series, a nonnegative function

$$\phi_u = \sum_{n=0}^{\infty} M^n u$$

in $C(I_2)$. It satisfies an abstract integral equation

$$(I - M)\phi_u = u$$

where I is the identity operator of $C(I_2)$. Observe that $K\phi_u$ is a solution of $\Delta u = Pu$ on B and $u + M\phi_u$ is a solution of $\Delta u = Pu$ on A with the same boundary values as those of $K\phi_u$ on $\partial(A \cap B)$ since $\phi_u - M\phi_u = u$ on I_2, $u + M\phi_u = K\phi_u$ on J_1, and $u = K\phi_u = M\phi_u = 0$ on $\partial(A \cap B) - (J_1 \cup I_2)$. Then we can define the function

$$v(z) = \begin{cases} (K\phi_u)(z) & z \in B \\ u(z) + (M\phi_u)(z) & z \in A \end{cases}$$

on $\Omega_{P,m}$ which belongs to the class $PP(\Omega_{P,m};\partial\Omega_{P,m})$. It satisfies

$$|v(z)| = |(K\phi_u)(z)| \leqq \|\phi_u\|_2 \leqq \sum_{n=0}^{\infty} \|u\|_2\rho^n = \frac{1}{1-\rho}\|u\|_2 \quad (z \in B),$$

$$|v(z) - u(z)| = |(M\phi_u)(z)| \leqq \sum_{n=1}^{\infty} \|u\|_2\rho^n = \frac{\rho}{1-\rho}\|u\|_2 \quad (z \in A).$$

8. We have seen in no.6 that the class $PP(A_j;\partial A_j)$ $(j = 1,\cdots,m)$ has a single generator k_j and also in no.7 that there exists a v_1 in $PP(\Omega_{P,m};\partial\Omega_{P,m})$ such that v_1 and $v_1 - k_1$ are bounded on B_1 and A_1, respectively. By the entirely same argument as applied to A_1 and B_1 we can see that there exists a v_j in $PP(\Omega_{P,m};\partial\Omega_{P,m})$ such that v_j and $v_j - k_j$ are bounded on B_j and A_j, respectively $(j = 1, \cdots,m)$, where $B_j = \Omega_{P,m} - (A_j \cap \{|z| \leqq a_2\})$. Since v_j is unbounded on A_j and bounded on any component of $\Omega_{P,m} \cap \{|z| < a_1\}$ except A_j $(j = 1,\cdots,m)$, $\{v_j\}_1^m$ is a linearly independent subclass of $PP(\Omega_{P,m};\partial\Omega_{P,m})$ so that $\dim(P,\Omega_{P,m}) \geqq m$. On the other hand for any function v in $PP(\Omega_{P,m};\partial\Omega_{P,m})$ there exists a nonnegative constant c_1 such that $v - P_v^{A_1} = c_1k_1$ on A_1 and hence $v - c_1v_1 = P_v^{A_1} + c_1(k_1 - v_1)$ is bounded on A_1. Similarly there exist nonnegative constants c_j $(j = 2,\cdots,m)$ such that $v - c_jv_j$ is bounded on A_j. Since $u = v - \sum_{j=1}^{m} c_jv_j$ is bounded on $\Omega_{P,m}$ and vanishes on $\partial\Omega_{P,m}$, it is identically zero. This can be seen as follows. First observe that u^2 is subharmonic on $\Omega_{P,m}$ and hence on Ω by defining $u^2 = 0$ on $\Omega - \Omega_{P,m}$ since $\Delta u^2 = 2(Pu^2 + |\nabla u|^2) \geqq 0$ on $\Omega_{P,m}$. For any

positive number ε, the function $-\varepsilon\log|z| - u^2(z)$ is superharmonic on Ω and has relative boundary values zero on $\partial\Omega$ and ideal boundary values $+\infty$ on $\delta\Omega$. Hence it is nonnegative on Ω or $u^2(z) \leqq -\varepsilon\log|z|$ on Ω for every ε. On making $\varepsilon \to 0$ we conclude that $u \equiv 0$. Now $\dim(P,\Omega_{P,m}) \leqq m$ and thus we have $\dim(P,\Omega_{P,m}) = m$.

The proof of the theorem 1 is herewith complete.

2. RELATIVE PICARD DIMENSIONS DETERMINED BY $\mathcal{Y}$-SEQUENCES

9. A sequence $\{\overline{Y}_n\}_1^\infty$ of closures $\overline{Y}_n$ of Jordan regions Y_n in Ω will be referred to as a *$\mathcal{Y}$-sequence* in Ω if $\overline{Y}_i \cap \overline{Y}_j = \emptyset$ $(i \neq j)$ and $\{\overline{Y}_n\}$ converges to $\delta\Omega$, i.e. there exists only a finite number of $\overline{Y}_n$ such that $\overline{Y}_n \cap \{\varepsilon \leqq |z| \leqq 1\} \neq \emptyset$ for any $\varepsilon > 0$. If every ∂Y_n is a smooth curve of class C^p $(p = 1,2,\cdots,\infty)$, then we say that $\{\overline{Y}_n\}$ is a C^p $\mathcal{Y}$-sequence $(p = 1,2,\cdots,\infty)$. For any $\mathcal{Y}$-sequence $\{\overline{Y}_n\}_1^\infty$ in Ω we assume that $\overline{Y}_n \subset \Omega$ $(n = 1,2,\cdots)$. The set

$$W = W(\{\overline{Y}_n\}) = \Omega - \bigcup_{n=1}^{\infty} \overline{Y}_n$$

is connected and thus a subregion of Ω. It is an admissible subregion of Ω in the sense of no.2.

We prove the following

THEOREM 2. *For any density* P *on* Ω *and any positive integer* m *there exists a* C^∞ *$\mathcal{Y}$-sequence* $\{\overline{Y}_n\}$ *in* Ω *such that the relative Picard dimension* $\dim(P,W(\{\overline{Y}_n\}))$ *is* m.

10. The required $\mathcal{Y}$-sequence $\{\overline{Y}_n\}$ is constructed by the region $\Omega_{P,m}$ obtained in the proof of Theorem 1 in nos.2-8 and two sequences $\{r_n\}$, $\{s_n\}$ as is described in the sequel. In particular notations used to determine or related to $\Omega_{P,m}$ appeared in nos.2-8 will also be used here.

First we choose $\{r_n\}_1^\infty$ as an arbitrary sequence in the interval $(0,1)$ convergent to zero such that

$$q_n < r_n < p_n \quad (n = 1,2,\cdots).$$

Although $\{r_n\}$ is entirely arbitrary except the above restriction, we will fix this throughout our construction once this is chosen.

The sequence $\{s_n\}_1^\infty$ is to be chosen in $(0,1)$ with

$$(q_n + r_n)/2 < s_n < r_n \quad (n = 1,2,\cdots).$$

We do not require the sequence $\{s_n\}$ anything more to satisfy the above inequalities for a while but it will be specified later so that the sequence $\{r_n - s_n\}_1^\infty$ converges to zero rapidly in a manner depending on $\Omega_{P,m}$ and $\{r_n\}$.

A closed Jordan regions $\overline{S}_{jn}$ $(j = 1,\cdots,m;\ n = 1,2,\cdots)$ in Ω bounded by a Jordan curve Γ_{jn} will be determined as follows. Consider simple arcs Λ_{jn}^+ and Λ_{jn}^- $(j = 1,\cdots,m;\ n = 1,2,\cdots)$ given by

$$\Lambda_{jn}^+ = \Gamma_j^+ \cap \{r_n \leqq |z| \leqq s_{n-1},\ 2(j-1)\pi/m \leqq \arg z \leqq (2j-1)\pi/m\},$$

$$\Lambda_{jn}^- = \Gamma_{j+1}^- \cap \{r_n \leqq |z| \leqq s_{n-1},\ (2j-1)\pi/m \leqq \arg z \leqq 2j\pi/m\}$$

where we set $\Gamma_{m+1}^- = \Gamma_1^-$ and $s_0 = a_1$. Take simple arcs λ_{jn}, ν_{jn} $(j = 1,\cdots,m;\ n = 1,2,\cdots)$ in

$$\{(r_n + s_n)/2 < |z| < r_n,\ |\arg z - (2j-1)\pi/m| < \pi/2m\},$$

$$\{s_n < |z| < (r_n + s_n)/2,\ |\arg z - (2j-1)\pi/m| < \pi/2m\},$$

respectively which joins two points $r_n e^{(4j-3)\pi i/2m}$ and $r_n e^{(4j-1)\pi i/2m}$, $s_n e^{(4j-3)\pi i/2m}$ and $s_n e^{(4j-1)\pi i/2m}$, respectively. We denote by Γ_{jn} $(j = 1,\cdots,m;\ n = 1,2,\cdots)$ the Jordan curve formed by four arcs $\nu_{j,n-1}$, Λ_{jn}^+, λ_{jn}, and Λ_{jn}^-, where we set $\nu_{j,0} = \emptyset$. We can and hence will choose $\nu_{j,n-1}$, λ_{jn} suitably so as to make Γ_{jn} a smooth curve of class C^∞. We thus obtain Jordan regions S_{jn} $(j = 1,\cdots,m;\ n = 1,2,\cdots)$ in Ω bounded by Γ_{jn} and hence a $\mathcal{Y}$-sequence $\{\overline{Y}_n\}_1^\infty = \{\overline{S}_{jn}\}_{1 \leqq j \leqq m,\, n \geqq 1}$ (see Fig. 3).

11. We now describe the *specification* of $\{s_n\}$. Consider a subregion U_{jn} $(j = 1,\cdots,m;\ n = 1,2,\cdots)$ of Ω given by

$$U_{jn} = \{(3q_n + r_n)/4 < |z| < (r_n + p_n)/2,$$

$$(2j-1)\pi/m - 3\pi/4m < \arg z < (2j-1)\pi/m - \pi/2m\}$$

$$\cup \{s_n < |z| < r_n,\ |\arg z - (2j-1)\pi/m| < \pi/2m\}$$

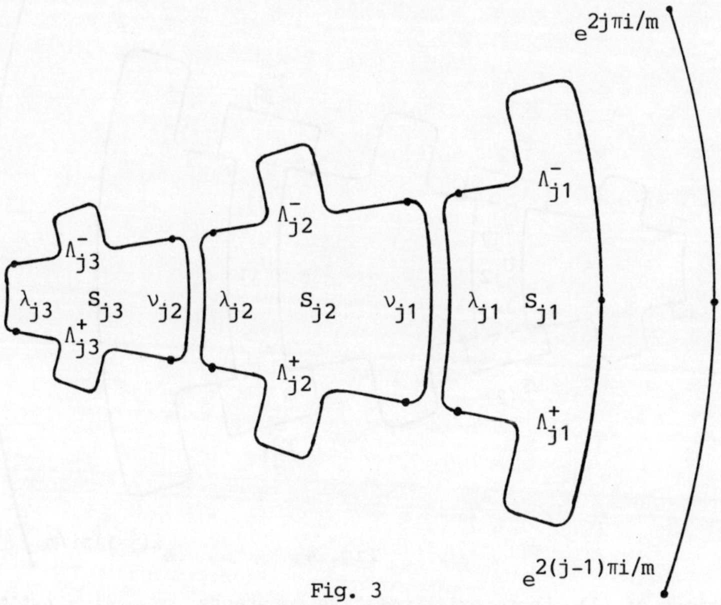

Fig. 3

$$\cup \{(3q_n + r_n)/4 < |z| < (r_n + p_n)/2,$$
$$(2j-1)\pi/m + \pi/2m < \arg z < (2j-1)\pi/m + 3\pi/4m\}.$$

We also consider a subset γ_{jn} of U_{jn} and a subset F_{jn} of ∂U_{jn} $(j = 1,\cdots,m;\ n = 1,2,\cdots)$ given by

$$\gamma_{jn} = (U_{jn} \cap \{\arg z = (2j-1)\pi/m - \pi/2m\})$$
$$\cup (U_{jn} \cap \{\arg z = (2j-1)\pi/m + \pi/2m\}),$$
$$F_{jn} = \partial U_{jn} - \{|\arg z - (2j-1)\pi/m| \leqq \pi/2m\} \quad \text{(see Fig. 4)}.$$

We denote by ω_{jn} $(j = 1,\cdots,m;\ n = 1,2,\cdots)$ the elliptic measure of F_{jn} considered on U_{jn} for the equation (1), i.e. the solution of $\Delta u = Pu$ on U_{jn} with boundary values 1 on F_{jn} and 0 on $\partial U_{jn} - F_{jn}$. By the similar argument as that of the lemma 1 in Nakai-Tada[5], ω_{jn} converges to 0 on γ_{jn} as $s_n \to r_n$ $(j = 1,\cdots,m;\ n = 1,2,\cdots)$:

(9) $$\lim_{s_n \to r_n} \sup_{\gamma_{jn}} \omega_{jn} = 0.$$

By the boundary Harnack inequality (Ancona[1]) for positive

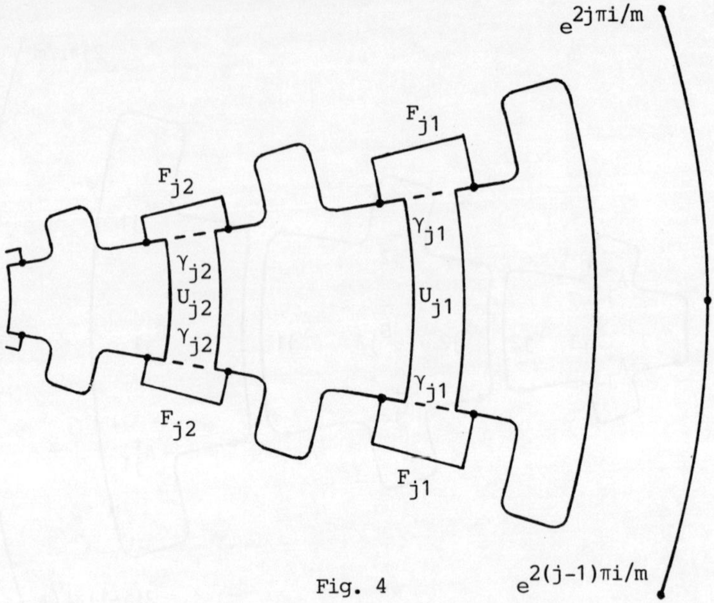

Fig. 4

solutions of (1), there exist positive constants c_{jn} $(j = 1,\cdots,m;\ n = 1,2,\cdots)$ such that

$$u(z) \leqq c_{jn}u(z_1) \tag{10}$$

for any z in F_{jn} and any u in $PP(\Omega_{P,m};\partial\Omega_{P,m})$, where z_1 was a fixed point in the interval (a_2,b_1) (cf. no.5). Here, in view of (9), we can and thus will choose s_n $(n = 1,2,\cdots)$ successively in $((q_n + r_n)/2, r_n)$ so small that

$$\sup_{\gamma_{jn}} \omega_{jn} < 1/2c_{jn} \quad (j = 1,\cdots,m). \tag{11}$$

This is the specification of $\{s_n\}_1^\infty$ we announced in no.10.

12. We will proceed to the analysis of the space $PP(W;\partial W)$ for

$$W = W(\{\overline{Y}_n\}) = W(\{\overline{S}_{jn}\}) = \Omega - \bigcup_{j=1}^{m}\bigcup_{n=1}^{\infty}\overline{S}_{jn}.$$

Consider the mapping Φ of $PP(W;\partial W)$ to $PP(\Omega_{P,m};\partial\Omega_{P,m})$ given by

$$\Phi u = u - P_u^{\Omega_{P,m}} \quad (u \in PP(W;\partial W)).$$

We also consider the mapping Φ_1 of $PP_1(W;\partial W)$ to $PP_1(\Omega_{P,m};\partial\Omega_{P,m})$ defined by

$$\Phi_1 u = \frac{\Phi u}{\ell(\Phi u)} \quad (u \in PP_1(W;\partial W))$$

where

$$\ell(\Phi u) = -\frac{1}{2\pi}\int_{\partial\Omega} \frac{\partial}{\partial|z|}(\Phi u)(z)\,|dz|.$$

By the similar argument as that of the proof of the theorem 1 in Nakai-Tada[5], we can prove by (10) and (11) that the mapping Φ is bijective. It is then easy to check that Φ_1 is also bijective. This means

$$\dim(P,W(\{\overline{S}_{jn}\})) = \dim(P,\Omega_{P,m}) = m$$

and completes the proof of the theorem 2.

3. THE PROOF OF THE MAIN THEOREM

13. Consider a density P on Ω and a $\mathcal{Y}$-sequence $\{\overline{Y}_n\}_1^\infty$ in Ω. Consider another density Q on Ω with $Q \geqq P$ and $\mathrm{supp}(Q - P) \subset \bigcup_1^\infty \overline{Y}_n$. Solutions u of $\Delta u = Qu$ on Ω are also solutions of $\Delta u = Pu$ on $W = W(\{\overline{Y}_n\}) = \Omega - \bigcup_1^\infty \overline{Y}_n$. Then we can define a mapping Ψ of $QP(\Omega;\partial\Omega)$ to $PP(W;\partial W)$ by

$$\Psi u = u - P_u^W \quad (u \in QP(\Omega;\partial\Omega)).$$

In general Ψ may or may not be injective and also surjective. If the mapping Ψ happens to be bijective, then the density Q is said to be *canonically associated* with the $\mathcal{Y}$-sequence $\{\overline{Y}_n\}$ for the density P. In this case we have $\dim Q = \dim(P,W(\{\overline{Y}_n\}))$ (Nakai-Tada[6]). It is shown in Nakai-Tada[6] that there always exists a density Q on Ω canonically associated with an arbitrary given $\mathcal{Y}$-sequence $\{\overline{Y}_n\}$ on Ω for an arbitrary density P on Ω. If P and $\{\overline{Y}_n\}$ are of class C^p, then Q can be chosen as being of class C^p $(p = 1,2,\cdots,\infty)$. Now we construct a density Q in the main theorem as follows. Let P be an arbitrary given density on Ω and m an arbitrary positive integer. We can assume that P is of class C^∞ by replacing it with a dominating C^∞ density if neccessary. Take a C^∞ $\mathcal{Y}$-sequence $\{\overline{Y}_n\}_1^\infty$ in the theorem 2 for P and m. This $\mathcal{Y}$-sequence $\{\overline{Y}_n\}$ satisfies

$\dim(P, W(\{\bar{Y}_n\})) = m$. Consider a canonically associated C^∞ density $Q_{P,m}$ with the $\mathcal{Y}$-sequence $\{\bar{Y}_n\}$ for the density P. The density $Q_{P,m}$ is the required density in the main theorem.

REFERENCES

1] Ancona, A., "Principe de Harnack a la frontière et théorème de Fatou pour un opérateur elliptique dans un domaine lipschitzien", Ann. Inst. Fourier 28, 169-213(1978).

2] Brelot, M., "Étude des l'équation de la chaleur $\Delta u = c(M)u(M)$, $c(M) \geq 0$, au voisinage d'un point singulier du coefficient", Ann. École Norm. Sup. 48, 153-246(1931).

3] Brelot, M., "Sur le principe des singularités positives et la notion de source pour l'équation (1) $\Delta u(M) = c(M)u(M)$ $(c \geq 0)$", Ann. Univ. Lyon Sci. Math. Astro. 11, 9-19(1948).

4] Nakai, M., "Picard principle for finite densities", Nagoya Math. J. 70, 7-24(1978).

5] Nakai, M. and Tada, T., "Nonmonotoneity of Picard principle", Trans. Amer. Math. Soc. 292, 629-644(1985).

6] Nakai, M. and Tada, T., "Extreme nonmonotoneity of Picard principle", NIT Sem. Rep. 26, 1-28(1987).

7] Phelps, R., "Lectures on Choquet's Theorem", Math. Studies No.7, Van Nostrand, Princeton, N.J., 1965.

GEOMETRY AND TOPOLOGY (pp. 251–258)
edited by G. M. Rassias & G. M. Stratopoulos

NEW AND OLD RESULTS AND PROBLEMS IN TOPOLOGY

George M. Rassias
The American College of Greece
Deree College
Aghia Paraskevi GR 153 10
Athens, Greece

ABSTRACT:

The purpose of this paper is: 1. to give an affirmative answer to Ulam's-Banach's problem, and 2. to give an example of a homogeneously invertible, Hausdorff topological space that is not normal.

1. ON A PROBLEM OF S.M. ULAM AND S. BANACH.

Let K be a closed subset of the n-dimensional sphere S^n; S.M. Ulam and S. Banach [10] posed the problem: "Does there exist a sequence of homeomorphic mappings of the sphere S^n onto itself, converging to a mapping of the sphere onto K?" An affirmative answer to this problem is given by the following theorem.

THEOREM: Let K be a closed subset of the n-dimensional sphere S^n. Then, there exists a sequence of homeomorphic mappings of the sphere S^n onto itself converging to a mapping of S^n onto K.

PROOF: Let X be a vector field on S^n that is nonzero everywhere except at one (or more) point in K, and such that for each $x \in S^n$, the flow of X, $a(t,x)$ converges to a singularity (or zero) of X for $t \to \infty$. (S^n-point$\approx R^n$).

Let f_m be a sequence of C^∞ functions converging to the characteristic function of $S^n - K$:

$$f_n \longrightarrow \begin{cases} 0 \text{ on } K \\ 1 \text{ on } S^n - K \end{cases}$$

uniformly on compact subsets on S^n-K (and interior of K). (C^∞ is dense in L^∞). Let $X_m = f_m X$.

Let $a_m(t,x)$, $t \geq 0$, $x \in S^n$ be the flow of X_m

$$\partial/\partial t \; a_m = X_m \circ a_m, \quad a_m(0,x) = x$$

For fixed t, $x \longrightarrow a(t,x)$ is a diffeomorphism $S^n \longrightarrow S^n$

$$h_m(x) = a_m(m,x)$$

For fixed m, $a_m(t,x)$ converges to zero set of X_m as $t \longrightarrow \infty$, which does not contain any given point of S^n-K if m is large enough.

Pointwise convergence $h_m(x)$ on S^n-K should then be true. Q.E.D.

II. ON THE INVERTIBILITY OF TOPOLOGICAL SPACES

P.H. Doyle and J.G. Hocking introduced the concept of a homogeneously invertible space and did important work on it [2-5]. J. Chew [1] proved that every homogeneously invertible, Hausdorff topological space is regular and posed the following problem: "Is a homogeneously invertible, Hausdorff topological space, normal?" In the sequel, an example is given of a homogeneously invertible, Hausdorff topological space that is <u>not</u> normal.

Definition: A topological space X is <u>homogeneously invertible</u> iff given a nonempty open set U and $a \in U$, $b \in X-U$ there is a self-homeomorphism h of X such that $h(b)=a$ and $h(X-U) \subset U$.

Counterexample:

Let S be the topological space whose points are points of the half-open interval $[0,1)$, and whose basic open sets are half-open intervals open on right. Because the space S is totally disconnected almost any function is continuous therefore the easiest chosen bijection proves that the space S is homogeneously invertible. In addition S is normal. However, the Cartesian product SxS is Hausdorff, homogeneously invertible but <u>not</u> normal.

We claim that the two closed sets $X_1=\{(x,1-x)/x$ is rational$\}$ and $X_2 = \{(x,1-x)/x$ is irrational$\}$ cannot be separated by open sets in SxS. Assume that there exist disjoint open sets V_1 and V_2 in SxS separating X_1 and X_2 i.e. $V_1 \supset X_1$ and $V_2 \supset X_2$. Every point (x,y) of SxS has a neighborhood base that consists of

$[x,x+\frac{1}{n}) \times [y,y+\frac{1}{n})$, $n \in N$.

So, if we denote by

$$I_n = \{(x,y) \in X_2 \,/\, [x,x+\tfrac{1}{n}) \times [y,y+\tfrac{1}{n}) \subset V_2\}$$

Then $X_2 = \bigcup_{n=1}^{\infty} I_n$. Consider $I = \{(x,1-x) \,/\, 0 \leqslant x < 1\}$

as a line with ordinary topology. Then, I_n is nowhere dense in I. Therefore,

$$I = \bigcup_{n=1}^{\infty} I_n \cup X_1,$$

which is a countable union of nowhere dense sets, that is impossible because of Baire's theorem.

Hence, SxS in <u>not</u> normal.

Q.E.D.

Research Problems:

Problem 1. Let X be a given topological space. We define that $A+B = (A \cup B) - (A \cap B)$ for every subsets A and B of X. Then is there a set function $f: 2^X \rightarrow 2^X$ such that $f(A+B) = f(A)+f(B)$ for every $A, B \subset X$ and such that f describes the topology of X?

Remark: $cl(A\cup B) = cl(A) \cup cl(B)$, $int(A\cap B) = int(A) \cap int(B)$, and both cl and int describe the topology of X.

Problem 2. Let o(X) denote the cardinality of the family of all open sets of a topological space X. Then is it true that $o(X) = 2^a$ for every Hausdorff space X and for some cardinal number a?

Problem 3. (G.M. Rassias [7])

(a) Determine which closed n-dimensional manifolds can be covered with n open disks for $n\geq 4$.

(b) Determine the minimal number of open disks that can cover an n-manifold for any n.

Remark: For part (b) of this problem some results have been obtained in G.M. Rassias [8].

Problem 4. (On a conjecture of René Thom)

René Thom made the following conjecture which appeared in Saunders Maclane [6]. The purpose of the conjecture was to find a topological characterization of degenerate critical points by the number of (non-degenerate) critical points and their indices in the non-degenerate approximation.

Conjecture of R. Thom: Let F be a C^2 differentiable function on a closed differentiable manifold M, $F: M\longrightarrow R$, having one degenerate critical point p. Perturb F to get a new function $G=F+\delta F$ which may be assumed to be a Morse function. The new function G has a finite number of nondegenerate critical points q_i corresponding to the original degenerate critical point p. Suppose also that the perturbation is such that the stable and unstable manifolds of the critical points meet transversally. The gradient trajectories between two of the relevant critical points form a set Ω. The conjecture asserts: (i) that Ω is contractible in itself, and (ii) that the the degenerate case arises by collapsing Ω (i.e., that the sets W_p^- and W_p^+ of trajectories entering and leaving p are determined by the sets $W_{q_i}^-$ and $W_{q_i}^+$ for the q_i, perhaps by taking the disjoint union of these and identifying the points qi). Then degenerate critical points could be characterized by the number of points and their indices in the nondegenerate approximation.

Two counterexamples to the above conjecture have been constructed by G.M. Rassias [10].

Problem 4a. (G.M. Rassias [7])

Determine the range of validity of Thom's conjecture for closed, differentiable manifolds of dimensions $n \geq 4$. Also, determine the homotopy type of Ω for any n=dimM.

Problem 5 (G.M. Rassias [7])

Let $\mu(M)$ be the Morse-Smale characteristic of a closed, C^∞ differentiable manifold. (a) Give necessary and sufficient conditions on a given homotopy class C of (closed, C^∞ differentiable) manifolds M such that

$$v(M) = \min_{f \in \Omega(M)} \sum_{i=0}^{m} c_i(M,f) - \sum_{i=0}^{m} \min_{f \in \Omega(M)} c_i(M,f)$$

is constant in C, for $n \geq 4$. (b) when there exists $M \in C$ such that $v(M)=0$? (c) Determine when $v \equiv 0$ on C. Note that $v(M)=0$ for every closed, C^∞ differentiable manifold M of dimension $n<4$. See G.M. Rassias [9].

Acknowledgement: I am deeply grateful to the Organizing committee of the British Mathemat. colloquium. Exeter (England), April 11-15, 1988, for inviting me to give this lecture.

REFERENCES

1. J. Chew, Invertibility of Function Spaces, Amer. Math. Monthly 82(1975), 618-621.

2. P.H. Doyle and J.G. Hocking, A Characterization of Euclidean n-space, Mich. Math. J., 7(1960), 199-200

3. P.H. Doyle and J.G. Hocking, Invertible Spaces, Amer. Math. Monthly 68(1961), 959-965.

4. P.H. Doyle and J.G. Hocking, Continously Invertible Spaces, Pacific J. of Math. 12(1962), 499-503.

5. P.H. Doyle and J.G. Hocking, Local Invertibility, Fundamenta Mathematicae, 54(1964), 15-25.

6. S. Maclane, Geometrical Mechanics, part II, Lect. Notes, Dept. of Math., Univ. of Chicago, 1968.

7. G.M. Rassias, New Results and Research Problems in Morse-Smale and Stallings Theories. Algebraic and differential Topology - Global differential Geometry (ed: G.M. Rassias), Teubner-Texte zur Mathematik, Band 70, Leipzig 1984, 7-16.

8. G.M. Rassias, Covering of Differentiable Manifolds with Open Disks, Archiv der Math. 38(1982), 366-367.

9. G.M. Rassias, On the Morse-Smale Characteristic of a Differentiable Manifold. Bull. Australian Math. Soc., vol. 20, 1979.

10. G.M. Rasias. Counterexamples to a Conjecture of René Thom. Geometry and Diff. Geometry Conference held in Haifa, Israel 1979. Lect. Notes in Math.. Springer-Verlag no. 792, 414-421.

11. S.M. Ulam. The Scottish Book, Problem 100 (Ulam-Banach), p. 28, 1977.

GEOMETRY AND TOPOLOGY (pp. 259–280)
edited by G. M. Rassias & G. M. Stratopoulos

DIFFERENTIAL-GEOMETRIC STRUCTURES ASSOCIATED WITH FAMILIES OF r-PARAMETER POINT TRANSFORMATIONS

Hanno Rund
Program in Applied Mathematics, University of Arizona
Tucson, Arizona 85721

ABSTRACT: With any r-parameter family of point transformations on a manifold M one can associate three distinct sets of vector fields on M, each set consisting of r elements. The successive imposition of algebraic conditions on these vector fields gives rise to certain differential-geometric structures by means of which the existence of an underlying Lie group may be discerned.

1. Vector Fields Associated with r-Parameter Point Transformations

One of the most basic results in the theory of transformation groups is a theorem, due to Sophus Lie, which establishes necessary and sufficient conditions that must be satisfied by an r-parameter family of point transformations on a manifold in order that these transformations constitute an r-parameter group ([2], p. 54). In the analysis that leads to Lie's theorem it is usually assumed *ab initio* that the r-parameter family of transformations is in fact a group, which gives rise in the first instance to the determination of necessary conditions, while the sufficiency of the latter is subsequently verified almost automatically by a direct appeal to the theorem of Frobenius. This may account for the fact that the situation in which there is given an r-parameter family of point transformations, not neccessarily a group, but nevertheless subject to certain weaker conditions, may have been partly overlooked, although an examination of such a state of affairs may not be entirely devoid of interest. It is the purpose of the present note to analyze a situation of this kind.

The approach presented below is based on the recognition that with any r-parameter family of differentiable point transformations on a given manifold one can associate three sets of vector fields on that manifold, each set consisting of

r elements. The successive imposition of conditions on these vector fields gives rise to differential-geometric structures by means of which the existence of an underlying Lie group may ultimately be determined. This in turn allows for a striking geometrical interpretation of those vector fields. However, at the same time it should be stressed that, from a purely practical point of view, the criteria of Lie's theorem are far more useful than those presented here: in fact, the primary objective of the latter is the construction of a geometrical background that is appropriate to this complex of ideas.

Let x^j $(j = 1, \cdots, n)$ denote local coordinates on an n-dimensional differentiable manifold M, and let u^α $(\alpha = 1, \cdots, r)$ represent the points of the r-dimensional number space $\mathbf{R}^r$. It is assumed that a set of n distinct real-valued C^∞ functions f^j is prescribed on $M \times \mathbf{R}^r$. For a given point p on M with coordinates x^h $(h = 1, \cdots, n)$ and for some $\{u^\alpha\} \in \mathbf{R}^r$ the values

$$y^j = f^j(x^h, u^\alpha) \tag{1.1}$$

are regarded as the coordinates of a point q on M; thus (1.1) is interpreted as a map $f: M \times \mathbf{R}^r \to M$, that is, an r-parameter family of point transformations on M. As usual, it is supposed that the values $\{u^\alpha\} = \{0\} \in \mathbf{R}^r$ in (1.1) imply the identity transformation:

$$x^j = f^j(x^h, 0). \tag{1.2}$$

[Here, and in the sequel, lower case Latin and Greek indices respectively range from 1 to n and from 1 to r; the summation convention applies to both sets of indices.]

It is also assumed that the parameters u^α in (1.1) are essential, that is, there do not exist $r-1$ functions v^a $(a = 1, \cdots, n-1)$ on $\mathbf{R}^r$ such that f^j admits the representation

$$f^j(x^h, u^\alpha) = F^j(x^h, v^a).$$

For this it is necessary and sufficient [2, p. 10] that there do not exist nonvanishing functions Φ^α on $\mathbf{R}^r$ for which

$$\Phi^\alpha \frac{\partial f^j}{\partial u^\alpha} = 0. \tag{1.3}$$

We shall write

$$f^j_h(x, u) = \frac{\partial f^j(x, u)}{\partial x^h}, \qquad f^j_\alpha(x, u) = \frac{\partial f^j(x, u)}{\partial u^\alpha}, \tag{1.4}$$

together with

$$\xi^j_\alpha(x) = f^j_\alpha(x, 0). \tag{1.5}$$

[Here, and in the sequel, arguments such as $(x^1, \cdots, x^n, u^1 \cdots u^r) = (x^h, u^\alpha)$ of f^j and its derivatives are denoted by (x, u).] From (1.2) it follows that

$$f^j_h(x, 0) = \delta^j_h, \tag{1.6}$$

so that the rank of the matrix $(f^j_h(x, u))$ is n on a neighborhood of $\{0\} \in \mathbf{R}^r$.

The map f defined by (1.1) induces corresponding maps of tangent spaces:

$$f_*: T_{(p,u)}(M \times \mathbf{R}^r) = T_p(M) \oplus T_u(\mathbf{R}^r) \to T_q(M),$$

where $T_p(M)$ denotes as usual the tangent space of M at the point p. In terms of the notation (1.4) we have, in particular,

$$f_*(\frac{\partial}{\partial x^h}) = f^j_h(x, u)\frac{\partial}{\partial y^j}, \qquad f_*(\frac{\partial}{\partial u^\alpha}) = f^j_\alpha(x, u)\frac{\partial}{\partial y^j}. \tag{1.7}$$

The given functions f^j give rise to three sets of vectors in $T_q(M)$, each set consisting of r elements. The first results directly from (1.5), namely

$$\xi_\alpha(y) = \xi^j_\alpha(y)\frac{\partial}{\partial y^j} = f^j_\alpha(y, 0)\frac{\partial}{\partial y^j}, \tag{1.8}$$

while the second coincides with the second member of (1.7), namely

$$B_\alpha(y) = f_*(\frac{\partial}{\partial u^\alpha}) = f^j_\alpha(x, u)\frac{\partial}{\partial y^j}. \tag{1.9}$$

For $u^\alpha = 0$ the point q coincides with p, and (1.8) assumes the form

$$\xi_\alpha(x) = \xi^j_\alpha(x)\frac{\partial}{\partial x^j} = f^j_\alpha(x, 0)\frac{\partial}{\partial x^j}, \tag{1.10}$$

which represents a set of r vectors in $T_p(M)$. The third set of vectors in $T_q(M)$ is represented by the images of (1.10) by f_*:

$$\eta_\alpha(y) = f_*(\xi_\alpha(x)) = \xi^h_\alpha(x) f_*(\frac{\partial}{\partial x^h}) = f^j_h(x, u)\xi^h_\alpha(x)\frac{\partial}{\partial y^j}. \tag{1.11}$$

Clearly $\eta_\alpha(x) = \xi_\alpha(x)$ in $T_p(M)$, while by (1.9) and (1.10) one has $B_\alpha(x) = \xi_\alpha(x)$, so that

$$\xi_\alpha(x) = B_\alpha(x) = \eta_\alpha(x). \tag{1.12}$$

However, at points other than p, these vectors will in general be distinct. For future reference we also note that (1.9) implies

$$[B_\alpha, B_\beta](y) = [B_\alpha(y), B_\beta(y)] = [f_*(\frac{\partial}{\partial u^\alpha}), f_*(\frac{\partial}{\partial u^\beta})] = f_*[\frac{\partial}{\partial u^\alpha}, \frac{\partial}{\partial u^\beta}] = 0. \tag{1.13}$$

Because of (1.6) the matrix (f_j^h) has a local inverse $(\hat{f}_h^j)$ by means of which we construct

$$b_\alpha^j(x,u) = -\hat{f}_m^j(x,u) f_\alpha^m(x,u). \tag{1.14}$$

This gives rise to the vectors

$$b_\alpha = b_\alpha^j(x,u)\frac{\partial}{\partial x^j} + \frac{\partial}{\partial u^\alpha} \tag{1.15}$$

in the tangent space $T_{(p,u)}(M \times \mathbf{R}^r)$, together with the 1-forms

$$C^j = dx^j - b_\alpha^j(x,u) du^\alpha \tag{1.16}$$

in the cotangent space $T^*_{(p,u)}(M \times \mathbf{R}^r)$, with

$$C^j(b_\alpha) = 0. \tag{1.17}$$

From (1.15), (1.14) and (1.7) it follows that

$$f_*(b_\alpha) = -\hat{f}_m^h(x,u) f_\alpha^m(x,u) f_h^j(x,u)\frac{\partial}{\partial y^j} + f_\alpha^j(x,u)\frac{\partial}{\partial y^j} = 0, \tag{1.18}$$

which shows that the vectors (1.15) span the kernel in $T_{(p,u)}(M \times \mathbf{R}^r)$ of the map f_*. Finally, it is seen that for the pull-back $f^*: T_q^*(M) \to T^*_{(p,u)}(M \times \mathbf{R}^r)$, one has

$$f^*(dy^j) = f_h^j(x,u)dx^h + f_\alpha^j(x,u)du^\alpha = f_h^j(x,u)C^h. \tag{1.19}$$

2. The First Two Assumptions

The Lie algebra of vector fields on the manifold M is denoted by $\mathcal{X}(M)$, while $\Lambda^s(M)$ is supposed to represent the vector space of s-forms on M. The r vector fields ξ_α on M as defined by (1.8) generate a module V_r over $\Lambda^\circ(M)$: $V_r = \{X \in \mathcal{X}(M): X = X^\alpha \xi_\alpha,\ X^\alpha \in \Lambda^\circ(M)\}$. If $X = X^\alpha\xi_\alpha$, $Y = Y^\alpha \xi_\alpha$, are arbitrary elements of V_r, then, for any $f \in \Lambda^\circ(M)$,

$$XY(f) = X^\alpha \xi_\alpha(Y^\beta \xi_\beta(f)) = X^\alpha \xi_\alpha(Y^\beta)\xi_\beta(f) + X^\alpha Y^\beta \xi_\alpha \xi_\beta(f),$$

from which it follows that

$$[X,Y] = (X(Y^\alpha) - Y(X^\alpha))\xi_\alpha + X^\alpha Y^\beta[\xi_\alpha, \xi_\beta]. \tag{2.1}$$

From this it is immediately evident that V_r is closed under composition by Lie brackets if and only if $[\xi_\alpha, \xi_\beta] \in V_r$, which is tantamount to the requirement that there exist 3-index symbols, possibly dependent on positional coordinates on M, such that

$$[\xi_\alpha, \xi_\beta] = e_\alpha{}^\epsilon{}_\beta \xi_\epsilon. \tag{2.2}$$

Under these circumstances V_r *is a Lie subalgebra of* $\mathcal{X}(M)$: *this is our first assumption.*

For the purposes of the present discussion we shall keep the point $p \in M$ fixed, and, as before, denote the representative of the field ξ_α in $T_q(M)$ by $\xi_\alpha(y)$, the coordinates y^j of $q \in M$ being given by (1.1). The 3-index coefficients that appear on the right-hand side of (2.2) may appear on y^j and hence on u^α; accordingly we shall write

$$E_\beta{}^\alpha{}_\gamma(u) = e_\beta{}^\alpha{}_\gamma(y). \tag{2.3}$$

In $T_q(M)$ we therefore have in consequence of (2.2):

$$[\xi_\alpha, \xi_\beta](y) = E_\alpha{}^\epsilon{}_\beta(u)\xi_\epsilon(y). \tag{2.4}$$

It is now stipulated that each of the r-vector fields B_α as defined by (1.9) be in the module V_r also, which requires the existence of a matrix $(\hat{X}^\beta_\alpha)$ with y^j-dependent entries such that $B_\alpha = \hat{X}^\beta_\alpha \xi_\beta$. In view of (1.1) we shall write

$$\hat{\chi}^\beta_\alpha(u) = \hat{X}^\beta_\alpha(y), \tag{2.5}$$

so that, in $T_q(M)$,

$$B_\alpha(y) = \hat{\chi}^\beta_\alpha(u)\xi_\beta(y). \tag{2.6}$$

Because of (1.12) we have

$$\hat{\chi}^\alpha_\beta(0) = \delta^\alpha_\beta. \tag{2.7}$$

Consequently the matrix $(\hat{\chi}^\alpha_\beta)$ has an inverse (χ^β_α) on a neighborhood of $\{0\} \in \mathbf{R}^r$, and (2.6) is equivalent to

$$\xi_\alpha(y) = \chi^\beta_\alpha(u)B_\beta(y). \tag{2.8}$$

This is our second assumption.

We then have

$$\xi_\alpha(y)\xi_\beta(y) = \chi^\epsilon_\alpha(u)B_\epsilon(y)(\chi^\gamma_\beta(u))B_\gamma(y) + \chi^\epsilon_\alpha(u)\chi^\gamma_\beta(u)B_\epsilon(y)B_\gamma(y), \tag{2.9}$$

in which, because of (1.9), (2.5), and (1.1),

$$B_\epsilon(y)(\chi^\gamma_\beta(u)) = f^j_\epsilon(x,u)\frac{\partial}{\partial y^j}(X^\gamma_\beta(y)) = \frac{\partial X^\gamma_\beta(y)}{\partial y^j}\frac{\partial y^j}{\partial u^\epsilon} = \frac{\partial \chi^\gamma_\beta(u)}{\partial u^\epsilon}. \tag{2.10}$$

This is substituted in (2.9). In view of (1.13) the resulting Lie bracket is then given by

$$[\xi_\alpha, \xi_\beta](y) = (\chi^\epsilon_\alpha(u)\frac{\partial \chi^\gamma_\beta(u)}{\partial u^\epsilon} - \chi^\epsilon_\beta(u)\frac{\partial \chi^\gamma_\alpha(u)}{\partial u^\epsilon})B_\gamma(y). \tag{2.11}$$

Also, with the aid of (2.8) the relation (2.4) can be expressed as

$$[\xi_\alpha, \xi_\beta](y) = E_\alpha{}^\epsilon{}_\beta(u)\chi^\gamma_\epsilon(u)B_\gamma(y).$$

This is compared with (2.11), the remarks preceding (1.3) being taken into account, together with (1.9). It is thus found that

$$\chi^\epsilon_\alpha(u)\frac{\partial \chi^\gamma_\beta(u)}{\partial u^\epsilon} - \chi^\epsilon_\beta(u)\frac{\partial \chi^\gamma_\alpha(u)}{\partial u^\epsilon} = E_\alpha{}^\epsilon{}_\beta(u)\chi^\gamma_\epsilon(u). \tag{2.12}$$

This is a system of first order partial differential equations that must be satisfied by the matrices $\chi^\beta_\alpha(u)$ that occur in the assumption (2.8). An equivalent representation of (2.12) in terms of the inverse matrices is given by

$$\frac{\partial \hat\chi^\alpha_\beta(u)}{\partial u^\epsilon} - \frac{\partial \hat\chi^\alpha_\epsilon(u)}{\partial u^\beta} = E_\mu{}^\alpha{}_\nu(u)\hat\chi^\mu_\beta(u)\hat\chi^\nu_\epsilon(u). \tag{2.13}$$

The structure of these equations suggests the introduction of the following set of 1-forms on $\mathbf{R}^r$:

$$\omega^\alpha(u) = \hat\chi^\alpha_\beta(u)du^\beta, \tag{2.14}$$

for which

$$d\omega^\alpha = \frac{1}{2}(\frac{\partial \hat\chi^\alpha_\beta}{\partial u^\epsilon} - \frac{\partial \hat\chi^\alpha_\epsilon}{\partial u^\beta})du^\epsilon \wedge du^\beta.$$

Consequently (2.13) can be expressed as

$$d\omega^\alpha + \frac{1}{2}E_\beta{}^\alpha{}_\gamma(u)\omega^\beta \wedge \omega^\gamma = 0, \tag{2.15}$$

which has the structure of Maurer-Cartan equations. (We shall return to this point in Section 4.) Exterior differentiation of (2.15) yields

$$\frac{\partial E_\beta{}^\alpha{}_\gamma}{\partial u^\sigma} du^\sigma \wedge \omega^\beta \wedge \omega^\gamma + 2E_\beta{}^\alpha{}_\gamma d\omega^\beta \wedge \omega^\gamma = 0,$$

to which we apply (2.15) and (2.14) to obtain

$$(\frac{\partial E_\beta{}^\alpha{}_\gamma}{\partial u^\sigma}\chi^\sigma_\epsilon + E_\gamma{}^\alpha{}_\sigma E_\epsilon{}^\sigma{}_\beta)\omega^\epsilon \wedge \omega^\beta \wedge \omega^\gamma = 0.$$

It is evident from (2.14) that the set $\{\omega^\alpha : \alpha = 1, \cdots, r\}$ constitutes a basis of 1-forms on $\mathbf{R}^r$; it therefore follows that

$$\begin{aligned}\frac{\partial E_\beta{}^\alpha{}_\gamma}{\partial u^\sigma}\chi^\sigma_\epsilon + \frac{\partial E_\epsilon{}^\alpha{}_\beta}{\partial u^\sigma}\chi^\sigma_\gamma + \frac{\partial E_\gamma{}^\alpha{}_\epsilon}{\partial u^\sigma}\chi^\sigma_\beta \\ + E_\gamma{}^\alpha{}_\sigma E_\epsilon{}^\sigma{}_\beta + E_\beta{}^\alpha{}_\sigma E_\gamma{}^\sigma{}_\epsilon + E_\epsilon{}^\alpha{}_\sigma E_\beta{}^\sigma{}_\gamma = 0.\end{aligned} \tag{2.16}$$

An alternative form of this relation follows directly from (2.3), since according to the latter, (1.4), and (1.9)

$$\frac{\partial E_\beta{}^\alpha{}_\gamma}{\partial u^\sigma} = f^j_\sigma(x,u)\frac{\partial e_\beta{}^\alpha{}_\gamma}{\partial y^j} = B_\sigma(y)e_\beta{}^\alpha{}_\gamma,$$

so that by (2.8)

$$\frac{\partial E_\beta{}^\alpha{}_\gamma}{\partial u^\sigma}\chi^\sigma_\epsilon = \xi_\epsilon(y)e_\beta{}^\alpha{}_\gamma. \tag{2.17}$$

Thus (2.16) is equivalent to

$$\xi_\epsilon e_\beta{}^\alpha{}_\gamma + \xi_\gamma e_\epsilon{}^\alpha{}_\beta + \xi_\beta e_\gamma{}^\alpha{}_\epsilon + e_\gamma{}^\alpha{}_\sigma e_\epsilon{}^\sigma{}_\beta + e_\beta{}^\alpha{}_\gamma e_\gamma{}^\sigma{}_\epsilon + e_\epsilon{}^\alpha{}_\sigma e_\beta{}^\sigma{}_\gamma = 0. \tag{2.18}$$

[This is consistent with a standard result concerning Lie subalgebras ([1], p. 57).]

We shall now indicate briefly how the above construction gives rise directly to a differential geometric setting. By means of the r 1-forms (2.14) on $\mathbf{R}^r$ we construct r^2 1-forms

$$\omega_\beta{}^\alpha(u) = E_\epsilon{}^\alpha{}_\beta(u)\omega^\epsilon(u), \tag{2.19}$$

which are regarded as *connection 1-forms* on $\mathbf{R}^r$. As usual, the associated *curvature 2-forms* are defined as

$$\Omega_\beta{}^\alpha(u) = d\omega_\beta{}^\alpha(u) + \omega_\epsilon{}^\alpha(u) \wedge \omega_\beta{}^\epsilon(u). \tag{2.20}$$

In order to evaluate the latter, we note that exterior differentiation of (2.19) yields with the aid of (2.14) and (2.15):

$$d\omega_\beta{}^\alpha = (\frac{\partial E_\nu{}^\alpha{}_\beta}{\partial u^\gamma}\chi^\gamma_\mu + \frac{1}{2}E_\beta{}^\alpha{}_\sigma E_\mu{}^\sigma{}_\nu)\omega^\mu \wedge \omega^\nu,$$

while

$$\omega_\epsilon{}^\alpha \wedge \omega_\beta{}^\epsilon = \frac{1}{2}(E_\mu{}^\alpha{}_\sigma E_\nu{}^\sigma{}_\beta - E_\nu{}^\alpha{}_\sigma E_\mu{}^\sigma{}_\beta)\omega^\mu \wedge \omega^\nu,$$

so that (2.20) assumes the form

$$\Omega_\beta{}^\alpha = [\frac{\partial E_\nu{}^\alpha{}_\beta}{\partial u^\gamma}\chi^\gamma_\mu + \frac{1}{2}(E_\beta{}^\alpha{}_\sigma E_\mu{}^\sigma{}_\nu + E_\mu{}^\alpha{}_\sigma E_\nu{}^\sigma{}_\beta - E_\nu{}^\alpha{}_\sigma E_\mu{}^\sigma{}_\beta)]\omega^\mu \wedge \omega^\nu.$$

Consequently, if we express the curvature 2-forms as

$$\Omega_\beta{}^\alpha(u) = -\frac{1}{2}K_\beta{}^\alpha{}_{\mu\nu}(u)\omega^\mu(u) \wedge \omega^\nu(u), \tag{2.21}$$

we must put

$$K_\beta{}^\alpha{}_{\mu\nu} = \frac{\partial E_\mu{}^\alpha{}_\beta}{\partial u^\gamma}\chi^\gamma_\nu - \frac{\partial E_\nu{}^\alpha{}_\beta}{\partial u^\gamma}\chi^\gamma_\mu + E_\beta{}^\alpha{}_\sigma E_\nu{}^\sigma{}_\mu + E_\mu{}^\alpha{}_\sigma E_\beta{}^\sigma{}_\nu + E_\nu{}^\alpha{}_\sigma E_\mu{}^\sigma{}_\beta.$$

By means of (2.16) this is readily reduced to

$$K_\beta{}^\alpha{}_{\mu\nu}(u) = \chi^\gamma_\beta(u)\frac{\partial E_\mu{}^\alpha{}_\nu(u)}{\partial u^\gamma}. \tag{2.22}$$

Consequently (2.16) may be expressed as

$$K_\epsilon{}^\alpha{}_{\beta\gamma} + K_\beta{}^\alpha{}_{\gamma\epsilon} + K_\gamma{}^\alpha{}_{\epsilon\beta} + E_\epsilon{}^\alpha{}_\sigma E_\beta{}^\sigma{}_\gamma + E_\beta{}^\alpha{}_\sigma E_\gamma{}^\sigma{}_\epsilon + E_\gamma{}^\alpha{}_\sigma E_\epsilon{}^\sigma{}_\beta = 0. \tag{2.23}$$

Moreover, since the matrix $(\chi^\gamma_\beta(u))$ is nonsingular on a neighborhood of $\{0\} \in \mathbf{R}^r$ by virtue of (2.7), it may also be inferred from (2.22) that *the 3-index symbols* $E_\mu{}^\alpha{}_\nu$ *are constants if and only if the curvature 2-forms vanish.*

3. The Third Assumption

The two assumptions that were imposed in the previous section are concerned with only two of the three sets of vector fields that had been identified in Section 1 in connection with the transformation equations (1.1). We now turn to the third set, namely (1.11), and demand that the elements of this set be also in the module V_r. As in the case of (2.8) this requirement is equivalent to the existence of a matrix with u^α-dependent entries $\hat{G}^\beta_\alpha(u)$ such that, in $T_q(M)$,

$$\eta_\alpha(y) = \hat{G}^\beta_\alpha(u)\,\xi_\beta(y). \tag{3.1}$$

Again, because of (1.12), this entails that

$$\hat{G}^\beta_\alpha(0) = \delta^\beta_\alpha, \tag{3.2}$$

so that $(\hat{G}^{\beta}_{\alpha}(u))$ has an inverse $(G^{\alpha}_{\beta}(u))$ on a neighborhood of $\{0\} \in \mathbf{R}^r$. By means of (1.1) and (1.11) the relation (3.1) can therefore be expressed as

$$\xi^j_\alpha(y) = \xi^j_\alpha(f(x,u)) = f^j_h(x,u)\xi^h_\epsilon(x)G^\epsilon_\alpha(u). \tag{3.3}$$

This is our third assumption. It is a relation that involves two sets of independent variables, namely $\{x^k : k = 1, \cdots, n\}$ and $\{u^\gamma : \gamma = 1, \cdots, r\}$.

First, differentiation of (3.3) with respect to x^k gives

$$\frac{\partial \xi^j_\alpha(y)}{\partial y^h} f^h_k(x,u) = \frac{\partial^2 f^j(x,u)}{\partial x^h \partial x^k}\xi^h_\epsilon(x)G^\epsilon_\alpha(u) + f^j_h(x,u)\frac{\partial \xi^h_\epsilon(x)}{\partial x^k}G^\epsilon_\alpha(u).$$

This is multiplied by $\xi^k_\gamma(x)G^\gamma_\beta(u)$, after which (3.3) is applied to the resulting expression on the left-hand side to yield

$$\begin{aligned}\xi^h_\beta(y)\frac{\partial \xi^j_\alpha(y)}{\partial y^h} &= \frac{\partial^2 f^j(x,u)}{\partial x^h \partial x^k}\xi^h_\epsilon(x)\xi^k_\gamma(x)G^\epsilon_\alpha(u)G^\gamma_\beta(u)\\ &\quad + f^j_h(x,u)\xi^k_\gamma(x)\frac{\partial \xi^h_\epsilon(x)}{\partial x^k}G^\epsilon_\alpha(u)G^\gamma_\beta(u).\end{aligned}$$

Since the first term on the right-hand side is symmetric in α and β, it is seen that

$$[\xi_\beta, \xi_\alpha]^j(y) = f^j_h(x,u)[\xi_\gamma, \xi_\epsilon]^h(x)G^\epsilon_\alpha(u)G^\gamma_\beta(u).$$

The relation (2.4) is now applied to the Lie bracket on each side: thus

$$E_\alpha{}^\epsilon{}_\beta(u)\xi^j_\epsilon(y) = f^j_h(x,u)\,\overset{\circ}{E}_\epsilon{}^\lambda{}_\gamma\xi^h_\lambda(x)G^\epsilon_\alpha(u)G^\gamma_\beta(u), \tag{3.4}$$

where we have put

$$\overset{\circ}{E}_\epsilon{}^\lambda{}_\gamma = E_\epsilon{}^\lambda{}_\gamma(0). \tag{3.5}$$

Since (3.3) is equivalent to

$$f^j_h(x,u)\xi^h_\lambda(x) = \xi^j_\epsilon(y)\hat{G}^\epsilon_\lambda(u), \tag{3.6}$$

it follows with the aid of (2.8) that (3.5) implies the relation

$$E_\alpha{}^\sigma{}_\beta(u) = \hat{G}^\sigma_\lambda(u)\,\overset{\circ}{E}_\epsilon{}^\lambda{}_\gamma G^\epsilon_\alpha(u)G^\gamma_\beta(u), \tag{3.7}$$

which is an explicit representation of the values of $E_\alpha{}^\sigma{}_\beta$ for arbitrary values of its arguments in terms of its values at $\{0\} \in \mathbf{R}^r$.

Second, differentiation of (3.3) with respect to u^γ gives

$$\frac{\partial \xi^j_\alpha(y)}{\partial y^h} f^h_\gamma(x,u) = \frac{\partial^2 f^j(x,u)}{\partial x^h \partial u^\gamma}\xi^h_\epsilon(x)G^\epsilon_\alpha(u) + f^j_h(x,u)\xi^h_\epsilon(x)\frac{\partial G^\epsilon_\alpha(u)}{\partial u^\gamma}. \tag{3.8}$$

It is now observed that (2.8) can be expressed as

$$f^j_\gamma(x,u) = \xi^j_\sigma(y)\hat{\chi}^\sigma_\gamma(u). \tag{3.9}$$

Thus, if (3.8) is mulitplied by $\chi^\gamma_\beta(u)$, it is found that

$$\begin{aligned}\xi^h_\beta(y)\frac{\partial \xi^j_\alpha(y)}{\partial y^h} &= \frac{\partial^2 f^j(x,u)}{\partial x^h \partial u^\gamma}\xi^h_\epsilon(x)\chi^\gamma_\beta(u)G^\epsilon_\alpha(u) \\ &\quad + f^j_h(x,u)\xi^h_\epsilon(x)\frac{\partial G^\epsilon_\alpha(u)}{\partial u^\gamma}\chi^\gamma_\beta(u).\end{aligned} \tag{3.10}$$

In order to analyze the first term on the right-hand side we note that, by (3.9),

$$\frac{\partial^2 f^j(x,u)}{\partial x^h \partial u^\gamma}\xi^h_\epsilon(x) = \frac{\partial f^j_\gamma(x,u)}{\partial x^h}\xi^h_\epsilon(x) = \frac{\partial \xi^j_\sigma(y)}{\partial y^h}f^h_k(x,u)\xi^k_\epsilon(x)\hat{\chi}^\sigma_\gamma(u).$$

With the aid of (3.3) this can be rewritten as

$$\frac{\partial^2 f^j(x,u)}{\partial x^h \partial u^\gamma}\xi^h_\epsilon(x) = \frac{\partial \xi^j_\sigma(y)}{\partial y^h}\xi^h_\tau(y)\hat{G}^\tau_\epsilon(u)\hat{\chi}^\sigma_\gamma(u),$$

so that

$$\frac{\partial^2 f^j(x,u)}{\partial x^h \partial u^\gamma}\xi^h_\epsilon(x)\chi^\gamma_\beta(u)G^\epsilon_\alpha(u) = \frac{\partial \xi^j_\beta(y)}{\partial y^h}\xi^h_\alpha(y). \tag{3.11}$$

Consequently (3.10) assumes the form

$$\xi^h_\beta(y)\frac{\partial \xi^j_\alpha(y)}{\partial y^h} - \xi^h_\alpha(y)\frac{\partial \xi^j_\beta(y)}{\partial y^h} = f^j_h(x,u)\xi^h_\epsilon(x)\frac{\partial G^\epsilon_\alpha(u)}{\partial u^\gamma}\chi^\gamma_\beta(u),$$

that is, again by (3.3),

$$[\xi_\beta, \xi_\alpha]^j(y) = \hat{G}^\sigma_\epsilon(u)\frac{\partial G^\epsilon_\alpha(u)}{\partial u^\gamma}\chi^\gamma_\beta(u)\xi^j_\sigma(y).$$

Because of (2.4) this gives

$$E_\beta{}^\sigma{}_\alpha(u) = \hat{G}^\sigma_\epsilon(u)\frac{\partial G^\epsilon_\alpha(u)}{\partial u^\gamma}\chi^\gamma_\beta(u),$$

so that

$$\frac{\partial G^\alpha_\beta(u)}{\partial u^\gamma} = G^\alpha_\sigma(u)E_\nu{}^\sigma{}_\beta(u)\hat{\chi}^\nu_\gamma(u). \tag{3.12}$$

But, according to (3.7),

$$G^\alpha_\sigma(u)E_\nu{}^\sigma{}_\beta(u) = \overset{\circ}{E}_\sigma{}^\alpha{}_\tau G^\sigma_\nu(u)G^\tau_\beta(u). \tag{3.13}$$

Consequently (3.12) can be expressed as

$$\frac{\partial G^{\alpha}_{\beta}(u)}{\partial u^{\gamma}} = \overset{\circ}{E}_{\sigma}{}^{\alpha}{}_{\tau}\lambda^{\sigma}_{\gamma}(u)G^{\tau}_{\beta}(u), \tag{3.14}$$

where we have put

$$\lambda^{\sigma}_{\gamma}(u) = G^{\sigma}_{\nu}(u)\hat{\chi}^{\nu}_{\gamma}(u). \tag{3.15}$$

It will be seen almost immediately that the system (3.14) is crucial to all subsequent developments. However, before an interpretation of (3.14) is given, it is necessary to derive the partial differential equations that must be satisfied by matrices (3.15).

From (3.15) it follows with the use of (3.14) that

$$\frac{\partial \lambda^{\alpha}_{\beta}(u)}{\partial u^{\epsilon}} = \overset{\circ}{E}_{\nu}{}^{\alpha}{}_{\mu}\lambda^{\nu}_{\epsilon}(u)\lambda^{\mu}_{\beta}(u) + G^{\alpha}_{\nu}(u)\frac{\partial \hat{\chi}^{\nu}_{\beta}(u)}{\partial u^{\epsilon}},$$

and hence, by (2.13)

$$\frac{\partial \lambda^{\alpha}_{\beta}(u)}{\partial u^{\epsilon}} - \frac{\partial \lambda^{\alpha}_{\epsilon}(u)}{\partial u^{\beta}} = 2\,\overset{\circ}{E}_{\nu}{}^{\alpha}{}_{\mu}\lambda^{\nu}_{\epsilon}(u)\lambda^{\mu}_{\beta}(u) + G^{\alpha}_{\nu}(u)E_{\sigma}{}^{\nu}{}_{\tau}(u)\hat{\chi}^{\sigma}_{\beta}(u)\hat{\chi}^{\tau}_{\epsilon}(u).$$

But from (3.13) and (3.15) it is evident that

$$G^{\alpha}_{\nu}(u)E_{\sigma}{}^{\nu}{}_{\tau}(u)\hat{\chi}^{\sigma}_{\beta}(u)\hat{\chi}^{\tau}_{\epsilon}(u) = -\,\overset{\circ}{E}_{\nu}{}^{\alpha}{}_{\mu}\lambda^{\nu}_{\epsilon}(u)\lambda^{\mu}_{\beta}(u),$$

so that

$$\frac{\partial \lambda^{\alpha}_{\beta}(u)}{\partial u^{\epsilon}} - \frac{\partial \lambda^{\alpha}_{\epsilon}(u)}{\partial u^{\beta}} = \overset{\circ}{E}_{\nu}{}^{\alpha}{}_{\mu}\lambda^{\nu}_{\epsilon}(u)\lambda^{\mu}_{\beta}(u). \tag{3.16}$$

This is the system that we have been seeking: in essence it represents the integrability conditions of the equations (3.14). The latter will presently be regarded as evolution equations for the matrices $(G^{\alpha}_{\beta}(u))$, subject to the initial conditions (3.2), for given matrices $(\lambda^{\alpha}_{\beta}(u))$ that satisfy (3.16). It should be noted already at this stage that the occurrence of the systems (3.14) and (3.16) suggests the existence of a Lie group structure on a neighborhood of $\{0\} \in \mathbf{R}^{r}$ ([5], p. 233, Theorem 1), which will in fact be verified directly in the next section.

The conclusions reached thus far may be summarized provisionally as follows: If the vectors $\{\xi_{\alpha}, B_{\alpha}, \eta_{\alpha} \colon \alpha = 1, \cdots, r\}$ as defined respectively by (1.8), (1.9) and (1.11) are such that the conditions (2.4), (2.8), and (3.1) hold, then the matrices $(\chi^{\alpha}_{\beta}(u))$ and $(G^{\alpha}_{\beta}(u))$ that occur in (2.8) and (3.1) satisfy the partial differential equations (2.12) and (3.14) respectively, the integrability conditions of the latter being ensured by (3.16).

4. Existence of a Group Structure: The Geometrical Background

In order to show that the three conditions imposed in Sections 2 and 3 entail the existence of a local Lie group structure we return to the connection 1-forms (2.19). With the aid of (2.14) these can be expressed as

$$\omega^{\alpha}_{\beta}(u) = E_{\epsilon}{}^{\alpha}{}_{\beta}(u)\hat{\chi}^{\epsilon}_{\gamma}(u)du^{\gamma},$$

in which we substitute from (3.7) and apply (3.15):

$$\omega^{\alpha}_{\beta}(u) = \hat{G}^{\alpha}_{\sigma}(u)\,\overset{\circ}{E}_{\nu}{}^{\sigma}{}_{\mu}\lambda^{\nu}_{\gamma}(u)G^{\mu}_{\beta}(u)du^{\gamma}.$$

When this is compared with (3.14) it is seen that

$$\omega^{\alpha}_{\beta}(u) = \hat{G}^{\alpha}_{\epsilon}(u)\frac{\partial G^{\epsilon}_{\beta}(u)}{\partial u^{\gamma}}du^{\gamma} = \hat{G}^{\alpha}_{\epsilon}(u)dG^{\epsilon}_{\beta}(u). \tag{4.1}$$

But an expression of this type is characteristic of ***zero curvature connections***. In fact, exterior differentiation of (4.1) yields

$$\begin{aligned} d\omega^{\alpha}_{\beta}(u) &= d\hat{G}^{\alpha}_{\epsilon}(u)\wedge dG^{\epsilon}_{\beta}(u) = -\hat{G}^{\alpha}_{\sigma}(u)\hat{G}^{\gamma}_{\epsilon}(u)dG^{\sigma}_{\gamma}(u)\wedge dG^{\epsilon}_{\beta}(u) \\ &= -(\hat{G}^{\alpha}_{\sigma}(u)dG^{\sigma}_{\gamma}(u))\wedge(\hat{G}^{\gamma}_{\epsilon}(u)dG^{\epsilon}_{\beta}(u)) = -\omega^{\alpha}_{\gamma}(u)\wedge\omega^{\gamma}_{\beta}(u), \end{aligned}$$

where, in the last step, we have used (4.1) once more. Thus (2.20) reduces to $\Omega^{\alpha}_{\beta}(u) = 0$, that is, by (2.21),

$$K_{\beta}{}^{\alpha}{}_{\mu\nu}(u) = 0. \tag{4.2}$$

Thus, according to the concluding remarks of Section 2,

$$\frac{\partial E_{\beta}{}^{\alpha}{}_{\epsilon}(u)}{\partial u^{\gamma}} = 0, \tag{4.3}$$

while (2.23) yields the standard Jacobi identity:

$$E_{\epsilon}{}^{\alpha}{}_{\sigma}E_{\beta}{}^{\sigma}{}_{\gamma} + E_{\beta}{}^{\alpha}{}_{\sigma}E_{\gamma}{}^{\sigma}{}_{\epsilon} + E_{\gamma}{}^{\alpha}{}_{\sigma}E_{\epsilon}{}^{\sigma}{}_{\beta} = 0. \tag{4.4}$$

Since the 3-index symbols $E_{\beta}{}^{\alpha}{}_{\gamma}$ are now constants, skew-symmetric in their subscripts, and satisfy (4.4), the converse of Lie's Third Theorem implies the existence of an abstract local Lie group whose structure constants coincide with these symbols.

For our present purposes it is necessary to derive an explicit expression for the composition functions of this group. Since the various steps of this derivation are well known ([3], Chapter 10), only a few particularly relevant features of this process are recalled here. First, it can be shown by means of the theorem of

Frobenius that the system (3.16) is tantamount to the integrability condition of the following system of first order partial differential equations

$$\lambda^\alpha_\epsilon(\psi^\lambda)\frac{\partial\psi^\epsilon}{\partial u^\beta} = \lambda^\alpha_\beta(u^\lambda) \tag{4.5}$$

for a set of r functions ψ^ϵ, the validity of (3.16) being a consequence of our assumptions. Thus there exists a neighborhood U of $\{0\} \in \mathbf{R}^r$ on which, for given u^λ_0, $\psi^\lambda_0 \in U$, there is prescribed a unique solution

$$\psi^\alpha = \psi^\alpha(u^\lambda, \psi^\lambda_0, u^\lambda_0) \tag{4.6}$$

of (4.5) that is subject to the initial conditions

$$\psi^\alpha_0 = \psi^\alpha(u^\lambda_0, \psi^\lambda_0, u^\lambda_0). \tag{4.7}$$

We now define a set of r functions on $U \times U$ by setting

$$\phi^\alpha(u^\lambda, v^\lambda) = \psi^\alpha(u^\lambda, v^\lambda, 0). \tag{4.8}$$

Then, by (4.7), with $u^\lambda = 0$,

$$\phi^\alpha(0, v^\lambda) = \psi^\alpha(0, v^\lambda, 0) = v^\alpha.$$

Second, a product may now be defined on U by means of the functions (4.8) as follows. Let u, $v \in U$ have coordinates u^λ, v^λ, respectively, and suppose that the point w with coordinates $w^\alpha = \phi^\alpha(u^\lambda, v^\lambda)$ is also in U. Under these circumstances the product uv of the points u, v is identified with w, which is expressed symbolically as

$$w = uv \Longleftrightarrow w^\alpha = \phi^\alpha(u^\lambda, v^\lambda). \tag{4.10}$$

The associativity of this composition can be established by known techniques ([3], p. 406). Also, if we denote $\{0\} \in U$ by e it follows from (4.9) that $ev = v$, which shows that e is a left identity. Finally, it follows directly from (2.7), (3.2), and (3.15) that

$$\lambda^\alpha_\beta(0) = \delta^\alpha_\beta, \tag{4.11}$$

so that, if we put $u^\lambda_0 = 0$, $\psi^\lambda_0 = 0$ in (4.7), we have $\psi^\alpha(0,0,0) = 0$: consequently (4.5) implies that $\partial\psi^\alpha(0,0,0)/\partial u^\beta = \delta^\alpha_\beta$ and hence, by (4.8), that $\partial\phi^\alpha(0,0)/\partial u^\beta = \delta^\alpha_\beta$. Thus the equation $\phi^\alpha(u^\lambda, v^\lambda) = w^\alpha$ with given v^α, w^α can be solved for u^α on a neighborhood $\tilde{U} \subset U$ of e. This ensures the existence of a right inverse, and the existence of a left inverse together with that of a right identity follows as usual. The neighborhood $\tilde{U}$ of $\mathbf{R}^r$, endowed with the composition (4.10), is therefore a local Lie group G with identity $e = \{0\} \in \mathbf{R}^r$.

From (4.8) and (4.5) it follows that the functions ϕ^ϵ are solutions of the system

$$\lambda^\alpha_\epsilon(\phi^\nu(u^\lambda, v^\lambda))\frac{\partial\phi^\epsilon(u^\lambda, v^\lambda)}{\partial u^\beta} = \lambda^\alpha_\beta(u^\lambda),$$

that is, in terms of the composition (4.10),

$$\lambda^\alpha_\epsilon(w^\lambda)\frac{\partial\phi^\epsilon(u^\lambda, v^\lambda)}{\partial u^\beta} = \lambda^\alpha_\beta(u^\lambda). \tag{4.12}$$

This holds for all u, $v \in \tilde{U}$. In particular, with $v = u^{-1}$, we have $w = e$, and thus, according to (4.11),

$$\lambda^\alpha_\beta(u) = \frac{\partial\phi(u^\lambda, v^\lambda)}{\partial u^\beta}\Big|_{v=u^{-1}}. \tag{4.13}$$

This shows that the functions λ^α_β are nothing other than the auxiliary functions of the Lie group G ([3], p. 403; [4], p. 128). But it is known that the matrices of the (r-dimensional) adjoint representation of G satisfy a system of first order partial differential equations whose coefficients are determined by the structure constants and the auxiliary functions: in fact, this system is (3.14), subject to the initial condition (3.2) ([4], p. 138). The structure of this system guarantees the local uniqueness of the solutions: consequently the matrices $(G^\alpha_\beta(u))$ that are defined implicitly by the condition (3.1) must necessarily refer to the adjoint representation of G. Moreover, because of (3.2) and (4.11) the equation (3.14) with $u^\lambda = 0$ reduces to

$$\frac{\partial G^\alpha_\beta(0)}{\partial u^\gamma} = E_\gamma{}^\alpha{}_\beta, \tag{4.14}$$

which provides a direct confirmation of the fact that the structure constants of G are identical with the constants $E_\gamma{}^\alpha{}_\beta$.

Our conclusions may be summarized in the

Theorem 1. *If the vector fields $\{\xi_\alpha, B_\alpha, \eta_\alpha\colon \alpha = 1, \cdots, r\}$ as defined by (1.8), (1.9), and (1.11) respectively satisfy the conditions*

$$[\xi_\alpha, \xi_\beta] = E_\alpha{}^\epsilon{}_\beta(u)\xi_\epsilon, \tag{i}$$

$$B_\alpha = \hat{\chi}^\beta_\alpha(u)\xi_\beta, \tag{ii}$$

and

$$\xi_\alpha = G^\beta_\alpha(u)\eta_\beta, \tag{iii}$$

then the 3-index symbols that occur in (i) are the structure constants of a local Lie group G. Moreover, under these circumstances the matrices $(G^\alpha_\beta(u))$ that occur in (iii) are identical with the matrices of the adjoint representation of G.

Remark. In this context the relation (3.7) is nothing other than the identification of the structure constants of G as type (1,2) objects in the adjoint representation.

From (4.13) and subsequent remarks it is evident that the matrices $(\lambda^\alpha_\beta(u))$ and $(G^\alpha_\beta(u))$ are completely determined by the composition functions of G. Because of (3.15) the same must be true of the matrices $(\chi^\alpha_\beta(u))$; indeed, according to the general theory ([4], p. 132) these matrices can be expressed as

$$\chi^\alpha_\beta(u) = \frac{\partial\phi^\alpha(u,v)}{\partial v^\beta}\Big|_{v=e}. \tag{4.15}$$

The r vector fields

$$E_\alpha = \chi^\beta_\alpha(u)\frac{\partial}{\partial u^\beta} \tag{4.16}$$

on $\tilde{U}$ constitute a basis of the set $\mathcal{G}$ of left-invariant vector fields of G, which, by definition, is the Lie algebra of G. Thus the 1-forms (2.14) are left-invariant 1-forms that are subject to the standard Maurer-Cartan equations (2.15). From (4.16) and (1.9) it also follows that

$$f_*(E_\alpha) = \chi^\beta_\alpha(u)f_*(\frac{\partial}{\partial u^\beta}) = \chi^\beta_\alpha(u)B_\beta,$$

and hence, by (2.6)

$$\xi_\alpha = f_*(E_\alpha). \tag{4.17}$$

Similarly, it is readily verified that the r vector fields

$$\tilde{E}_\alpha = \hat{\lambda}^\beta_\alpha(u)\frac{\partial}{\partial u^\beta} \tag{4.18}$$

(in which $(\hat{\lambda}^\beta_\alpha(u))$ denotes the matrix inverse of $(\lambda^\alpha_\beta(u))$) constitute a basis of the set $\hat{\mathcal{G}}$ of the right-invariant vector fields of G, with which are associated the right-invariant 1-forms

$$\tilde{\omega}^\alpha = -\lambda^\alpha_\beta(u)du^\beta. \tag{4.19}$$

In consequence of (3.16) these also satisfy a system of Maurer-Cartan equations

$$d\tilde{\omega}^\alpha + E_\beta{}^\alpha{}_\gamma\tilde{\omega}^\beta \wedge \tilde{\omega}^\gamma = 0. \tag{4.20}$$

Again, it follows from (4.18), (1.9), and (2.6) that

$$f_*(\tilde{E}_\alpha) = \hat{\lambda}^\beta_\alpha(u) B_\beta = \hat{\chi}^\epsilon_\beta(u)\hat{\lambda}^\beta_\alpha(u)\xi_\epsilon,$$

to which we apply (3.15) and (3.1) to obtain

$$\eta_\alpha = \hat{G}^\beta_\alpha(u)\xi_\beta = f_*(\tilde{E}_\alpha). \tag{4.21}$$

According to (4.17) and (4.21) *the vector fields* $\{\xi_\alpha\}$, $\{\eta_\alpha\}$ *now emerge as the images under the map* f_* *of the bases* (4.16) *and* (4.18) *of the left- and right-invariant vector fields of the group* G *respectively.*

In conclusion let us return briefly to the r-parameter family of transformations (1.1) that was our point of departure. With the aid of (1.9) the condition (2.8) can be expressed as

$$\frac{\partial y^j(x,u)}{\partial u^\alpha} = f^j_\alpha(x,u) = \xi^j_\beta(y)\hat{\chi}^\beta_\alpha(u). \tag{4.22}$$

In the present context the 3-index symbols that occur in (2.4) are constants. It is therefore possible to apply the fundamental theorem on transformation groups ([2], p. 54) according to which the solutions $y^j = f^j(x,u)$ of (4.22), with $x^j = f^j(x,0)$, define a transformation group on the manifold M. Thus, as expected, the three conditions enunciated in the theorem above, are sufficient to ensure that the equations (1.1) represent a transformation group.

5. Absolute Invariants

A function $F : M \to \mathbf{R}$ is said to be an *invariant* of the r-parameter family of transformations (1.1) whenever

$$F(y^j) = F(f^j(x,u)) = F(x^j) \tag{5.1}$$

for all admissible values of the parameters u^α. Such an invariant must necessarily satisfy the partial differential equations

$$\frac{\partial F(y)}{\partial y^j} f^j_\alpha(x,u) = 0, \tag{5.2}$$

together with

$$\frac{\partial F(y)}{\partial y^j} f^j_h(x,u) = \frac{\partial F(x)}{\partial x^h}. \tag{5.3}$$

With $u^\alpha = 0$ in (5.2), one obtains

$$\frac{\partial F(x)}{\partial x^j}\xi^j_\alpha(x) = 0 \tag{5.4}$$

in terms of the notation (1.5), so that $\xi_\alpha F = 0$ at the point p on M with coordinates x^h. The function F is called an *absolute invariant* ([2], p. 61) if, in addition to (5.1), it satisfies the conditions

$$\xi_\alpha(y)F(y) = \xi^j_\alpha(y)\frac{\partial F(y)}{\partial y^j} = 0 \tag{5.5}$$

for all admissible values of the parameters u^α. This stipulation is suggested by the relations (5.4); the latter also indicate that there are restrictions on the number of independent invariants (if any) of the transformations (1.1). Accordingly we shall have to deal with two distinct cases.

Case I: It is assumed that $r \leq n$, and that the rank of the matrix (ξ^j_α) is maximal, namely r.

From (5.5) it follows directly that under these circumstances there can exist at most $n - r$ independent absolute invariants. We shall now suppose that such a set of invariants $\{F^A : A = r+1, \cdots, n\}$ does in fact exist, and we shall endeavor to determine the conditions that must be satisfied by the transformations (1.1) in order that this be possible.

Let $\{\sigma^A : A = r+1, \cdots, n\}$ denote a set of real parameters. The equations

$$F^A(x^h) = \sigma^A \tag{5.6}$$

describe an $(n-r)$-parameter family of r-dimensional subspaces $N(\sigma)$ of M. If the values of F^A at the fixed point $p \in M$ are given by σ^A_0, then, by (5.1), the values of F^A at the point $q \in M$ are also equal to σ^A_0, which implies that both p and q are on the subspace $N(\sigma_0)$. Moreover, it is possible to construct a parametric representation of $N(\sigma_0)$ in terms of a set of r parameters $\{v^\alpha : \alpha = 1, \cdots, r\}$ of the form

$$y^j = g^j(v, \sigma_0), \tag{5.7}$$

which is such that the coordinates x^j of p are given by

$$x^j = g^j(0, \sigma_0), \tag{5.8}$$

while the coordinates y^j of q as given by (1.1) are represented directly by (5.7). The derivatives

$$b^j_\alpha(v, \sigma_0) = \frac{\partial g^j(v, \sigma_0)}{\partial v^\alpha} \tag{5.9}$$

define a set of r vectors

$$b_\alpha(y) = \frac{\partial g^j(v, \sigma_0)}{\partial v^\alpha}\frac{\partial}{\partial y^j} \tag{5.10}$$

in $T_q(M)$: these vectors obviously constitute a basis of the tangent space $T_q(N(\sigma_0))$. The substitution of (5.7) in (5.6) results in an identity in the parameters v^α, σ_0^A, differentiation of which yields

$$\frac{\partial F^A(y)}{\partial y^j}\frac{\partial g^j(v,\sigma_0)}{\partial v^\alpha} = \frac{\partial F^A(y)}{\partial y^j} b^j_\alpha(v,\sigma_0) = 0. \tag{5.11}$$

By hypothesis, the $n-r$ functions F^A satisfy (5.5), while, because of their independence, the rank of the matrix $(\partial F^A/\partial y^j)$ is $n-r$. Thus a comparison of (5.5) with (5.11) yields

$$\xi^j_\alpha(y) = A^\beta_\alpha b^j_\beta(v,\sigma_0), \tag{5.12}$$

in which (A^β_α) denotes a nonsingular $r\times r$ matrix whose entries may depend on (v^α,σ_0^A). The vectors $\{\xi_\alpha : \alpha = 1,\cdots,r\}$ therefore represent a basis of $T_q(N(\sigma_0))$. Also, since (5.5) implies that $[\xi_\alpha,\xi_\beta]F^A = 0$, that is,

$$\frac{\partial F^A(y)}{\partial y^j}[\xi_\alpha,\xi_\beta]^j(y) = 0, \tag{5.13}$$

it follows similarly from (5.11) that each Lie bracket $[\xi_\alpha,\xi_\beta]$ is in $T_q(N(\sigma_0))$, which is tantamount to the condition (2.4). Thus, as expected, the system (5.5) is complete.

Let us now turn to equation (5.2). As before, a comparison of (5.2) with (5.5) yields

$$f^j_\alpha(x,u) = B^\beta_\alpha b^j_\beta(v,\sigma_0), \tag{5.14}$$

in which (B^β_α) denotes a nonsingular $r\times r$ matrix whose entries may depend on (v^α,σ_0^A). This gives rise directly to two inferences. First, in terms of (1.9) and (5.10),

$$B_\alpha(y) = f^j_\alpha(x,u)\frac{\partial}{\partial y^j} = B^\beta_\alpha b^j_\beta(v,\sigma_0)\frac{\partial}{\partial y^j} = B^\beta_\alpha b_\beta(y), \tag{5.15}$$

in which the vectors on the left-hand side represent a basis of the tangent space at q of the r-dimensional subspace $L(p)$ of M that is represented parametrically by the equations $y^j = f^j(x^h,u^\alpha)$ for fixed x^h, while the right-hand side consists of a linear combination of the basis vectors of $T_q(N(\sigma_0))$. *This implies that the subspaces $L(p)$ and $N(\sigma_0)$ coincide, at least locally.* (Since this conclusion does not depend on (5.5), it is evidently valid also for invariants that are not absolute.) Second, the relations (5.12) and (5.14) can be combined to give

$$\xi^j_\alpha(y) = A^\epsilon_\alpha \hat{B}^\beta_\epsilon f^j_\beta(x,u), \tag{5.16}$$

where $(\hat{B}^\beta_\epsilon)$ denotes the matrix that is inverse to (B^ϵ_β). But this is nothing other than the condition (2.8).

Finally, when (5.3) is multiplied by $\xi^h_\alpha(x)$, the relation (5.4) being taken into account, it is seen that

$$\frac{\partial F^A(y)}{\partial y^j} f^j_h(x,u)\xi^h_\alpha(x) = \frac{\partial F^A(x)}{\partial x^h}\xi^h_\alpha(x) = 0.$$

Because of (1.11) this can be expressed as

$$\eta^j_\alpha(y)\frac{\partial F^A(y)}{\partial y^j} = 0, \tag{5.17}$$

which, when taken in conjunction with (5.11), shows that the vectors $\{\eta_\alpha : \alpha = 1, \cdots, r\}$ are in $T_q(N(\sigma_0))$ also, and this in turn entails the condition (3.1). Consequently all three hypotheses of Theorem 1 are satisfied. The latter therefore gives rise to

Theorem 2. *Let the r-parameter family of point transformations (1.1) in n independent variables, with $r \leq n$, be such that the rank of the matrix defined by the elements (1.5) is r. In order that the transformations (1.1) admit the maximal number $n - r$ of independent absolute invariants, it is necessary that these transformations constitute a group.*

The converse, of course, is well known: any transformation *group* that satisfies the hypothesis of the theorem, admits $n - r$ absolute invariants.

Case II. It is assumed that $r > n$.

We shall denote the rank of the matrix (ξ^j_α) by s, so that $s \leq n < r$, and only s elements of $\{\xi_\alpha : \alpha = 1, \cdots, r\}$ are independent. (See the example cited at the end of this section.) It is supposed that $\{\xi_\alpha\}$ is labelled such that $\{\xi_a : a = 1, \cdots, s\}$ are independent, so that the remaining $\{\xi_d : d = s+1, \cdots, r\}$ can be expressed as

$$\xi_d = C^a_d\,\xi_a; \tag{5.18}$$

however, these elements are not necessarily independent. The system (5.4) is therefore equivalent to the reduced system

$$\xi_a F = \xi^j_a \frac{\partial F}{\partial x^j} = 0, \qquad (a = 1, \cdots, s) \tag{5.19}$$

which can have at most $n - s$ independent solutions.

The existence of $n - s$ absolute invariants $\{F^A : A = s+1, \cdots, n\}$ is now presupposed: these satisfy the equations (5.2) and (5.3), namely

$$\frac{\partial F^A(y)}{\partial y^j} f^j_\alpha(x,u) = 0, \qquad (A = s+1, \cdots, n) \tag{5.20}$$

and

$$\frac{\partial F^A(y)}{\partial y^j} f_h^j(x,u) = \frac{\partial F^A(x)}{\partial x^h}, \qquad (A = s+1, \cdots, n) \tag{5.21}$$

while the application of (5.18) to (5.5) yields

$$\frac{\partial F^A(y)}{\partial y^j} \xi_a^j(y) = 0. \tag{5.22}$$

The equation (5.6) [in which the index A now ranges from $s+1$ to n] describes an $(n-s)$-parameter family of s-dimensional subspaces $\tilde{N}(\sigma)$ of M. A parametric representation such as (5.7) of $\tilde{N}(\sigma)$ in terms of s parameters $\{v^a : a = 1, \cdots, s\}$ again gives rise to basis elements

$$c_a = c_a^j(v, \sigma_0)\frac{\partial}{\partial y^j} \tag{5.23}$$

of the tangent space $T_q(\tilde{N}(\sigma_0))$ of $\tilde{N}(\sigma_0)$, where

$$c_a^j(v, \sigma_0) = \frac{\partial g^j(v, \sigma_0)}{\partial v^a}. \tag{5.24}$$

As for (5.11) it follows that

$$\frac{\partial F^A(y)}{\partial y^j} c_a^j(v, \sigma_0) = 0. \tag{5.25}$$

Since the absolute invariants F^A are assumed to be independent, the rank of the matrix $(\partial F^A/\partial y^j)$ is maximal, namely $n-s$, and a comparison of (5.22) with (5.25) reveals the existence of a nonsingular $s \times s$ matrix (A_a^b) in terms of which we may write

$$\xi_a^j(y) = A_a^b c_b^j(v, \sigma_0). \tag{5.26}$$

This shows that the vectors $\{\xi_a : a = 1, \cdots, s\}$ constitute a basis of $T_q(\tilde{N}(\sigma_0))$. Moreover, since (5.25) implies that $[\xi_\alpha, \xi_\beta](y)F^A = 0$, we deduce as in the case of (5.14) that the Lie brackets $[\xi_\alpha, \xi_\beta](y)$ are in $T_q(\tilde{N}(\sigma))$, which again implies condition (2.4).

In terms of the notation (1.9) the equations (5.20) can be expressed as $B_\alpha(y)F^A = 0$; consequently the r vectors $B_\alpha(y)$ are in $T_q(\tilde{N}(\sigma_0))$, and, since this is also the case of the r vectors $\xi_\alpha(y)$ by virtue of (5.5), it follows that there exists an $r \times r$ matrix $\chi_\alpha^\beta(u)$ such that (2.8) holds, together with (2.7), the latter being a consequence of the initial conditions (1.12). Finally, as for Case I, it is evident that (5.21) implies the condition (3.1). Consequently the hypotheses of Theorem 1 are satisfied, which gives rise to

Theorem 3. *Let (1.1) represent an r-parameter family of point transformations in n independent variables, with $r > n$, and let s denote the rank of the matrix defined by the elements (1.5), so that $s \leq n < r$. In order that the transformation (1.1) admit the maximal number $n - s$ of independent absolute invariants, it is necessary that these transformations constitute a group.*

Remark. The conditions of the theorem are not sufficient to guarantee the existence of nontrivial invariants. This is demonstrated by the example of the Euclidean group in the plane, for which (1.1) assumes the form

$$f^1(x,y) = x^1 cosu^3 - x^2 sinu^3 + u^1, \qquad f^2(x,y) = x^1 sinu^3 + x^2 cosu^3 + u^2.$$

Here $r = 3$, $n = 2$, and since

$$\xi^j_\alpha(x) = \begin{pmatrix} 1 & 0 \\ 0 & 1 \\ -x^2 & x^1 \end{pmatrix},$$

it is seen that $s = 2$. Also

$$\xi_1 = \frac{\partial}{\partial x^1}, \quad \xi_2 = \frac{\partial}{\partial x^2}, \quad \xi_3 = -x^2\frac{\partial}{\partial x^1} + x^1\frac{\partial}{\partial x^2}, \tag{5.27}$$

so that $\xi_3 = x^1\xi_2 - x^2\xi_1$ [which is a counterpart of (5.18)], while

$$[\xi_\alpha, \xi_\beta] = E_\alpha{}^\epsilon{}_\beta \xi_\epsilon \qquad \text{with} \qquad E_2{}^1{}_3 = E_3{}^2{}_1 = 1,$$

all other E's being zero. The condition (2.8) holds with

$$\chi^\alpha_\beta(u) = \begin{pmatrix} 1 & 0 & -u^2 \\ 0 & 1 & u^1 \\ 0 & 0 & 1 \end{pmatrix},$$

and the matrix that appears in (3.3) is given by

$$G^\alpha_\beta(u) = \begin{pmatrix} \cos u^3 & \sin u^3 & u^1 \sin u^3 - u^2 \cos u^3 \\ -\sin u^3 & \cos u^3 & u^1 \cos u^3 + u^2 \sin u^3 \\ 0 & 0 & 1 \end{pmatrix}.$$

However, from (5.27) it is immediately evident that this transformation group has no nontrivial invariants. This phenomenon is obviously related to the transitivity of the group.

A more typical example of the situation discussed above is perhaps represented by the orthogonal group in 4 dimensions. In this case $r = 6$, $n = 4$, and $s = 3$. The number of independent absolute invariants is $4 - 3 = 1$, this invariant being $F(R)$, where $R = [(x^1)^2 + (x^2)^2 + (x^3)^2 + (x^4)^2]^{1/2}$.

References

[1] D. G. B. Edelen. *Applied Exterior Calculus.* Wiley- Interscience, New York, 1985.

[2] L. P. Eisenhart. *Continuous Groups of Transformations.* Dover, New York, 1961.

[3] L. S. Pontryagin. *Topological Groups.* Second edition, translated by A. Brown. Gordon & Breach, New York, 1966.

[4] H. Rund. Differential-geometric and variational background of classical gauge field theories. Aeq. Math. *24*, 121-174 (1982).

[5] H. Rund. Variational principles for field variables that are subject to group actions. *Algebraic and Differential Topology–Global Differential Geometry* (dedicated to Marston Morse; ed. G. M. Rassias), Teubner, Leipzig (1984), 224-251.

GEOMETRY AND TOPOLOGY (pp. 281–290)
edited by G. M. Rassias & G. M. Stratopoulos

THE CLASSIFICATION AND UNIQUENESS OF NEW LINE INVOLUTIONS

George Stratopoulos

Department of Mathematics
National Technical University of Athens
Athens, Greece

ABSTRACT:

In this paper we introduce a new series of line involutions in the projective space, S_3, of three dimensions. We investigate the various properties of the involutions and determine their characteristics. The characteristics of the invlutions are used to classify the new involution and prove their uniqueness (1), (2).

1. THE INVOLUTIONS AND THEIR CHARACTERISTICS

The involutions of our (2,4n)-series of line involutions, as we shall be refering to them, are defined by an involutory point transformmation of the points in which a general line, L, meets a nonsingular quadric surface, V, bearing a curve of symbols (2,4n) where n=1,2,... as follows:

Let V be a nonsingular quadric surface in a projective space, S_3, of three dimensions with reguli R and Q bearing the curve, C, of order 4n+2 and symbols (2,4n) for n=1,2,... Fig. 1.

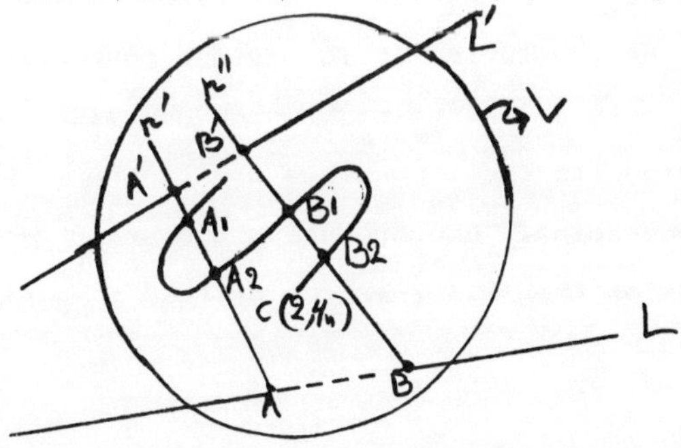

Fig. 1

Then, a general line, L, of S_3 meets V in the points A and B. We know (3) through each of the points A and B passes only one generator, r, of the R-regulus and that each such r is a secant of C. Thus, let r′, r″ be the generators of R on the points A and B, respectively (Fig. 1). further, let A′ and B′ be the harmonic conjugates of A and B relative to the pairs (A_1, A_2) and (B_1, B_2) intersections of the generators r′ and r″ with C, respectively. The transformations

$$W_n: \quad L \dashrightarrow L' = A'B' \qquad \text{(Fig. 1)}$$

induced by the involutory point transformations

$$W: \quad A \dashrightarrow A' \quad \text{and} \quad B \rightarrow B' \quad \text{on } V$$

establish a series of line involutions (W^2_n (L)=L) for each n=1,2,...The characteristics of the involutions are obtained as we prove the following propositions:

PROPOSITION 1: The singular lines of the (2,4n)-series of line involutions form a complex of lines of order k=4n+2.

Proof: It is clear that no line, L, can meet its image, L′ at points other than one of the intersections of L and V. Because if there were in fact a line, L, and its image, L′, meeting at some point (not A or B) then the plane of L and L′ would contain two generators of the same regulus R which is absurd. Moreover, because of the harmonic tetrads on generators of R used in defining the involution it is obvious that L and its image L′ can meet only on points of V which are also on C. Thus, the complex of singular lines consists of all the lines of S_3 which meet C at one point. Consequently, the order of the complex of singular lines of the (2,4n)-series of line involutions is 4n+2.

PROPOSITION 2: The order of the complex of invariant lines is $i=0$.

Proof: Since a line, L, and its image L' can meet, if they do, only on a point of C, all the secants of C are invariant and there are no other such lines. That is, there is no complex of invariant lines and hence $i=0$.

There is however a congruence of invariant lines, namely the secants of C whose symbols are (2,4n). Since

$$a^2+b^2-a-b/2$$

secants of a curve with symbols (a,b) on V pass through an arbitrary point and

$$(a+b)!\ /\ 2!(a+b-2)!$$

secants of such a curve lie in an arbitrary plane, the order and the class of the congruence of invariant lines of the involution are

$$(2)^2+(4n)^2-2-4n/2 = 8n^2-2n+1$$

$$(2+4n)!\ /\ 2!(2+4n-2)! = 8n^2+6n+1, \text{ respectively.}$$

Now, since the characteristics i, k, m and j of an involution satisfy the equations

$$m+1 = 2j-k = j+i \qquad [2]$$

and since $i=0$ and $k=4n+2$, we obtain $m=4n+1$ for the order of the involution and $j=4n+2$ for the order of the image regulus. For the classification process of the line involutions in a projective space, we have shown the existance of this (2,4n)-series of line involutions whose characteristics fill infinitely many cells of the first row ($i=0$) of the classification table 1, [1].

	j=1	j=2	j=3	j=4	j=5	j=6 .....
i=0	k=1 m=0	k=2 m=1	k=3 m=2			
i=1	k=0 m=1	k=1 m=2				
i=2		k=0 m=3				
i=3						
i=4						

. . . .

Table 1.

2. SPECIAL LINES

As in all line involutions there exist sets of special lines. i.e. lines whose images are not unique. In our involutions the set of special lines contains the following subsets of such lines:

a) The generators, g, of the G-regulus

and b) The tangents to V.

In the following we investigate the images of the abve sets of special lines.

a) let C_g be the image of a general generator, g, of the G-Regulus under the point involution on V. Then, since r-generators are invariant and meet each g-generator at only one point, C_g has symbols (1,4n). This in fact is so because r meets C_g in one point, A , the image of A·which is the only intersection of r and g. Further, a g-generator meets C_g in 4n points namely the invariant intersections of C and g. Therefore, under our involution, each g-generaor is transformed into the congruences of secants of its image curve C_g. The order and class of these congruences are

$$1^2+(4n)^2-1-4n/2 = 8n^2-2n$$

and

$$(4n+1)! / 2!(4n+1-2)! = 8n^2-2n, \text{ respectively.}$$

b) The tangents to the nonsingular quadric, V, form a quadratic complex of lines all of which are special. The image of one such tangent to V, at the point A is the pencil of tangents to V at A′ (image of A under the point involution on V). Thus, the pencil of tangent lines to V at A is transformed into a pencil of tangenst to V at A′. In fact, we can assign a one-to-one correspondence on the lines of the two tangent pencils.

PROPOSITION 3: The image of a conic on V is a curve with symbols (1, 4n+1) on V.

Proof: A conic, C_2, on V is a (1,1) curve, and it meets the image curve (a (1,4n)-curve) of a g-generator in 1+4n points. Consequently, the image C_2' of C_2 meets any g-generator in 4n+1 points. Since any r-generator meets C_2' at only one point, C_2' must be a curve of symbols (1,4n+1) on V.

PROPOSITION 4: The images of plane-fields and bundles are congruences with characteristics $(8n^2+2n, 8n^2+6n+1)$ and $(8n^2-2n+1, 8n^2+6n+1)$, respectively.

Proof: Since a plane-field, p, meets V in a conic, the lines of p are secants of the conic, and hence their images must be secants of the image-curve of the conic. Since the image-curve of a conic has symbols (1,4n+1) on V, the class of th)e image congruence of p is

$$1^2+(4n+1)^2-1-(4n+1) \;/\; 2 = 8n^2+2n$$

and the order is

$$(4n+2)! \;/\; (4n+2-2)! = 8n^2+6n+1$$

Now, since a bundle has for its image, under any line involution, a congruence with class the order of the image-congruence of a plane field and order the order of the congruence of invariant lines, we have $8n^2-2n+1$ and $8n^2+6n+1$ for the order and class of the image-conoruence of a bundle, respectively.

3. THE UNIQUENESS OF THE LINE INVOLUTION

In the first part of this paper we have stablished the existence of line involutions of order m=4n+1 for n=1,2,... with no complex of singular lines, i=0, and a complex of singular lines, namely those meeting a twisted curve C of order 4n+2 and symbols (2,4n) on V. Subsequently, we shall show that any inivolution with these characteristics is described (defined) as the one presented in the first part of this paper. We do this by showing first the following propositions:

PROPOSITION 5: In an involution with order m=4n+1, i=0, and a complex of singular lines, those meeting a twisted curve K on V, every line meeting K at a point O meets its image at O.

Proof: If we consider a general pencil, p, of lines containing a general secant of K. then the image of p is a ruled surface of order 4n+1 met by the plane, P. of the pencil. p, in a curve, C, of order 4n+1. On the curve C there is a 4n+1:1 correspondence, in which 4n+1 points cut from C by a general line, L, of the pencil p correspond to the point on C which is the intersection of the image L′ of L and the plane of the pencil. This correspondance has 4n+2 coincidences 4 for each of which we have a line of the pencil meeting its image. This means that each pencil, p. contains 4n+2 singular lines. But the pencil contains 4n+2-1 = 4n+1 singular lines which can be valid only if C is composite consisting of a secant of K and a curve of order 4n. Therefore, the secants of K are all invariant. When this is the case, however. an arbitrary pencil of lines having a point, P on K as vertex is transformed into a ruled surface of order 4n+1 having 4n+1 generators concurrent at the point P. Now, since a ruled surface of order 4n+1 with 4n+1 concurrent generatqors is necessarily a cone, it follows that every line through a point, P, on K meets its image at P. This completes the proof of our proposition.

From the above conclusions and remarks it follows that any bundle with vertex P on K is transformed into itself in an involutorial fashion with a cone of order 4n+1 of invariant lines. This cone is in fact the one of secants of K passing through P. But, then, the involution within the bundle is necessarily a perspective de Jonquieres involution of order 4n+1, and the invariant locus must have a multiple line of multiplicity either 4n or 4n-1. The first case requires that there be a line through P which meets K in 4n+1 points, and the second requires there be a line through P meeting K in 4n points. In each of the two cases, lines of the bundles are transformed by involutions within the pencils they determine with the multiple secant. In the first case, the fixed elements within each pencil are the multiple secant and the line joining the vertex P to the intersection of K and the plane of the pencil which does not lie on the multiple secant. In the second case, the fixed elements are the lines which join the vertex, P, to the two in-intersections of K and the plane of the pencil which does not lie on the multiple secants which are exceptional and in each case are transformed into cones of order 4n. These remarks hold true for each point of K, and thus K must have either a regulus of (4n+1)-fold secants or a regulus of 4n-fold secants. Further, since 4n>1, no two of the multiple secants can intersect. If they did intersect, either the plane of the two lines would meet K in more than 4n+2 points or alternatively, the order of the image regulus of the pencil determined by the two lines would be too high. But if no two lines of the regulus of multiple secants of K can intersect, then the regulus must be quadratic or K must be either a (1,4n+1) or a (2,4n) curve n a nonsingular quadric surface.

In what follows, we examine the two possible types of K mentioned above.

a. Suppose K is a (1,4n+1) curve on a nonsingular quadric and that the complex of singular lines are those meeting K, then any pencil in a plane containing a (4n+1)-fold secant of K has an image regulus which meets the plane of the pencil in 4n+1 lines, namely those which are images of the lines of the pencil which pass through the intersection of K and the multiple secant, plus an additional component to account for the intersections of the images of the general lines of the pencil. But if there is no additional complex of singular lines, the order of the image regulus of a pencil is 4n+1. This contradicts the preceding observation, and the case of K being a (1,4n+1) curve on a nonsingular quadric surface must be rejected. Thus, K must be a (2,4n) curve on a nonsingular quadric surface.

b. For K a (2,4n) curve on a nonsingular quadric V, we observe that the second regulus of V consists of the lines which join the two free intersections of K and the planes through any one of the multiple secants. For each of these lines meets V in three points: two on K and one point on one of the multiple secants.

Let us consider an arbitrary line, L, meeting V in two points A and B. If r is the multiple secant of K which passes through B and g is the simple secant of K which passes through A while A_1, A_2, A_3, ... A_{4n} are the points in which r meets K, and A is the image of A on the generator g, it follows that the image of line AA_1 is $A A_1$. Moreover, the image of AA is the same line.

Hence, the pencil (A,r) is transformed into a ruled surface of order 4n+1 containing 4n+1 cuncurrent generators. This image surface must be a cone, and hence every line of the pencil (A,r), including L, is transformed into a line through A , the harmonic conjugate of A, with respect to the two intersections of g and K. By a similar argument, interchanging the roles of B and A, we conclude that the other intersection of L is similarly transformed on the generator which passes through B. This proves our assertion on the uniqueness of this series of line involutions.

REFERENCES

1. Stratopoulos, G., "A New Line Involution and the Classification Process". Bull. Greek Math Soc., Vol. 21, 1980.

2. Stratopoulos, G., and Wylie G.R., "The Uniqueness of a Certain Certain Line Involution". Proc. Amer. Math. Soc., Vol. 19, 1968 pp. 1283-1289.

3. Snyder, V., and Sisam, C.H., "Analytic Geometry of Space", Henry Holt and Co., New York, 1924.

4. Semple, J.G., and Roth, L., "Introduction to Algebraic Geometry", Oxford, The Clarendon Press, 1949.

NATIONAL TECHNICAL UNIVERSITY OF ATHENS

GEOMETRY AND TOPOLOGY (pp. 291–305)
edited by G. M. Rassias & G. M. Stratopoulos

A negative answer to Hatcher's question

by Moto-o TAKAHASHI

Institute of Mathematics

University of Tsukuba

ABSTRACT

In [1], Hatcher raised the following question.

Question: For knot exteriors in S^3, must the boundary slopes of incompressible, ∂-incompressible surfaces always be integers ?

In this paper we shall give a negative answer to this question, that is:

Theorem. There exist a knot k and an incompressible and ∂-incompressible surface S in the exterior of k such that the slope of the boundary curves of S is not an integer.

In this paper we shall prove the following theorem.

Theorem. There exist a knot k and an incompressible and ∂-incompressible surface S in $S^3-\mathring{N}(k)$ such that the slope of the boundary curves of S is not an integer.

(Here $\mathring{N}(k)$ is the interior of a regular neighborhood of k.)

Examples of non-integer slopes are also obtained in [3], independently.

Before constructing k and S we shall prove a lemma. Let H be a handlebody of genus 2 and ℓ be a loop on ∂H. Let $N(\ell, \partial H)$ be a regular neighborhood of ℓ in ∂H and let $S_\ell=\partial H-\mathring{N}(\ell, \partial H)$. Moreover, let A and B be meridian disks of H and let $a=\partial A$, $b=\partial B$.

Lemma. Let ℓ be the loop as shown in the Figure 1, where $p>r>0$, $s>q>0$, $(p,r)=1$ and $(s,q)=1$. Then, S_ℓ is an incompressible and ∂-incompressible surface in H.

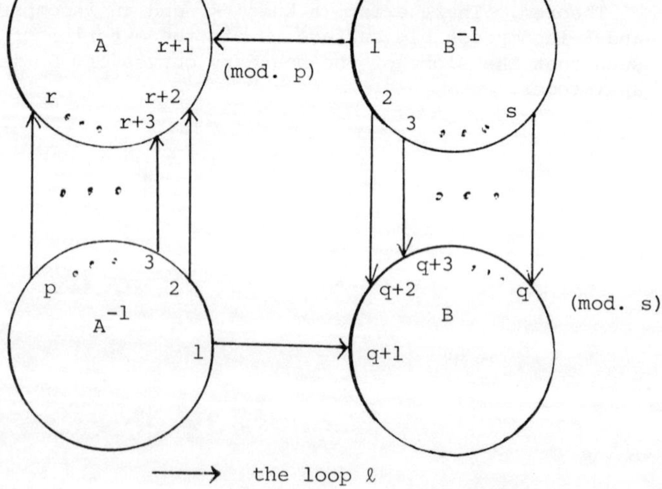

A and A^{-1} are glued together such that the same labels coincide, and similarly for B and B^{-1}.

Figure 1

Proof. S_ℓ is homeomorphic to a twice punctured torus $T-D^+-D^-$. Let $a_1,\cdots,a_n$ be the arcs in S_ℓ such that $\partial a_i=a_i\cap\partial S_\ell$ and $a\cap S_\ell=a_1\cup a_2\cup\cdots\cup a_n$. Similarly, let $b_1,\cdots,b_m$ be the arcs in S_ℓ such that $\partial b_i=b_i\cap\partial S_\ell$ and $b\cap S_\ell=b_1\cup b_2\cup\cdots\cup b_m$.

In the universal covering plane of T, the arcs $a_1,\cdots,a_n$ and $b_1,\cdots,b_m$ are drawn as in the Figure 2, where r', $p-r'$, q', $s-q'$ indicate the number of parallel arcs and $p>r'>0$, $rr'\equiv 1$ (mod. p), $s>q'>0$, $qq'\equiv 1$ (mod. s).

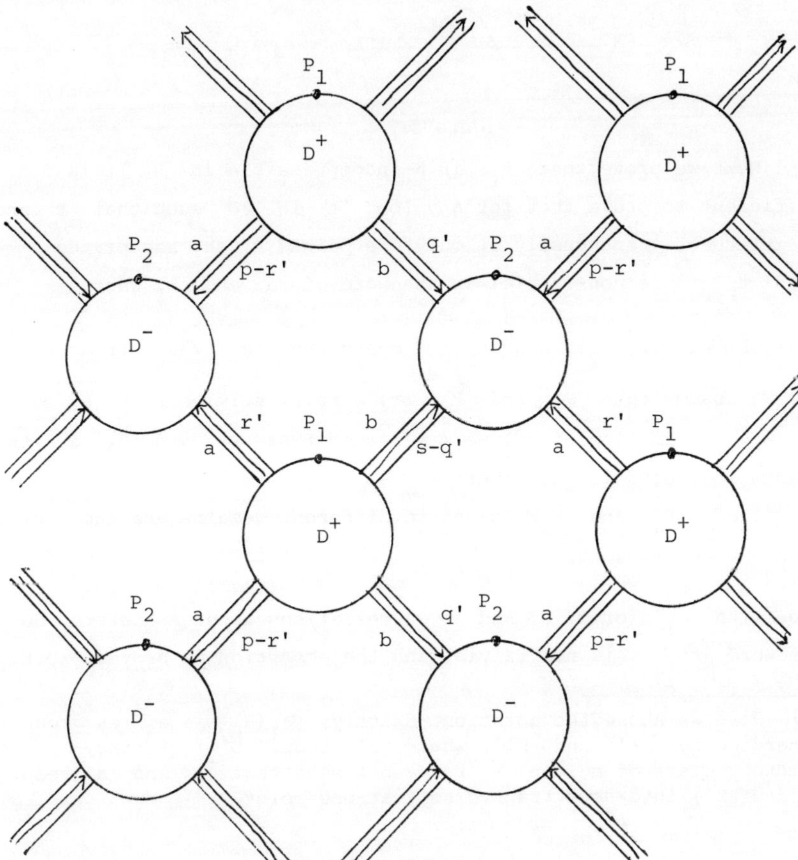

the reverse development of ℓ

Figure 2

We call this infinite graph, the reverse development of the loop ℓ on ∂H. (See [2].)

First we prove that S_ℓ is incompressible in H. For this it is sufficient to show that any non-trivial loop t in S_ℓ does not bound a 2-disk in H. Since $\pi_1(H)$ is a free group generated by a and b,*) it is sufficient to show that the word $\tilde{t}$ associated with the loop t is non-trival as the word of a and b. But this is obvious from the Figure 2, since no cancellations such as aa^{-1}, $a^{-1}a$, bb^{-1}, $b^{-1}b$ occur anywhere, after removing trivial intersections as in the Figure 3.

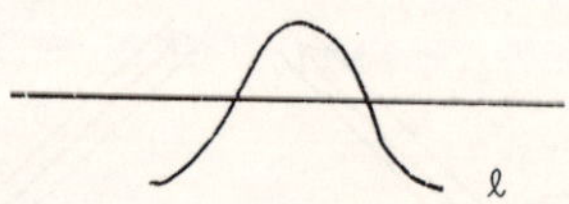

Figure 3

Next we prove that S_ℓ is ∂-incompressible in H. It is sufficient to prove that for any loop t in ∂H such that t and ℓ intersect transversely at only one point P, the associated word $\tilde{t}_{P_1P_2}$ is non-trivial as the word of a and b, where $\partial N(\ell,\partial H) \cap t = \{P_1, P_2\}$ and $t_{P_1P_2}$ means the arc $t \cap S_\ell$ from P_1 to P_2. We assume that P_1 and P_2 are sufficiently near P. By an isotopy deformation of t, we may assume that P_1 and P_2 are the points indicated in the Figure 2.

Then, P_1 and P_2 belong to different domains and the arc $t_{P_1P_2}$ connecting P_1 and P_2 in S_ℓ must intersect at least once with a_i's or b_i's. And no cancellations occur as before. So, the word $\tilde{t}_{P_1P_2}$ is non trivial and the assertion is proved. (Q.E.D.)

*) Here we abuse the notation. Exactly, $\pi_1(H)$ is a free group generated by a'' and b'', where a''(resp. b'') is the homotopy class of a loop a'(resp. b') such that a and a'(resp. b and b') intersect transversely at one point.

We are now in a position to prove the theorem. Let N be the 3-manifold obtained from a genus 2 handlebody H by attaching a 2-handle along the loop ℓ_1 shown in the Figure 4. The boundary of N is a torus. The fundamental group of N can be read from the diagram of the Figure 4:

$$\pi_1(N) \cong \langle a,b \mid a^5b^2a^2b^3a^2b^2=1 \rangle.$$

First we shall show that N is homeomorphic to a knot exterior. For this it is sufficient to show that the 3-sphere can be obtained by a Dehn surgery of N.

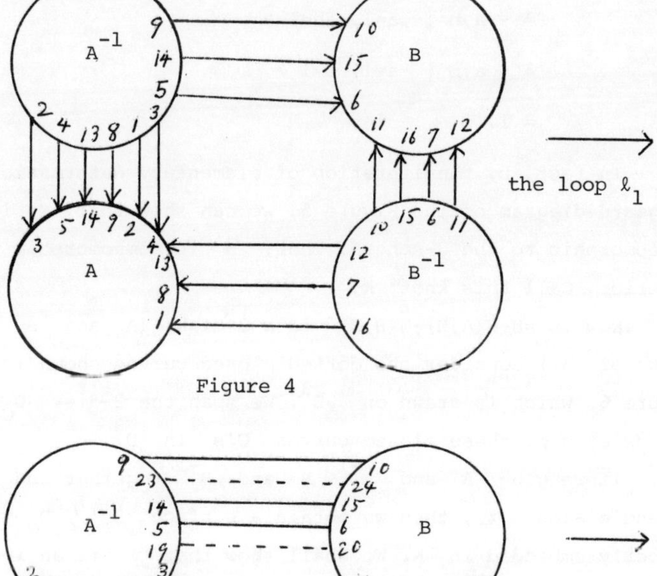

Figure 4

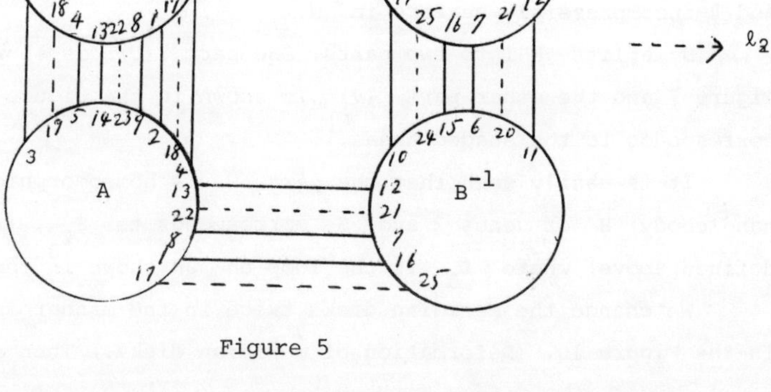

Figure 5

Let M' be the 3-manifold obtained from N by attaching a 2-handle along the loop ℓ_2 as shown in the Figure 5. The boundary of M' is a 2-sphere. Let M be the closed 3-manifold obtained from M' by attaching a 3-disk to $\partial M'$. Then, M is obtained by a Dehn surgery of N. Now,

$$\begin{aligned}
\pi_1(M) &\cong \pi_1(M') \\
&\cong \langle a,b \mid a^5b^2a^2b^3a^2b^2=1,\ a^3b^2a^2b^2=1 \rangle \\
&\cong \langle a,b \mid a^2ba^2b^2=1,\ a^3b^2a^2b^2=1 \rangle \\
&\cong \langle a,b \mid a^2ba^2b^2=1,\ ab=1 \rangle \\
&\cong \langle a,b \mid a=1,\ ab=1 \rangle \\
&\cong \langle a,b \mid a=1,\ b=1 \rangle \\
&\cong 1.
\end{aligned}$$

In fact, by the iteration of elementary deformation of Heegaard diagram of the Figure 5, we can show that M is homeomorphic to the 3-sphere. Thus, N is homeomorphic to a knot exterior. Call this knot k.

Now $D^3=H-\mathring{N}(A,H)-\mathring{N}(B,H)$ is a 3-disk. (A and B are meridian disks of H.) Consider the dotted closed curves shown in the Figure 6, which is drawn on ∂D^3. We span the 2-disk D_C which bounds each of these closed curves C's in D^3.

If we glue A and A^{-1}, B and B^{-1} together and attach a 2-handle along ℓ_1, then we obtain a surface $S=\bigcup_C D_C$, which is properly embedded in N. We shall show that S is an incompressible and ∂-incompressible surface in N.

S splits N into two parts. One part (U) is shown in the Figure 7 and the other part (V) is shown in the Figure 8. (S corresponds to the shaded areas.)

It is easily seen that the part U is homeomorphic to a handlebody H of genus 2 and S corresponds to S_{ℓ_3}, (S_ℓ is defined above) where ℓ_3 is the loop on ∂H shown in the Figure 9.

We change the meridian disks twice in the manner indicated in the Figure 10. (Deformation of meridian disks.) Then ℓ_4 is of

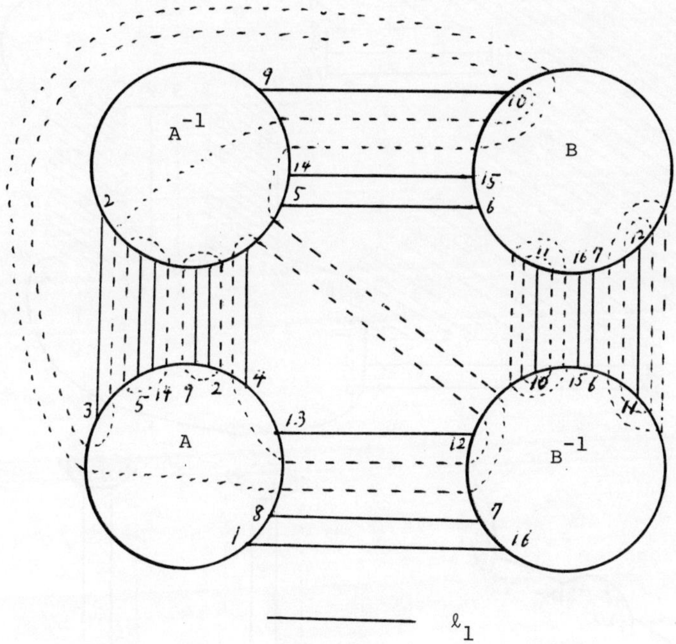

Figure 6

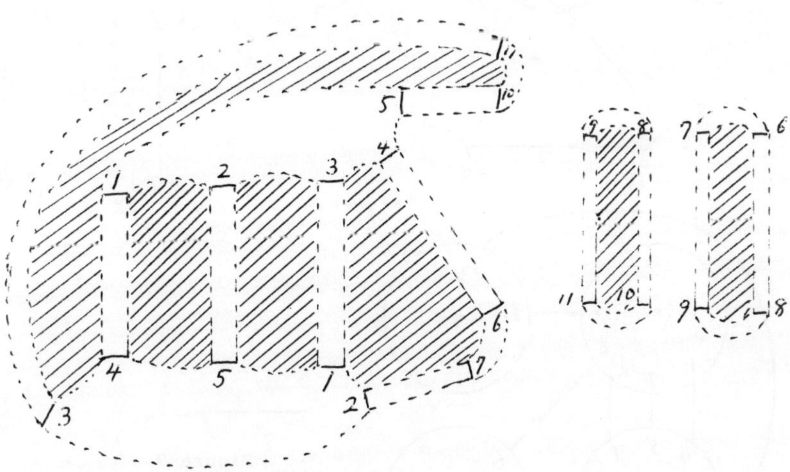

Figure 7

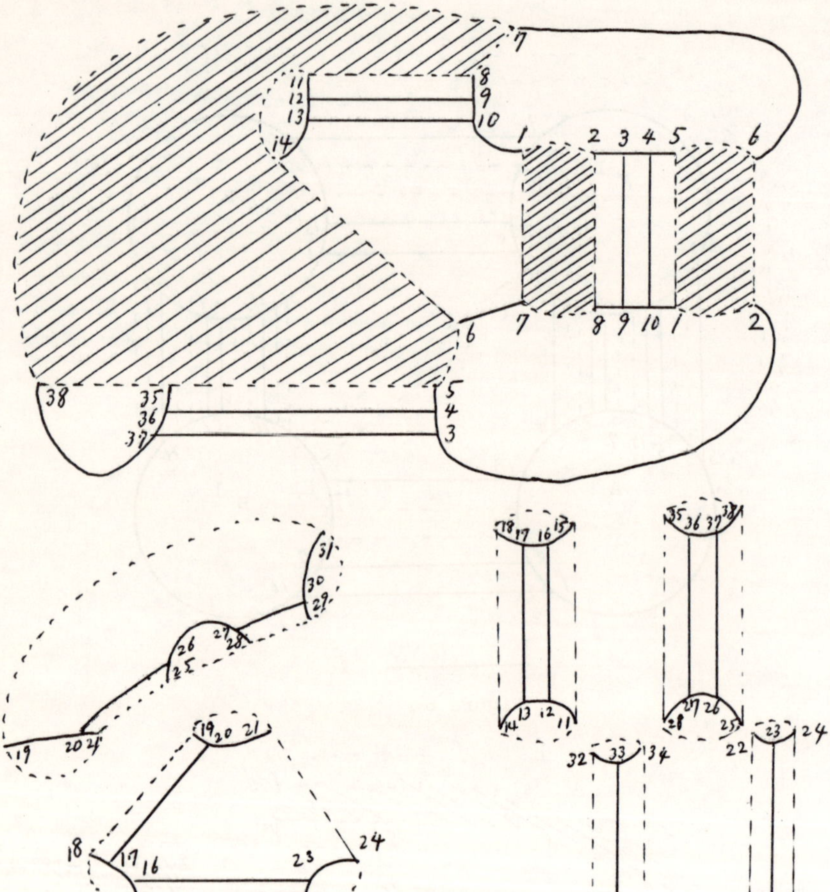

Figure 8

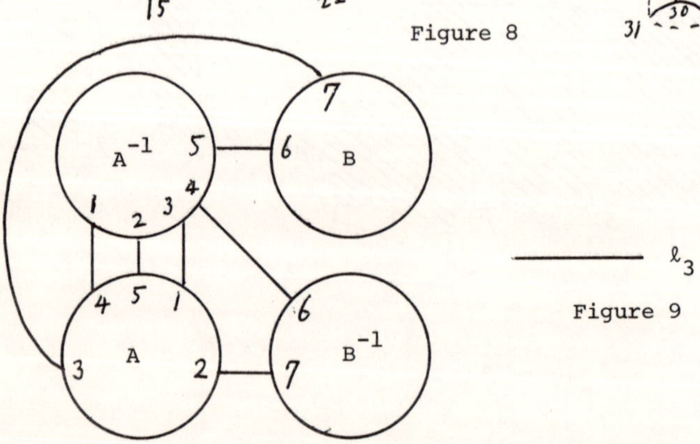

Figure 9

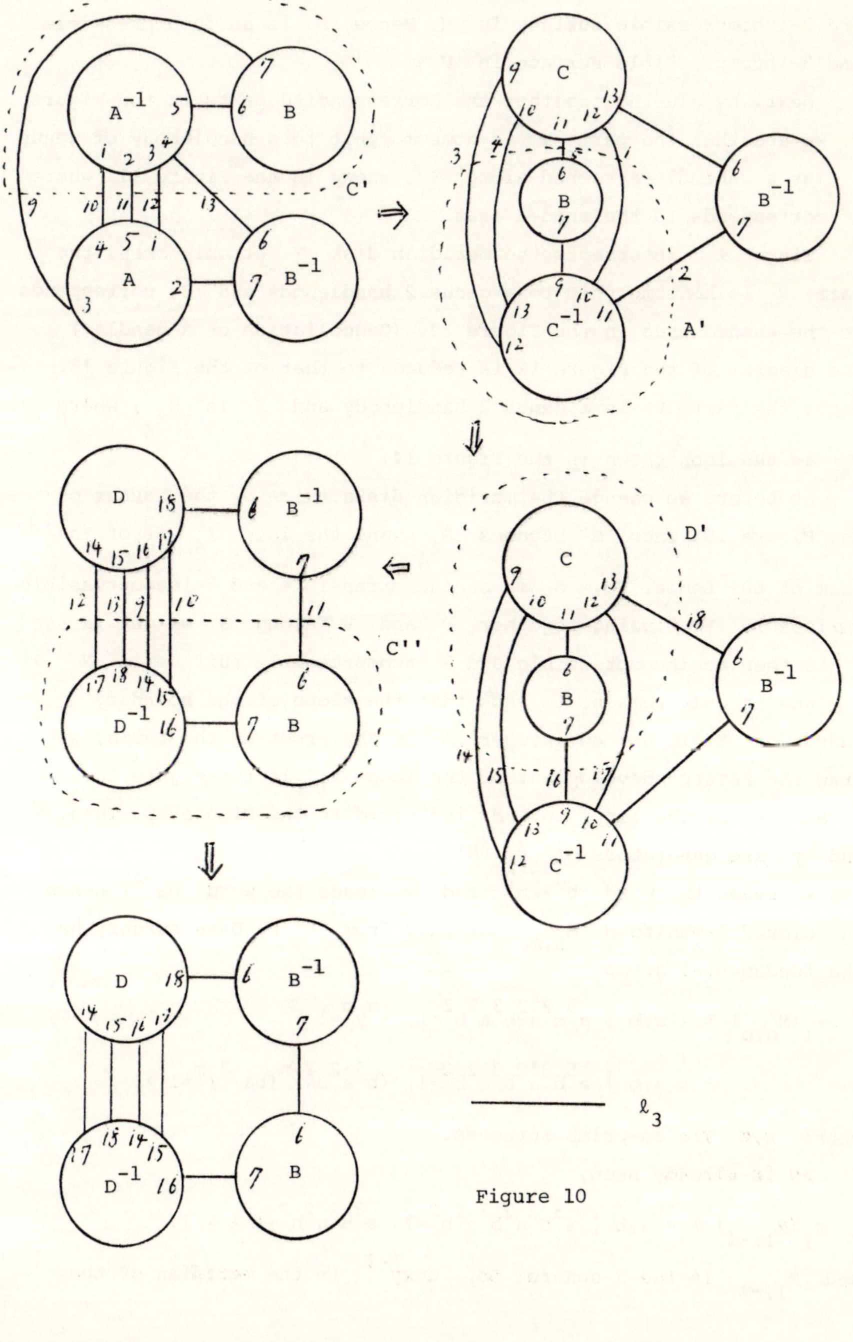

Figure 10

the form of ℓ in the above Lemma. Thus, S_{ℓ_4} is an incompressible and ∂-incompressible surface in H. Hence S is an incompressible and ∂-incompressible surface in U.

Next, by glueing together the corresponding parts in the Figure 8, we see that the part V is homeomorphic to a handlebody of genus 3 plus a 2-handle attached along ℓ_1, shown in the Figure 11, where S corresponds to the shaded area.

Since ℓ_1 intersects the meridian disk Y at only once, the part V is homeomorphic to a genus 2 handlebody and S corresponds to the shaded area in the Figure 12. (Cancellation of a handle.) The diagram of the Figure 12 is reduced to that of the Figure 13. Hence the part V is a genus 2 handlebody and S is S_{ℓ_4}, where ℓ_4 is the loop shown in the Figure 14.

As before we change the meridian disks twice by the manner of the Figure 15. Then S becomes S_{ℓ_4} and the loop ℓ_4 is of the form of the Lemma. So, S is an incompressible and ∂-incompressible surface in V. Glueing together U and V along S we get N and S is then an incompressible and ∂-incompressible surface in N.

Now it only remains to show that the slope of the boundary curves of S is not an integer. As in the proof of the Lemma, we draw the reverse development of the loop ℓ_1 in the Figure 16. Let x and y be the loops on ∂N indicated in the Figure 16. Then, x and y are generators of $\pi_1(\partial N)$.

x reads the word $b^3a^2b^2$ and y reads the word ba^{-3}. Hence the closed 3-manifold $M_{n,m}$ obtained from N by Dehn surgery has the fundamental group

$$\pi_1(M_{n,m}) \cong \langle a,b \mid a^5b^2a^2b^3a^2b^2=1,\ x^ny^m=1 \rangle$$

$$\cong \langle a,b \mid a^5b^2a^2b^3a^2b^2=1,\ (b^3a^2b^2)^n(ba^{-3})^m=1 \rangle$$

where n,m are co-prime integers.

As is already seen,

$$\pi_1(M_{1,-1}) \cong \langle a,b \mid a^5b^2a^2b^3a^2b^2=1,\ a^3b^2a^2b^2=1 \rangle \cong 1,$$

and $M_{1,-1}$ is the 3-sphere. So, $u=xy^{-1}$ is the meridian of the

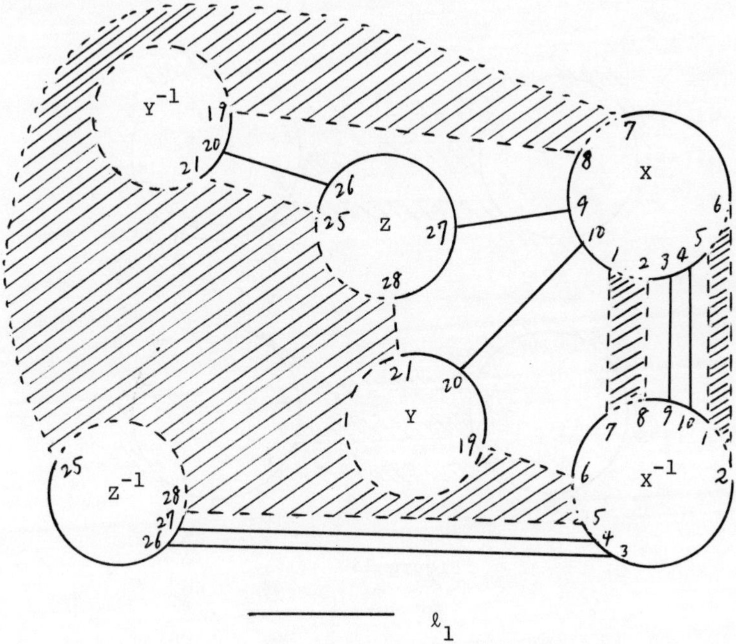

Figure 11

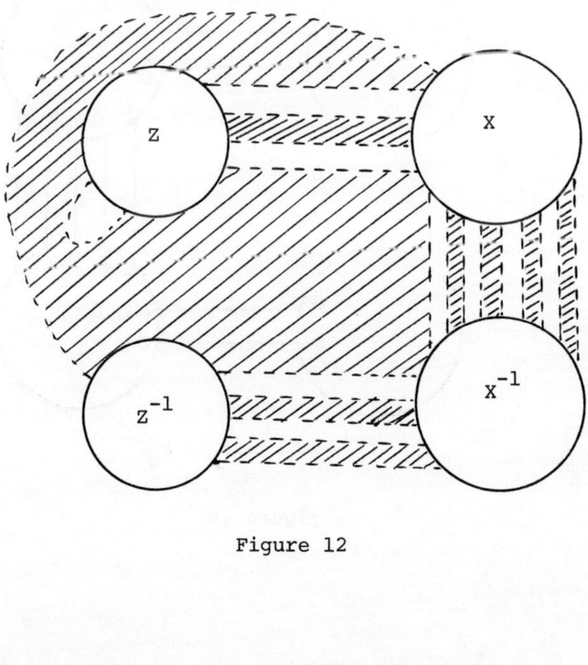

Figure 12

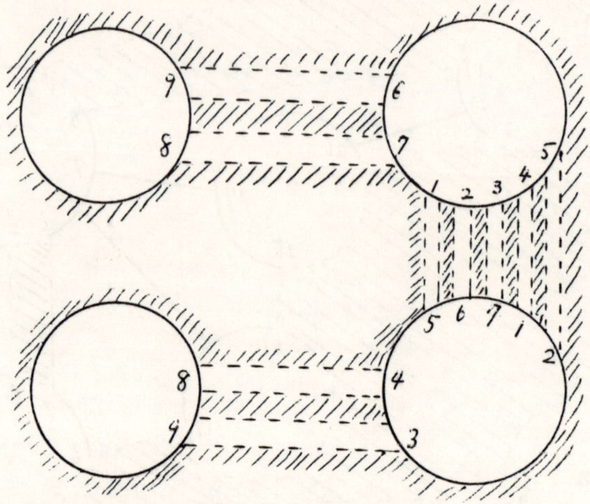

Figure 13

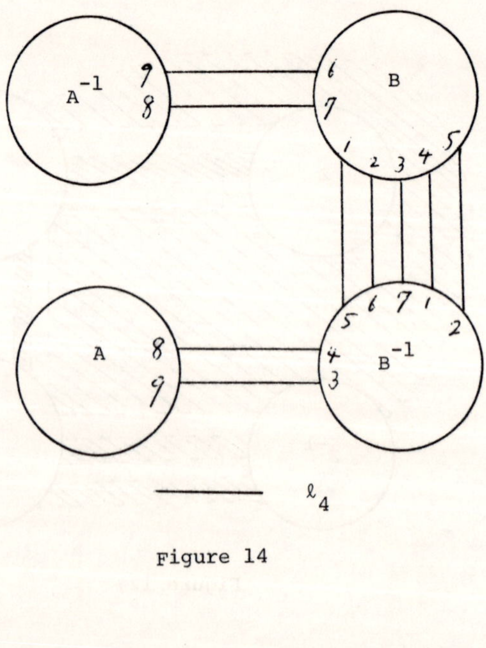

Figure 14

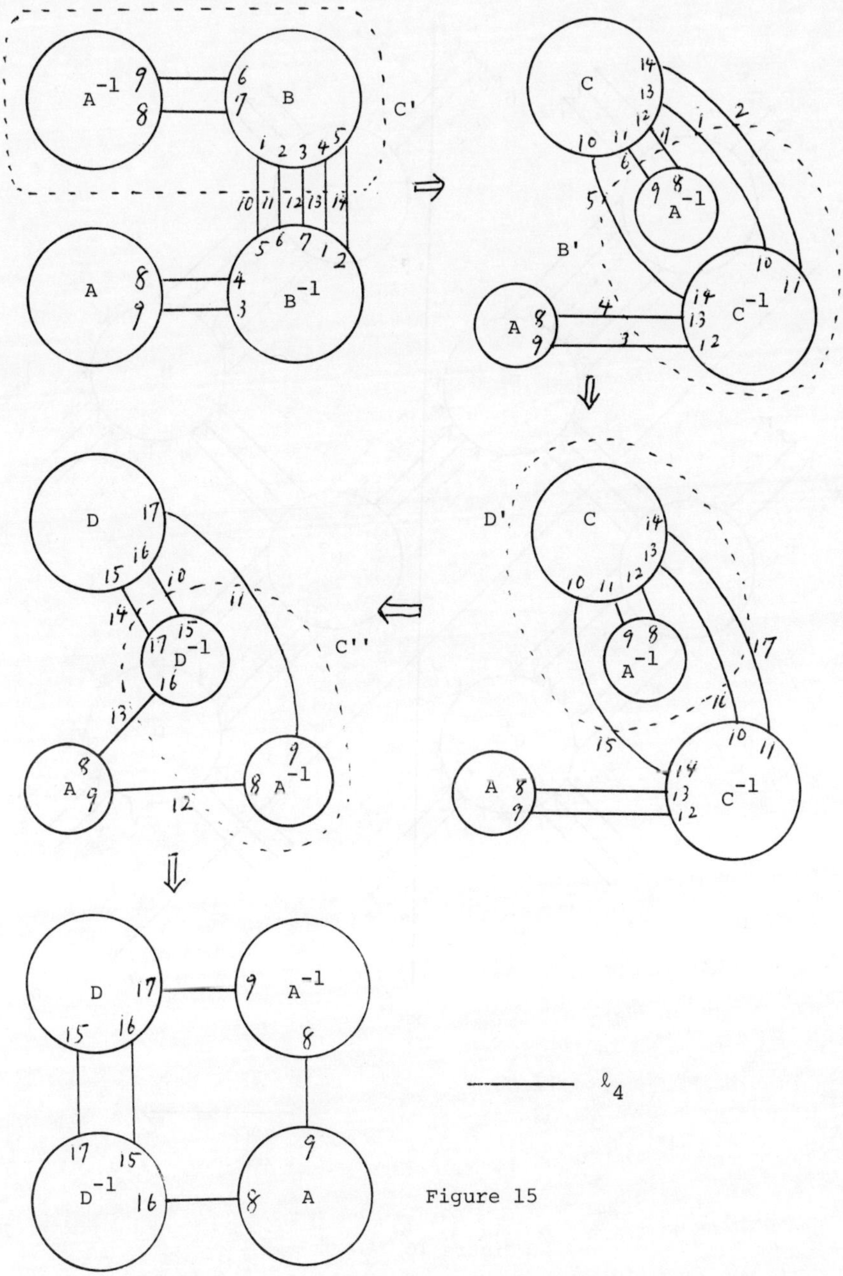

Figure 15

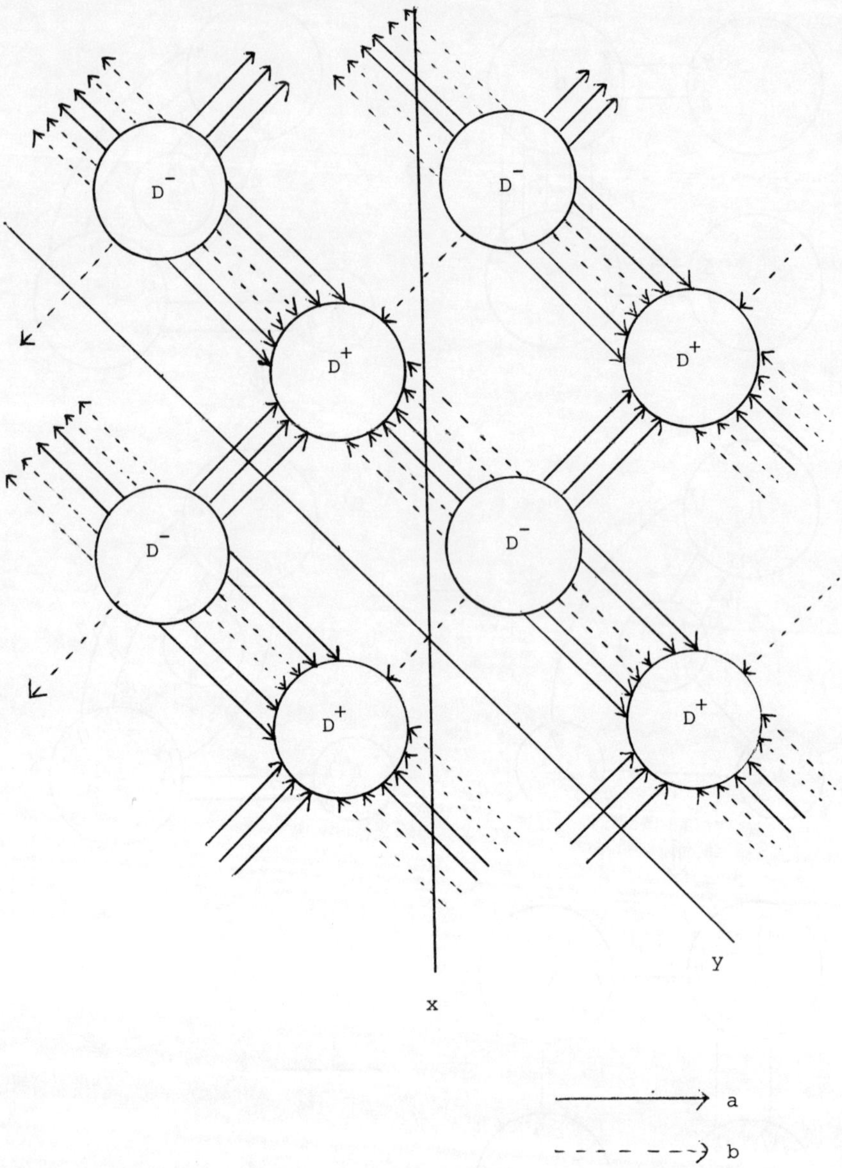

Figure 16

knot k, which we constructed. Let $v=x^{n_0}y^{m_0}$ be the longitude of k. Then $H_1(M_{n_0,m_0})$ is infinite. Since $H_1(M_{n_0,m_0})$ is the abelianization, we have

$$H_1(M_{n_0,m_0}) \cong \langle a,b \mid [a,b]=1,\ a^9b^7=1,\ a^{2n_0-3m_0}b^{5n_0+m_0}=1 \rangle .$$

If it is infinite, then we must have

$$9(5n_0+m_0)-7(2n_0-3m_0)=31n_0+30m_0=0.$$

Thus, the longitude of k is $v=x^{30}y^{-31}$. So, $x=u^{31}v^{-1}$ and $y=u^{30}v^{-1}$. Hence the slope of x^ny^m is $(31n+30m)/(n+m)$.

It can be easily observed that the boundary curves of S (see the Figure 6) are homotopic to xy. Thus the slope of them is 61/2, not an integer. This completes the proof of the Theorem.

Remark. By a similar method we can construct a knot k and an incompressible and ∂-incompressible surface S such that the slope of ∂S is p/q, where q is any integer and $(p,q)=1$.

References

[1] A.H.Hatcher: On the boundary curves of incompressible surfaces, Pacific Journal of Mathematics, vol. 99, No.2, (1982) 373-377.

[2] M.Takahashi: Some simple cases of Poincaré conjecture, Journal of Mathematical Society of Japan, vol. 32, No.2, (1980), 373-397.

[3] A.H.Hatcher and U.Oertel: Boundary slopes for Montesinos Knots, preprint.

GEOMETRY AND TOPOLOGY (pp. 306–314)
edited by G. M. Rassias & G. M. Stratopoulos

HARMONIC POLYNOMIAL MAPS BETWEEN SPHERES AND COMPLEX PROJECTIVE SPACES

GABOR TOTH*
RUTGERS UNIVERSITY
CAMDEN, NJ 08102 U.S.A.

ABSTRACT: We construct compact convex moduli spaces of harmonic maps between spheres and complex projective spaces.

§1. Introduction and generalities.

Let M be a compact oriented (isotropy) irreducible Riemannian homogeneous space. We write $M = G/K$, where G is a transitive Lie group of isometries of M and the isotropy subgroup K acts irreducibly on the tangent space $T_o(M)$ at the origin $o = \{K\}$. Let $\lambda \in Spec(M)$ and consider the associated (finite dimensional) eigenspace $V_\lambda \subset C^\infty(M)$. We endow V_λ with the normalized L_2-scalar product $\langle\, , \rangle$ defined by

$$\langle \mu , \mu' \rangle = \frac{n(\lambda)+1}{\int_M vol(M)} \int_M \mu \cdot \mu' \, vol(M),$$

where $\mu , \mu' \in V_\lambda$, and $n(\lambda) + 1 = \dim V_\lambda (=$ multiplicity of $\lambda)$. Precomposing eigenfunctions with isometries on M gives rise to an orthogonal G-module structure on V_λ [3].A map $f : M \to S^n$ into the Euclidean n-sphere is said to be a *λ-eigenmap* if the components f^i ,$i = 0, ..., n$, of f with respect to $S^n \subset \mathbf{R}^{n+1}$ belong to V_λ. Such maps are *harmonic* in the sense of J. Eells and J.H. Sampson [7] and, in fact, they can be characterized as harmonic maps of constant energy density $(= \frac{\lambda}{2})$ [8]. A map $f : M \to S^n$ is said to be *full* if the image of f is not contained in a proper linear subspace of $\mathbf{R}^{n+1}$. Two maps f ,$f' : M \to S^n$ are said to be *equivalent* if there exists an isometry $U \in O(n+1)$ such that $f' = U \circ f$. One of the fundamental problems in harmonic map theory posed by J. Eells and L. Lemaire [8] and R.T. Smith [15] in 1972 (for spherical domains) is to classify, for fixed $\lambda \in Spec(M)$, the equivalence classes of all full λ-eigenmaps $f : M \to S^n$.
The standard minimal immersion $f_\lambda : M \to S^{n(\lambda)}$ is the prototype of λ-eigenmaps

---

*Research carried out in part while the author was a member of the Institute for Advanced Study. 1980 Mathematics Subject Classification (1985 Revision). Primary 58E20, Secondary 53C42.

which is, in fact, a full (minimal) homothetic immersion. Its components comprise an orthonormal base of V_λ; different choices of the base give rise to equivalent standard minimal immersions. From here on we fix $f_\lambda : M \to S^{n(\lambda)}$ and thereby an isomorphism $\mathbf{R}^{n(\lambda)+1} \cong V_\lambda$.The G- module structure of V_λ then carries over to a G- module structure $\rho_\lambda : G \to SO(n(\lambda)+1)$ of $\mathbf{R}^{n(\lambda)+1}$. By construction, $f_\lambda : M \to S^{n(\lambda)}$ is equivariant with respect to the homomorphism ρ_λ.
A full λ-eigenmap $f : M \to S^n$ can then be written as $f = A \cdot f_\lambda$, where A is an $(n+1) \times (n(\lambda)+1)$-matrix of maximal rank. The symmetric matrix

$$\langle f \rangle = A^t \cdot A - I_{n(\lambda)+1} \in S^2(\mathbf{R}^{n(\lambda)+1})$$

(depends only on and) represents the equivalence class of f uniquely. (From here on, we use [16] as a standard reference, cf. also [19].) The condition $\langle f, f \rangle = 1$ translates into $\langle f \rangle \in E_\lambda$, where E_λ is the orthogonal complement of

$$span\{f_\lambda(x)^2 | x \in M\} \subset S^2(\mathbf{R}^{n(\lambda)+1}),$$

where $f_\lambda(x)^2$ is the symmetric square of the unit vector $f_\lambda(x) \in \mathbf{R}^{n(\lambda)+1}$ which is, in fact, orthogonal projection onto $\mathbf{R} \cdot f_\lambda(x)$. The orthogonal complement is taken with respect to the scalar product $\langle B, B' \rangle = trace\, B^t \cdot B', B, B' \in S^2(\mathbf{R}^{n(\lambda)+1})$. Clearly, E_λ is a G-submodule of $S^2(\mathbf{R}^{n(\lambda)+1})$ with respect to the induced module structure $Ad\rho_\lambda$ on $S^2(\mathbf{R}^{n(\lambda)+1})$.Setting

$$L_\lambda = \{C - I_{n(\lambda)+1} \in E_f | C \geq 0\}$$

(where $\geq$ stands for positive semidefinite) the correspondence $f \longmapsto \langle f \rangle$ gives rise to a parametrization of the equivalence classes of full λ- eigenmaps $f : M \to S^n$ by the compact convex body L_λ of E_λ. The G-module structure of E_λ leaves L_λ invariant and, on L_λ, the G-action is induced by precomposing λ-eigenmaps with isometries on M. For fixed $f : M \to S^n$, the isotropy subgroup $G_f = G_{\langle f \rangle} = \{g \in G | \exists U \in O(n+1)$ such that $f \circ g = U \circ f\}$ is nothing but the (maximal) *symmetry group* of f. The classification problem for λ-eigenmaps posed above can then be translated into the problem of understanding the geometry of L_λ (in particular, its boundary ∂L_λ) in E_λ.As a first step we introduce a natural cell structure of L_λ as follows. For a fixed full λ-eigenmap $f : M \to S^n$, define

$$E_f = (span\{f(x)^2) | x \in M\})^\perp \in S^2(\mathbf{R}^{n+1})$$

and

$$L_f = \{C - I_{n+1} \in E_f | C \geq 0\}.$$

Then the affine map $\phi : L_f \to L_\lambda$ defined by $\phi(C - I_{n+1}) = A^t \cdot C \cdot A - I_{n(\lambda)+1}$, injects L_f onto a compact convex set $\bar{I}_f$ containing $\langle f \rangle$. For a full λ-eigenmap

$f' : M \to S^{n'}$,$\langle f' \rangle \in \bar{I}_f$ iff $f' = A' \cdot f$ for some $(n'+1) \times (n+1)$-matrix A' of maximal rank, or equivalently, iff the components of f' are contained in the linear span of the components of f in V_λ. Denoting by $\mathcal{A}_f$ the affine subspace of E_λ spanned by $\bar{I}_f$, we have $\mathcal{A}_f \cap L_\lambda = \bar{I}_f$ and the interior I_f of $\bar{I}_f$ in $\mathcal{A}_f$ is a convex body containing $\langle f \rangle$. The convex sets I_f, for the various f, give rise to a cell decomposition of L_f. Clearly, $\mathrm{I}_{f_\lambda} = int\, L_\lambda$ and when passing to the boundary of a cell the range dimension of the corresponding full λ-eigenmaps decrease. Note also that the range dimension is constant on any I_f, in particular, it is $n(\lambda)$ on I_{f_λ}. For $g \in G$, we have $g \cdot I_f = I_{f \circ g^{-1}}$ so that the G-action on L_λ respects the cell structure. We now subdivide the classification problem introduced above into the following problems:

I. Compute $\dim L_\lambda = \dim E_\lambda$.

II. Decompose $E_\lambda(\otimes_{\mathbf{R}} \mathbf{C})$ into irreducible components and determine the highest weight vectors of the components.

III. Describe the cell structure of L_λ modulo the action of G.

Almost nothing is known for rank $M \geq 2$ (cf. [17]). For M rank 1 , E_λ is the sum of those irreducible G- submodules of $S^2(\mathbf{R}^{n(\lambda)+1})$ which are not class 1 with respect to (G, K) (i.e. which when restricted to K do not contain the trivial K-module) [17]. For $M = S^m$, I and II have completely been resolved [16]; III is known for the first nonrigid range ($m = 3$ and $\lambda = 8$) [19] while for $m \geq 5$ odd a cell on ∂L_λ is known corresponding to λ-eigenmaps arising from the Hopf- Whitehead construction applied to orthogonal multiplications [18].For $M = \mathbf{C}P^m$,I has been resolved in [1] using subtle representation theory while,for II,some components of E_λ have been determined previously [11,12,20].Some components of E_λ have also been discovered for $M = \mathbf{H}P^m$ whose dimensions thereby give a lower bound for I.

In §2 we treat the spherical case and select among the legion of examples some classical and recent ones of interest. The results are then applied in §3 to construct specific cells on ∂L_λ which give rise to a massive amount of new examples for *harmonic nonholomorphic maps between complex projective spaces*; an other fundamental question in harmonic map theory (cf. [8]). They include those discovered by A. Din and W. Zakrzewski [5,6] and further classified by J. Eells and J.C. Wood [9].

§2. Harmonic polynomial maps between spheres.

For $M = S^m$ with $(G, K) = (SO(m+1)\,, SO(m))$,we have $\lambda = \lambda_a = a(a+m-1) \in Spec(S^m)$ and the associated eigenspace V_{λ_a} is nothing but the irreducible $SO(m+1)$-module of spherical harmonics of order a on $S^m \subset \mathbf{R}^{m+1}$ [3]. To simplify the notation, from now on, we write $V_{\lambda_a} = V_a$, $E_{\lambda_a} = E_a$, etc. By the rigidity theorem of E. Calabi [4], for $m = 2$, we have $L_a = \{0\}$, i.e. the only full λ_a-eigenmap is $f_a : S^2 \to S^{2a}$, the standard minimal immersion, which is nothing but the classical Veronese surface in S^{2a} [3]. For $a = 1$ it is elementary that $L_a = \{0\}$. However, for $m \geq 3$ and $k \geq 2$, we have $\dim L_a > 0$. More precisely, for $m = 3$,

$$E_a \otimes_{\mathbf{R}} \mathbf{C} \cong \sum_{\substack{(b,c)\in\Delta_a \\ b,c\ even}} \left\{ V_3^{(b,c)} \oplus V_3^{(b,-c)} \right\}$$

and, for $m \geq 4$,

$$E_a \otimes_{\mathbf{R}} \mathbf{C} \cong \sum_{\substack{(b,c)\in\Delta_a \\ b,c\ even}} V_m^{(b,c,0,\ldots 0)},$$

where $\Delta_a \subset \mathbf{R}^2$ is the closed triangle with vertices $(2,2)$,(a,a) and $(2a-2,2)$ and V_m^ρ is the (complex) irreducible $SO(m+1)$ -module with highest weight $\rho = (\rho_1, \rho_2, \ldots, \rho_l) \in (1/2 \cdot \mathbf{Z})^l$, $l = [(m+1)/2]$ (cf. [16]). Applying the Weyl dimension formula, $\dim L_a$ can be computed.

The first nonrigid range $m = 3$ and $a = 2$, i.e. the structure of full quadratic eigenmaps $f : S^3 \to S^n$, is of particular interest [19]. Using coordinates (x,y,u,v) on $S^3 \subset \mathbf{R}^4$, let $f_n : S^3 \to S^n, 2 \leq n \leq 8, n \neq 3$, be defined by $f_n(x,y,u,v) =$

$$\begin{aligned}
&(x^2+y^2-u^2-v^2, 2(xu-yv), 2(xv+yu)), (Hopf\ map), n = 2\\
&(x^2+y^2-u^2-v^2, 2xu, 2xv, 2yu, 2yv), n = 4\\
&(x^2-y^2, u^2-v^2, 2xy, \sqrt{2}(xu+yv), \sqrt{2}(yu-xv), 2uv), n = 5\\
&(1/\sqrt{2}(x^2+y^2-u^2-v^2), 1/\sqrt{2}(x^2-y^2), 1/\sqrt{2}(u^2-v^2),\\
&\sqrt{2}xy, \sqrt{3}(xu+yv), \sqrt{3}(yu-xv), \sqrt{2}uv), n = 6\\
&(x^2-y^2, u^2-v^2, 2xy, \sqrt{2}xu, \sqrt{2}xv, \sqrt{2}yu, \sqrt{2}yv, 2uv), n = 7\\
&f_{\lambda_2}(x,y,u,v), (f_{\lambda_2}\ a\ standard\ minimal\ immersion), n = 8.
\end{aligned}$$

Then, $I_{f_n}, 2 \leq n \leq 8, n \neq 3$, comprise all cells of L_2 modulo the action of $O(4)$. Moreover, I_{f_2} = point, I_{f_4} = segment, I_{f_5} = 2-disk, I_{f_6} =(finite) solid cone, $\dim I_{f_7} = 5$ and $\dim I_{f8} = 10$. Note also that the $O(4)$-orbit of the point $\langle f \rangle$ corresponding to the Hopf map $f_2 : S^3 \to S^2$ has 2 components which arc imbedded in the appropriate 4-spheres of the 5-dimensional components $V_3^{(2,2)}$ and $V_3^{(2,-2)}$ of E_2 as *Veronese surfaces.*

An other discovery of cells in ∂L_2 is offered by the Hopf-Whitehead construction. Recall that a bilinear map $F : \mathbf{R}^{m+1} \times \mathbf{R}^{m+1} \to \mathbf{R}^n$ is called an *orthogonal multiplication* if $|F(x,y)| = |x| \cdot |y|, x, y \in \mathbf{R}^{m+1}$. For given F, the map $f_F : S^{2m+1} \to S^n$ defined by

$$f(x,y) = (|x|^2 - |y|^2, 2F(x,y)),$$
$$x, y \in \mathbf{R}^{m+1}, |x|^2 + |y|^2 = 1,$$

is a quadratic eigenmap (which is full iff F is surjective) [2]. Notice that f_2 and f_4 introduced above correspond to complex multiplication and real tensor product on $\mathbf{R}^2$, respectively. In general, we consider the cell $\bar{I}_\otimes = \bar{I}_{f_\otimes} \subset \partial L_2$, where $f_\otimes : S^{2m+1} \to S^{(m+1)^2}$ is associated with the tensor product $\otimes : \mathbf{R}^{m+1} \times \mathbf{R}^{m+1} \to \mathbf{R}^{(m+1)^2}$. Since $f_\otimes$ is equivariant with respect to $\rho_\otimes = \otimes : SO(m+1) \times SO(m+1) \to SO((m+1)^2)$, the point $\langle f_\otimes \rangle \in \bar{I}_\otimes$ is left fixed by $SO(m+1) \times SO(m+1)$ so that setting it as the origin of the affine span $\mathcal{A}_\otimes$ of $\bar{I}_\otimes$, we obtain an $SO(m+1) \times SO(m+1)$-module $\mathcal{A}_\otimes$. In fact, $\mathcal{A}_\otimes \cong so(m+1) \otimes so(m+1)$, where on the right hand side the module structure is given by $Ad \otimes Ad$ [18]. It follows that $\dim I_\otimes = (m(m+1)/2)^2$. Note that, for $m = 1, 2, \bar{I}_\otimes$ has been determined by M. Parker explicitly [13].

§3. Harmonic polynomial maps between complex projective spaces.

For $p > q \geq 0, p + q = a$, let $\mathcal{H}^{p,q}$ be the (complex) irreducible $U(m+1)$-module of complex harmonic polynomials on $\mathbf{C}^{m+1}$ of bidegree (p, q) [3]. A base $\{f^i_{p,q}\}_{i=0}^{n(p,q)} \subset \mathcal{H}^{p,q}, \dim_{\mathbf{C}} \mathcal{H}^{p,q} = n(p,q) + 1$, with respect to a normalized Hermitian L_2-scalar product on $\mathcal{H}^{p,q}$ induces a full λ_a-eigenmap $f_{p,q} : S^{2m+1} \to S^{2n(p,q)+1}$, where the components of $f_{p,q}$ are $\{Re(f^i_{p,q}), Im(f^i_{p,q})\}_{i=0}^{n(p,q)}$. Then, $f_{p,q}$ is equivariant with respect to the homomorphism $\rho_{p,q} : U(m+1) \to SO(2(n(p,q) + 1))$. Moreover as the central (diagonal) subgroup $S^1 \subset U(m+1)$ acts on $\mathcal{H}^{p,q}$ via $\rho_{p,q}$ by the single weight $p - q$, the map $f_{p,q}$ projects down to a map $\tilde{f}_{p,q} : \mathbf{C}P^m \to \mathbf{C}P^{n(p,q)}$ such that $\pi \circ f_{p,q} = \tilde{f}_{p,q} \circ \pi$, where π stands for the respective Hopf maps.

LEMMA. *The map $f_{p,q} : S^{2m+1} \to S^{2n(p,q)+1}$ is horizontal with respect to $\pi : S^{2m+1} \to \mathbf{C}P^m$, i.e. $(f_{p,q})_*(ker \pi_*)$ and $(f_{p,q})_*((ker \pi_*)^\perp)$ are orthogonal in $T(S^{2n(p,q)+1})$*

PROOF: For $z \in S^{2m+1} \subset \mathbf{C}^{m+1}$, the horizontal subspace $(ker \pi_*)^\perp_z$ is the orthogonal complement of $\mathbf{C} \cdot z$ in $\mathbf{C}^{m+1}$ (shifted to z). Given $w \in (ker \pi_*)^\perp_z$, we have to show that

$$\langle df_{p,q}(w)\, ,\, df_{p,q}(iz) \rangle = 0,$$

where we used Hermitian scalar product in $\mathbf{C}^{n(p,q)+1}$ and iz is considered in $T_z(S^{2m+1})$. By homogeneity, we have

$$\begin{aligned} df_{p,q}(iz) &= \frac{d}{dt} f_{p,q}(e^{it} z)|_{t=0} = \frac{d}{dt} e^{i(p-q)t}|_{t=0} \cdot f_{p,q}(z) \\ &= i(p-q) f_{p,q}(z) = i \frac{p-q}{a} df_{p,q}(z). \end{aligned}$$

On the other hand, differentiating $|f_{p,q}(\cos t \cdot z + \sin t \cdot w)|^2 = 1$ at $t = 0$, we obtain $\langle df_{p,q}(w)\, ,\, df_{p,q}(z) \rangle = 0$ and the proof is complete.

Applying the Reduction Theorem of R.T. Smith [15], we obtain that $\tilde{f}_{p,q} : \mathbf{C}P^m \to \mathbf{C}P^{n(p,q)}$ is a harmonic map.

REMARKS:
1.For $m = 1, n(p,q)+1 = (p+1)(q+1)-pq = p+q+1 = a+1$ we obtain the harmonic maps $\tilde{f}_{p,q} : \mathbf{C}P^1 \to \mathbf{C}P^a$. They and their conjugates comprise all harmonic maps of $\mathbf{C}P^1$ into $\mathbf{C}P^a$. (For a even we also have to add $\tilde{f}_{a/2,a/2} : \mathbf{C}P^1 \to \mathbf{C}P^a$ which is induced by $f_{a/2,a/2} : S^3 \to S^{2a+1}$ which is not full.) These are the harmonic maps which were discovered by A. Din and W. Zakrzewski [5,6] and classified by J. Eells and J.C. Wood [9].
2. For $q = 0$, the map $\tilde{f}_{a,0} = \tilde{v}_a : \mathbf{C}P^m \to \mathbf{C}P^{n(a,0)}, n(a,0) = \binom{m+a}{a}$ is nothing but the (holomorphic) Veronese map [14] induced by $v_a : S^{2m+1} \to S^{2n(a,0)+1}$, where

$$v_a(z_0, \dots, z_m) = ((a!/i_0!\dots i_m!)^{1/2} z_0^{i_0} \dots z_m^{i_m})_{\substack{i_0+\dots+i_m=a \\ i_0,\dots,i_m \geq 0}}.$$

We now take $\bar{I}_{p,q} = \bar{I}_{f_{p,q}}$ and intersect it with the linear subspace $Fix_{Ad\rho_a}(S^1, E_a)$ to obtain the convex set $Fix_{Ad\rho_a}(S^1, \bar{I}_{p,q})$. Given a full λ_a-eigenmap $f : S^{2m+1} \to S^N$ with $\langle f \rangle$ in the intersection, we have $f = A \cdot f_{p,q}$ for some $(N+1)\times(2(n(p,q)+1))$-matrix A of maximal rank. Moreover, as $\langle f \rangle \in Fix_{Ad\rho_a}(S^1, E_a)$, the map f is equivariant with respect to a homomorphism $\rho : S^1 \to SO(N+1)$ and $A : \mathbf{R}^{2(n(p,q)+1)} \to \mathbf{R}^{N+1}$ is intertwining between ρ_a and ρ. As ρ acts on $\mathbf{R}^{2(n(p,q)+1)}$ with the single weight $p - q (> 0)$ the same holds for ρ, in particular, $N = 2n + 1$ is odd and $A : \mathbf{C}^{n(p,q)+1} \to \mathbf{C}^{n+1}$ is complex linear. By equivariance, f projects down to a map $\tilde{f} : \mathbf{C}P^m \to \mathbf{C}P^n$ such that $\pi \circ f = \tilde{f} \circ \pi$. Repeating the proof of the previous lemma, it follows that f is horizontal with respect to $\pi : S^{2m+1} \to \mathbf{C}P^m$ so that $\tilde{f} : \mathbf{C}P^m \to \mathbf{C}P^n$ is harmonic. It follows that the convex set $Fix_{Ad\rho_a}(S^1, \bar{I}_{p,q})$ parametrizes the harmonic maps $\tilde{f} : \mathbf{C}P^m \to \mathbf{C}P^n$ obtained in the above manner.

REMARK: Examples are easy to construct. For instance, for $m = 2$ and $p = 2, q = 1$,

$$\begin{aligned} f(z,w,t) = (&\sqrt{7/8}(|z|^2 - 2|w|^2)z, \sqrt{1/8}(|w|^2 - 2|z|^2)w, \sqrt{7/8}(|w|^2 - 2|t|^2)w, \\ &\sqrt{1/8}(|t|^2 - 2|w|^2)t, \sqrt{7/8}(|t|^2 - 2|z|^2)t, \sqrt{1/8}(|z|^2 - 2|t|^2)z, \\ &\sqrt{6}z^2\bar{w}, \sqrt{6}w^2\bar{t}, \sqrt{6}t^2\bar{z}, \sqrt{6}\bar{z}wt) \end{aligned}$$

gives rise to a harmonic map $\tilde{f} : \mathbf{C}P^2 \to \mathbf{C}P^9$.

Returning to the general situation, we first note that $f_{p,q} : S^{2m+1} \to S^{2n(p,q)+1}$ is equivariant with respect to the homomorphism $\rho_{p,q} : U(m+1) \to SO(2(n(p,q)+1))$ corresponding to the $U(m+1)$-module structure on $\mathcal{H}^{p,q} \cong \mathbf{C}^{n(p,q)+1}$. Setting $\langle f_{p,q} \rangle$ as the origin of the affine span $\mathcal{E}_{p,q}$ of $Fix_{Ad\rho_a}(S^1, \bar{I}_{p,q})$ we obtain that $\mathcal{E}_{p,q}$ is a $U(m+1)$-module and

$$\mathcal{E}_{p,q} \cong Fix_{Ad\rho_a}(S^1, E_{p,q})$$

as $U(m+1)$-modules, where $E_{p,q} = E_{f_{p,q}}$ (cf.§1).
Let $\bar{E}_{p,q}$ be the sum of those irreducible $U(m+1)$- submodules of $S^2(\mathcal{H}^{p,q}_{\mathbf{R}})$ ($\mathcal{H}^{p,q}_{\mathbf{R}}$ = the realification of $\mathcal{H}^{p,q}$) which do not contain $U(m)$-fixed vectors.

PROPOSITION. $\bar{E}_{p,q} \subset E_{p,q}$.

PROOF: Let $p : S^2(\mathcal{H}^{p,q}_{\mathbf{R}}) \to W_0$ denote the orthogonal projection, where $W_0 = \mathbf{R} \cdot f_{p,q}(o)^2$ and $o = (1,0,...,0) \in S^{2m+1}$ is the base point left fixed by $U(m)$. Consider the induced representation

$$\begin{aligned} I = Ind_{U(m)}^{U(m+1)}(W_0) &= \{\psi : U(m+1) \to W_0 | \psi(u \cdot v) \\ &= u \cdot \psi(v), u \in U(m), v \in U(m+1), \psi \textit{ continuous}\}. \end{aligned}$$

For $\sigma \in S^2(\mathcal{H}^{p,q}_{\mathbf{R}})$ we define the map $\Psi(\sigma) : U(m+1) \to W_0$ by $\Psi(\sigma)(v) = p(v \cdot \sigma), v \in U(m+1)$. Then $\Psi(\sigma) \in I$ so that we obtain a homomorphism $\Psi : S^2(\mathcal{H}^{p,q}_{\mathbf{R}}) \to I$ of $U(m+1)$-modules. We have $\ker \Psi = (span\{U(m+1) \cdot W_0\})^{\perp} = (span\{f(x)^2 | x \in S^{2m+1}\})^{\perp} = E_{p,q}$ so that $im\Psi = E^{\perp}_{p,q} \subset I$ as $U(m+1)$-modules. By Frobenius reciprocity [21], we have $\dim \hom_{U(m+1)}(\bar{E}_{p,q}, E^{\perp}_{p,q}) \leq \dim \hom_{U(m+1)}(\bar{E}_{p,q}, I) = \dim \hom_{U(m)}(\bar{E}_{p,q}, W_0) = 0$ and the claim follows.

By the proposition above a lower estimate on $\dim Fix_{Ad\rho_a}(S^1, \bar{I}_{p,q}) = \dim \mathcal{E}_{p,q}$ is provided by

$$L(p,q) = \dim Fix_{Ad\rho_a}(S^1, \bar{E}_{p,q}).$$

To enumerate $L(p,q)$, we complexify

$$\begin{aligned} &Fix_{Ad\rho_a}(S^1, S^2(\mathcal{H}^{p,q}_{\mathbf{R}})) \otimes_{\mathbf{R}} \mathbf{C} \\ &= Fix_{Ad\rho_a}(S^1, S^2(\mathcal{H}^{p,q} \oplus \mathcal{H}^{q,p})) \\ &= \mathcal{H}^{p,q} \otimes \mathcal{H}^{q,p} \oplus \mathcal{H}^{q,p} \otimes \mathcal{H}^{p,q} \end{aligned}$$

and obtain that

$$Fix_{Ad\rho_a}(S^1, \bar{E}_{p,q} \otimes_{\mathbf{R}} \mathbf{C}) \subset \mathcal{H}^{p,q} \otimes \mathcal{H}^{q,p} \oplus \mathcal{H}^{q,p} \otimes \mathcal{H}^{p,q}$$

is the sum of those irreducible $U(m+1)$-submodules which do not contain $U(m)$-fixed vectors. This is just the condition for spherical harmonics $\mathcal{H}^{b,b}$ so that we obtain.

$$\begin{aligned} L(p,q) = &2 \dim_{\mathbf{C}} \mathcal{H}^{p,q} \otimes \mathcal{H}^{p,q} \\ &- 2 \sum_b m\left[\mathcal{H}^{b,b} : \mathcal{H}^{p,q} \otimes \mathcal{H}^{p,q}\right] \cdot \dim_{\mathbf{C}} \mathcal{H}^{b,b}. \end{aligned}$$

For $m, a \geq 4$, the Littlewood-Richardson rule [10,22] gives the multiplicity

$$m\left[\mathcal{H}^{b,b} : \mathcal{H}^{p,q} \otimes \mathcal{H}^{q,p}\right] = \min\{b+1, q+1, a-b+1\}$$

(cf. also [1]).

Since

$$\dim_{\mathbf{C}} \mathcal{H}^{p,q} = \binom{m+p}{p}\binom{m+q}{q} - \binom{m+p-1}{p-1}\binom{m+q-1}{q-1}$$

we finally get, for $m, a \geq 4$,

$$\dim Fix_{Ad\rho_a}(S^1, \bar{I}_{p,q}) \geq L(p,q) = 2\left[\binom{p+m}{p}\binom{m+q}{q} - \binom{m+p-1}{p-1}\binom{m+q-1}{q-1}\right]^2$$

$$-2\sum_{b=0}^{a} \min\{b+1, q+1, a-b+1\}\left[\binom{m+b}{b}^2 - \binom{m+b-1}{b-1}^2\right].$$

REMARKS:

1. Enumerating, for $m = a = 4$, we find

$$L(3,1) = 36{,}600.$$

2. For $q = 0$, $L(a,0) = 0$. In fact, as can be easily shown, $Fix_{Ad\rho_a}(S^1, \bar{I}_{a,0}) = \{\langle v_a \rangle\}$.

References

1. D.Barbasch, J. Glazebrook and G. Toth, *Harmonic maps of complex projective spaces into quotients of spheres*, Comm. Math. Helv. (to appear).

2. P. Baird, *Harmonic maps with Symmetry*, Pitman, 1983.

3. M. Berger, P. Gauduchon and E. Mazet, *Le spectre d'une variété riemannienne*, Lecture Notes in Math. No. 194, Springer, New York, 1971.

4. E. Calabi, *Minimal immersions of surfaces in Euclidean spheres*, J. Diff. Geom. 1(1974) 111-125.

5. A. Din and W. Zakrzewski, *General classical solutions in the $\mathbf{C}P^{n-1}$ model*, Nucl. Phys. B 174 (1980) 397- 406.

6. A. Din and W. Zakrzewski, *Properties of the general classical $\mathbf{C}P^{n-1}$ model*, Phys. Letters 95 B (1980) 419-422.

7. J. Eells and J. H. Sampson, *Harmonic mappings of Riemannian manifolds*, Amer. J. Math. 86(1964) 109-160.

8. J. Eells and L. Lemaire, *Selected topics in harmonic maps*, Reg. Conf. Ser. in Math. No.50, AMS, 1982.

9. J. Eells and J. C. Wood, *Harmonic maps from surfaces to complex projective spaces*, Adv. in Math. Vol. 49, No.3(1983) 217-263.
10. G.D. James, *The representation theory of the symmetric group*, Lecture Notes in Math. No. 682, Springer, New York, 1978.
11. K. Mashimo, *Degree of the standard isometric minimal immersions of complex projective spaces into spheres*, Tsukuba J. Math. Vol. 4 (1980) 133-135.
12. K. Mashimo, *Minimal immersions of quaternion projective spaces into spheres*, preprint.
13. M. Parker, *Orthogonal multiplications in small dimensions*, Bull. London Math. Soc. 15(1983) 368-372.
14. I. Shafarevich,*Basic algebraic geometry*, Springer, New York, 1977.
15. R.T. Smith, *Harmonic mappings of spheres*, Thesis, Warwick University, 1972.

16. G. Toth, *Harmonic and minimal maps*, E. Horwood Series, Halsted Press, John Wiley and Sons, New York, 1984.
17. G. Toth, *On nonrigidity of harmonic maps into spheres*, Proc. Amer. Math. Soc. Vol. 94, No.4 (1985) 711-714.
18. G. Toth, *On classification of orthogonal multiplications á la DoCarmo-Wallach*, Geom. Dedicata 22(1987) 251-254.
19. G. Toth, *Classification of quadratic harmonic maps of S^3 into spheres*, Indiana Univ. Math. J. Vol. 36, No.2 (1987) 231-239.
20.H. Urakawa, *Minimal immersions of projective spaces into spheres*, Tsukuba J. Math. Vol. 9,No.2 (1985) 321-347.
21. N.R. Wallach, *Minimal immersions of symmetric spaces into spheres*, in "Symmetric Spaces",Dekker, New York (1972) 1- 40.
22. H. Weyl, *The Classical groups*, Princeton Univ. Press 1946.

GEOMETRY AND TOPOLOGY (pp. 315–331)
edited by G. M. Rassias & G. M. Stratopoulos

TOPOLOGICAL TECHNIQUES IN GENERAL RELATIVITY

U.D. Vyas* and P.S. Joshi+
*Department of Mathematics and Computer Science
Winston-Salem State University
Winston-Salem, North Carolina 27110
+Tata Institute of Fundamental Research
Homi Bhabha Road, Bombay, 400005 India

ABSTRACT

The purpose of this paper is to illustrate some applications of Topology in General Theory of Relativity by surveying our work in this field. We prove some new results and also point out few open problems.

1. INTRODUCTION

The General Theory of Relativity has exhibited beautiful mathematical structures soon after its discovery. However, the development of modern technology has opened up many possibilities of experimental tests and the subject of 'Experimental Relativity' has fast evolved. On the other hand the mathematical developments have continued unabated and in the last twenty-five years more and more general forms of mathematics such as topology and manifold theory have been applied in relativity to discuss local and global problems of the spacetime universe. It is our purpose here to give an idea of some of these topological properties of certain physically significant subsets of spacetime by discussing several recently obtained results. It is quite natural after all that topology, which is a generalized form of geometry, should find profound applications in general relativity, which is, as Penrose[1)] has pointed out, a supremely beautiful description of the geometry of the world.

In the treatment presented here we have avoided giving most of the detailed proofs while providing ample references so that the interested reader can pursue the matter. Section 2 below discusses the mathematical modelling adopted for the universe in the general theory of relativity.

Higher order causality conditions are discussed in Section 3 whereas Section 4 introduces a special tool to measure volumes in spacetime, the causal functions in it. Automorphism or isomorphisms preserving causal structure are specially important in relativity and are dealt with in Section 5. Finally, Section 6 and 7 discuss respectively the physically important phenomenon of causality violation and spacetime singularities which may provide some introduction to the current research. The final section mentions some open problems.

2. MODELLING OF THE UNIVERSE

General Relativity requires a model of the Universe. The technical name of this model is spacetime and we shall denote it by M. Every point, called an event, has definite configuration and specific time giving us coordinates (x,y,z,t). These coordinates are not Euclidean but it is enough to keep them locally Euclidean. Thus one must map any 'small' region of a spacetime into a subset of Euclidean four plane by a bijection. This is being done by imposing a four dimensional manifold structure on M. To make the model physically reasonable we need differentiable, topological and causal structure on it.

In the differentiable structure of the spacetime, the theory requires covariance of the coordinates, C^∞ differentiability of coordinates at the overlapping of two charts, a Lorentzian metric g on it and orientability of the spacetime manifold M. We also grant the existence of a nowhere vanishing time like vector field on the spacetime which divides the non-spacelike vectors of each T_p into two classes, which we label arbitrarily as the future directed and past directed non-spacelike vectors. Here T_p is the collection of all tangent vectors to curves passing through a point p of M and vectors v of T_p are non-spacelike if $g(v,v) \leq 0$, see figure 1.

The atlas $\{U_\alpha, \phi_\alpha\}$ of the manifold will be assumed to be complete and by declaring any subset of M open if it is a union of such U_αs we obtain a Topology for our model called the manifold topology. We assume that our manifold is without boundary and also that the manifold topology is Hausdorff and connected. This topology is paracompact too because Geroch[2)] has proved that a T_2 - manifold with a Lorentzian metric is always paracompact. We further note that the Hausdorff nature and connectedness of M implies that M must have a countable basis because it is paracompact (Willard[3]).

The topology introduced above will help us in building causal structure of the spacetime. In special theory of relativity it is assumed that no physical information can travel faster than light; this is known as principle of causality in the special theory of relativity. This principle is equivalent to saying that thare are no closed non-spacelike curves in the spacetime. In the general theory of relativity the principle of causality is assumed locally only; i.e. there are no closed non-spacelike curves locally in M. More precisely, using the manifold topology, we say, $\forall$ p $\in$ M, $\exists$ a neighborhood U_p of p such that $\forall$ q $\in$ U_p there is a message (particle or light signal) from p to q if and only if there is a unique C^1 - non-spacelike curve from p to q lying entirely in U_p. If this condition is satisfied globally that is we do not allow closed non-spacelike curves globally in M is known as the chronology condition on M.

The incorporation of the principle of causality, or causality condition or chronology condition is not sufficient many times to bring in a reasonable causal structure which can match with the physical reality. Before we go to higher order causality conditions we state that the spacetime satisfies Einstein's equations

$$R_{ab} - \frac{1}{2} R g_{ab} + \Lambda g_{ab} = 8\pi T_{ab}$$

with usual notation. (e.g. Hawking and Ellis[4]).

3. OTHER CONDITIONS ON SPACETIME

Besides the non-occurance of closed non-spacelike curves, we need according to the physical reality, more conditions on the spacetime. In this section we state few such conditions and also give theorems in which topology plays an important role.

The chronological relation $\ll$ (causal relation $<$) between points of M are defined by saying $x \ll y$ ($x < y$) if and only if there is a future directed timelike (non-spacelike) C^2 curve from x to y. The chronological future $I^+(x)$ and causal future $J^+(x)$ of a point x in M are defined as under. (The definitions of the chronological past $I^-(x)$ and the causal past $J^-(x)$ are dual.)

$$I^+(x) = \{y \in M \mid x \ll y\}$$
$$J^+(x) = \{y \in M \mid x < y\}$$

The chronological comman future $\uparrow S$ of open set S of M is defined by

$$\uparrow S = \text{int}\{x \in M \mid \forall s \in S, \; s < x\}.$$

The definition of chronological comman past $\downarrow S$ is dual.

A spacetime M is distinguishing if $I^+(x) = I^+(y) \Rightarrow x = y$ and $I^-(x) = I^-(y) \Rightarrow x = y$. Further, M is future reflecting if $I^-(x) \subset I^-(y) \Rightarrow I^+(y) \subset I^+(x)$ and it is past reflecting if $I^+(x) \subset I^+(y) \Rightarrow I^-(y) \subset I^-(x)$; whereas M is reflecting if it is both past and future reflecting. We say M is causally continuous if it is distinguishing and reflecting.

Sometimes a condition stronger than distinguishing but weaker than the causal continuity is useful. Such a condition is strong causality. This property holds at a point $p \in M$ if given a neighbourhood N_p of p there is another neighbourhood $U_p \subset N_p$ of p such that no non-spacelike curve intersects U_p more than once. This physical condition can be characterized by a topology, called the Alexandroff topology of M. It is the topology generated by considering all sets of the type $I^+(x) \cap I^-(y)$, $\forall x, y \in M$ as the base of the topology. Penrose[5] and McWilliams[6] proved following theorem.

Theorem 1: In a spacetime M following are equivalent:

1) M strongly causal
2) The Alexandroff topology is T_2
3) The Alexandroff topology is T_1 - completely regular
4) The Alexandroff topology is metrizable.

Further, the whole space is non-compact with respect to the manifold topology otherwise the elementry condition of chronology is violated. But we need compactness in defining a strong condition, stronger than the causal continuity, called the global hyperbolicity. This holds if M is strongly causal and $J^+(p) \cap J^-(q)$ are compact $\forall p, q \in M$. We also need the concept of compactness in defining causal simplicity of M. The spacetime M is causally simple if for every compact set K in M, $J^+(K)$ and $J^-(K)$

are closed.

To observe the non-occurance of closed time like curves in the slight perterbation of the metric g we define stable causality. We say that the stable causality holds on M if the spacetime metric g has an open neighbourhood in the C^{o} open topology (refer Hawking and Ellis[4]) - a topology on the bundle of Lorentz metrics on M - such that there are no closed time curves in any metric belonging to the neighbourhood.

So far we have introduced few causality conditions on M, but in fact Carter[7] has pointed out that there exists an infinite heirarchy of causality conditions for a spacetime. At this juncture some one may ask if it is always necessary to deal with all these different causality conditions separately. We show below that if the sets $J^{\pm}(x)$ are closed in $M \,\forall x \in M$ then many of the important causality conditions are characterized by a single topological restriction on M in terms of the boundaries or frontiers of the sets $I^{\pm}(x)$. First we state the following theorem whose proof is given by Joshi[8],

Theorem 2: Let $J^{+}(x)$ be closed in $M \,\forall x \in M$. Then following are equivalent:

1) M is causal

2) $\partial I^{+}(x) \cap \partial I^{-}(x) = \{x\}$

3) M is distinguishing.

Now it can be verified that a spacetime in which the sets $J^{\pm}(x)$ are closed will be reflecting. We have also shown (Akolia, Joshi and Vyas[9]) that for such a spacetime the causality conditions like stable causality, strong causality and distinguishingness become equivalent. Thus we have essentially shown that under suitable conditions our topological boundary condition characterizes the causal regularity of M in a fairly complete manner.

Next two theorems characterize the chronology and reflectingness of the spacetime in terms of the Set Theory or Topology.

Theorem 3: M is Chronological

$\Leftrightarrow z \notin \uparrow I^{-}(z) \;\forall\; z \in M$

$\Leftrightarrow z \notin I^{+}(z) \;\forall\; z \in M$

Theorem 4:

$$M \text{ is reflecting} \Leftrightarrow p \in \partial I^+(z) \Leftrightarrow z \in \partial I^-(p), \ \forall\ p,z \in M.$$

The proofs of these are given in Joshi[8].

4. CAUSAL FUNCTIONS

Through the concept of causal functions we measure the 'size' or 'volume' of any subset of M and in particular that of $I^+(x)$ and $I^-(x)$ $\forall\ x \in M$.
Definition: Let $\sigma : \mathcal{P}(M) \to \mathbb{R}$ be a finitely subadditive set function with $\sigma(\emptyset) = 0$ and $\sigma(M) = 1$, where $\mathcal{P}(M)$ is the power set. Then the function $F^+ : M \to \mathbb{R}$ defined by

$$F^+(x) = \sigma[I^+(x)] \ \forall\ x \in M,$$

is called a future causal function on M. The definition of a past causal function is dual.

In this definition, the attempt is to attach a non-negative real number to each subset and particularly to $I^+(x)$ and $I^-(x)$. The usefulness of this view depends on the careful selection of σ. Later in this survey and also in our paper[10] we have suggested some choices of σ. Arbitrary choice of σ would provide a amthematical exercise but not a contribution to the theory of relativity. For example consider following σ:

$$\sigma(A) = 0 \text{ if A is countable.}$$

$$\sigma(A) = 1 \text{ if A is uncountable.}$$

This will bring $F^+(x) = 1$ for all x in M and such a causal function has no application in the literature so far.

Now, before we choose any particular σ to obtain a definite pair of causal functions. We see the nature of causal functions in two models of spacetimes; other models are discussed by Vyas and Joshi[10]. Further, when you think about the causal function in following two models remember that $F^+(x)$ has a numerical value which is proportional to its size.

Example: 1: (Klein-Minkowski Space). This space is formed by putting a Minkowski metric on $\mathbb{R}$xK, where K is the Klein Bottle (see e.g. Willard[3]). To obtain this space, top and bottom of a rectangle (figure 2) are identified giving a cylinder. Then the other sides of the rectangle are identified with a twist. Such a space is not time orientable. Consider a timelike curve through a point p and a null geodesic $(x^1, x^2, x^3) =$ constant, through a point x. Clearly both of them are closed curves. The route of γ is p-q-r-p. Along γ or η, F^+ and F^- both are constants with value equal to 1 because $I^+(p) = I^-(p) = I^+(x) = I^-(x) = M$.

Example:2: (Einstein's Universe)

Two sides of a rectangle are identified to obtain a cylinder called Einstein's Cylinder or Einstein's Universe. (See Figure 3). The light cones at bottom are vertical and gradually they bend and ultimately their null directions coincide with the null geodesic η. Again light cones tip back and obtain original position at top. One point is removed from η. There are no closed non-spacelike curves but for $p \neq q$ we have $I^+(p) = I^+(q)$. Here γ is timelike curve x-y-z and η is a null geodesic which is not closed.

F^+ is decreasing along γ but on η its value is constant and equal to ½; we note that the space is not future distinguishing. But by removing a strip, we can make the space past distinguishing and it can be seen that on η, F^- is not constant but infact monotone. In this space we do not have a single point x with $F^+(x)$ or $F^-(x)$ equal to unity, as it was the case in example 1.

As shown by Vyas and Joshi[10)] the choice of a reasonable causal functions $h^{\pm}$ can provide a unified description of the causal structure in terms of this tool. We point out here several interesting results while refering to the original paper for proofs and further detail. For example it is possible to show

Proposition 5: M is chronological if and only if causal functions $h^{\pm}$ are strictly monotone along each timelike curve.

We can understand the behaviour of $h^{\pm}$ in a distinguishing spacetime which is as follows:

Proposition 6: For a spacetime M, the following are equivalent:

1) M is future (past) distinguishing
2) h^+ (respectively h^-) is injective along all future (past) directed non-spacelike curves, and

3) h^+ (respectively h^-) is strictly decreasing (increasing) along all future (past) directed non-spacelike curves.

Proposition 7: A spacetime M is globally hyperbolic if and only if $h^{\pm}$ are strictly monotone and continuous along all timelike curves and $h^+ \to 0$ along future inextendible timelike curves and $h^- \to 0$ along past inextendible timelike curves.

5. CHRONAL ISOMORPHISMS AND AUTOMORPHISMS

When we deal with two models of the universe the isomorphisms play an important role in carrying properties of one model into the other one.

Definition: Let (M,g) and (M',g') be two spacetimes. A one-one and onto map $\emptyset$ from M to M' is a chronal isomorphism if for all $x,y \in M$, we have $x \ll y \Rightarrow \emptyset(x) \ll \emptyset(y)$.

This definition is same as that of Malaments'[11] causal isomorphism but since it preserves the chronological precedence only, we call it chronal isomorphism. Further, it is different from that of Budic and Sachs[12] who also assume $x < y$ implies $\emptyset(x) < \emptyset(y)$. If M = M' then this isomorphism called a chronal automorphism is same as automorphisms introduced by Zeeman[13] and Lester[14], however they call it a causal automorphism. The difference is that these two authors deal with particular forms of M and M' say Minkowskian, DeSitter or Einstein's cylinder spacetime where as we discuss here almost for a general spacetime.

Following basic lemma is proved in Vyas and Akolia[15].

Lemma 8: For a chronal isomorphism $\emptyset$, $\emptyset(I^{\pm}(x)) = I^{\pm}(\emptyset(x))$ for all x of the domain spacetime.

The mathematical interpretation of this is that $\emptyset$ and I commute and physical significance of this is that the chronal isomorphism maps a future (past) light cone $I^+(x)$ (respectively $I^-(x)$) at x into future (past) light cone $I^+(\emptyset(x))$ (respectively $I^-(x)$) at $\emptyset(x)$ and vice versa. That also means that it maps light rays to light rays.

Note: One can define a causal isomorphism Ψ by preserving the causal relationship $<$ and also causal automorphism Ψ_a. In that case one could obtain $\Psi(J^{\pm}(x)) = J^{\pm}(\Psi(x))$, which indicates preservation of non-spacelike curves in both ways.

Following lemma is very useful in proving many results. The proof is given in Vyas and Akolia[15].

Lemma 9: Let $\emptyset$: $(M,g) \rightarrow (M',g')$ be a chronal isomorphism. Then the Alexandroff topologies are homeomorphic.

One can see that without any extra condition on the underlying spacetime this $\emptyset$ is converted into a homeomorphism for the Alexandroff Topology. Now using this and theorem 1, we prove the following.

Proposition 10: Let $\emptyset$: $(M,g) \rightarrow (M',g')$ be a chronal isomorphism. Then M is strongly causal if and only if M' is strongly causal.

Proof: M strongly causal

$\Leftrightarrow$ The Alexandroff topology of M is T_2

$\Leftrightarrow$ The Alexandroof topology of M' is T_2

$\Leftrightarrow$ M' is strongly causal.

Further it can be proved that $\emptyset$ also preserves distinguishingness, W-Causality (Akolia, Joshi and Vyas[9]), reflectingness, causal continuity and the global hyperbolicity in both ways. But it is shown there by means of an example that $\emptyset$ cannot preserve the causality which means in general $\emptyset \neq \Psi$.

Following proposition is proved by Zeeman[13] for Minkowski spacetimes and Lester[14] for Minkowski and DeSitter spacetimes. Here we present an alternative proof of the same. Before we do this let us clarify the notations. The chronological ordering $\ll$ is denoted by $<$ in Zeeman's paper and by $\leftarrow$ in Lester's paper. They also use $\leftarrow\cdot$ notation for null ordering. That is $x \leftarrow\cdot\ y$ if and only if $x < y$ but $x \not\ll y$ or $x < y$ and $y \in J^+(x) - I^+(x)$.

Proposition 11: Let x and y be distinct points of M. Then

$$x \leftarrow\cdot\ y \Leftrightarrow \begin{cases} x \not\ll y \\ \forall z \in M,\ z \ll x \Rightarrow z \ll y. \end{cases}$$

Proof: $x \leftarrow\cdot\ y$ implies $x < y$ but $x \not\ll y$

Now let $z \in M$ be $\ni z \ll x$.

Then we have $z \ll x$ and $x < y$.

But Penrose[5] has proved in this case that $x \ll y$.

To prove the converse assume $x \not\leftarrow\cdot\ y$ and $x \ll y$.

Now,

$x \not<. \, y, \; x \not\ll y$

$\Rightarrow$ y is in the spacelike region of x

$\Rightarrow y \notin \overline{I^+(x)} \cup \overline{I^-(x)}$

$\Rightarrow x \notin \overline{I^+(y)} \cup \overline{I^-(y)}$

$\Rightarrow x \in \complement(\overline{I^+(y)} \cup \overline{I^-(y)})$

$\Rightarrow \exists$ an open set $O_x \ni x \in O_x \subset \complement(\overline{I^+(y)} \cup \overline{I^-(y)})$.

Now, let $z \in O_x \cap I^-(x)$ be a point.

i.e. z is in the local past of x.

Clearly $z \notin \overline{I^+(y)} \cup \overline{I^-(y)} \Rightarrow z \not\ll y$.

But we do have $z \ll x$.

This means $\exists\, z \in M, z \ll x$ with $z \ll y$ is false. Thus the negation of $x <. \, y$ implies the negation of the condition stated in the proposition.

For a Minkowski and DeSitter spacetimes it follows from the discussion given by Lester that the collection of all automorphisms of M is the same as the collection of all Lorentz transformations up to translation and dilatations. Further, if M = M' is Minkowski spacetime then it follows from the paper of Zeeman that the chronology group (the abstract group of all chronal automorphisms with binary operation as the composition of functions) is generated by (a) the orthochronal Lorentz group (b) translations of M and (c) dilatations of M.

6. VIOLATION OF CAUSAL CONTINUITY

It is found that the causal continuity is strong enough for most of the models of the universe. In many cases causal continuity is good substitute of even stronger conditions like global hyperbolicity. Looking into the importance of the causal continuity it is useful to discuss the sets where causal continuity breaks down. Now, causal continuity can be violated by not having reflectingness or distinguishingness in the spacetime. Thus we study the sets of non-reflecting points of M and also the sets of non-distinguishing points of M.

If R is the subset of M where reflecting condition is violated then Ishikawa[16)] has proved that the interior of R is empty, that the violation of reflectingness can not take place in 'big' part of the spacetime. We proved that at the same time such a violation is not an individual happening, or,

Proposition 12: No point of R is an isolated point.

Now suppose the distinguishingness fails on a subset Δ of M. Naturally the set Δ is the union of Δ^+ and Δ^- where Δ^+ (respectively Δ^-) is the set of points where future (past) distinguishing condition is violated. We also say that Δ-symmetric condition holds at a point $x \in M$ if $x \in \Delta^+ \cap \Delta^-$. Further, if $x \in \Delta^+$(respectively, Δ^-) then we define $\Delta^+(x)$ (respectively, $\Delta^-(x)$) as under.

$$\Delta^+(x) = \{ y \mid I^+(x) = I^+(y),\ y \neq x \}$$
$$\Delta^-(x) = \{ y \mid I^-(x) = I^-(y),\ y \neq x \}$$

In this connection following results are obtained in Vyas and Akolia[17].

Proposition 13: The formation of $\Delta^+(x)$ in Δ^+ gives rise to a partition of Δ^+.

Proposition 14: If Δ-symmetric condition holds at x then $\Delta^{\pm}(x)$ are closed in manifold topology.

Proposition 15: No point of Δ^+ is an isolated point.

If the spacetime M satisfies the condition of imprisonment (see Hawking and Ellis[4]) or causality it is shown that the interior of Δ is empty.

7. SINGULARITIES OF THE SPACETIME

The general theory of relativity covers study of two remarkable situations. First is the gravitational collapse of massive stars. This occurs when the nuclear fuel of the star is exhausted. In this case the star disappears from the view and leaves a 'black hole' in the spacetime. Second remarkable situation is the prediction of singularities in the spacetime. These singularities were at the beginning of the universe or will be at the end of the universe or it may be due to the break-down of physical laws somewhere in between in the gravitational collapse situations.

There are many forms of singularity theorems discussed (Hawking and Ellis[4]), predicting the existence of singularity. Very interesting situation is that all of them contain assumption of causality and predict that either causality is violated or spacetime is singular. On the other hand Tipler[18] has shown that causality violation is no longer an alternative to singularity. But what about a spacetime which is causal but violates one of the higher order causality conditions? In reply to this we have the following theorem:

Theorem 16: Let M be a causal general-relativistic spacetime. Then the violation of any one of strong causality, stable causality, causal continuity or causal simplicity in M implies that M is null geodesically incomplete, (i.e. singular) provided,

1) the weak energy condition (Hawking and Ellis[4]) holds in M.
2) the matter tensor satisfies on all null geodesics in M: $T_{ab}V^aV^b \geqslant K > 0$ where T_{ab} is the stress-energy tensor for the spacetime.

The proof of this theorem depends on the following lemma.

Lemma 17: Let M be causal. Let any one of strong causality, stable causality, causal continuity or causal simplicity in M be violated. Then there is a proper closed subset S of M such that $\partial I^+(S) \neq \emptyset$.

Proof: Violation of any one of the above mentioned strong causality conditions will imply the violation of causal simplicity. Thus $\exists$ a closed set $S \subset M \ni$ the causal simplicity is violated on S. So $J^+(S)$ is not closed by theorem 2.

Now if possible suppose S = M. Then $S = J^+(S) = M$ implies that $J^+(S)$ is closed. So S is proper subset of M. Further, $J^+(S) \subsetneq M \Rightarrow \overline{J^+(S)} \subseteq \bar{M} = M$.

Now since $J^+(S)$ is not closed $\exists\, p \in \overline{J^+(S)}$ with $p \notin J^+(S)$. This implies $p \in \partial J^+(S) = \partial I^+(S)$.

The complete proof and the explanation of the theorem 17 is given by Joshi[19] and it is predicted that the higher order causality violations in a causal spacetime imply a singularity in the spacetime.

8. OPEN PROBLEMS

The domain of intersection between topology and general relativity is indeed vast and important and interesting results can be found at many places in the literature. There are several important open problems as well and we list here a few relevant to the topics discussed in the present paper.

(1) Along with conditions of imprisonment or causality it is shown in Section 6 that the interior of the subset Δ of M, when distinguishingness is violated, is empty. It is still open to check the interior of Δ for a general spacetime.

(2) It has been pointed out in Section 7 that higher order causality violations give rise to singularities in the form of null geodesic incompleteness. However, one would expect such a singularity to be all-pervasive and material particle trajectories also should be incomplete in such a spacetime. Thus, the open problem is to see whether the timelike geodesic incompleteness is also associated with causality violations.

(3) An all important open problem is the nature of singularities associated with causality violations. It would be useful to show that these are curvature singularities. Further, it is also necessary to ascertain whether these accompanied by a horizon or they are naked singularities.

(4) New exact solutions of Einstein's equations which exhibit causality violating properties, however which have physically reasonable matter content would be most useful to provide insight in the phenomena of closed non-spacelike curves.

REFERENCES

1. Penrose, R. 'Singularities and Time-asymmetry', 'General Relativity' (Ed. Hawking S.W. and Israel W.) Cambridge Univ. Press 581 (1979).
2. Geroch, R. 'Spinor Structure of Spacetimes in General Relativity' J.Math. Phys.9 1739 (1968).
3. Willard, S. 'General Topology' Addison Wesley Publishing Co. (1970).
4. Hawking, S.W. and Ellis, G.F.R. 'The Large Scale Structure of Spacetime', Cambridge Univ. Press (1977).
5. Penrose, R. 'Techniques of Differential Topology in Relativity', A.M.S. Colloquim Publication SIAM, Philadelphia (1971).
6. McWilliams, K.J. 'Completeness of the Alexandrov Topology for Spacetime', Int.J.Theo.Phys. 20 905 (1981)

7. Carter, B. 'Causal Structure in Spacetime' Gen.Rel.Grav. 1 349 (1971).
8. Joshi, P.S. 'Topological Properties of Certain Physically Significant Subsets of Spacetime' 'A Random Walk in General Relativity and Causality' (Ed. N. Dadhich et al), John Weiley and Co. 128 (1984).
9. Akolia, G.M., Joshi, P.S. and Vyas, U.D. 'On Almost Causality' J.Math.Phys. 22 1243 (1981).
10. Vyas,U.D. and Joshi,P.S. 'Causal Functions in General Relativity' Gen.Rel.Grav. 15 553 (1983).
11. Malament, D.B. 'The Class of Continuous Timelike Curves Determines the Topology of Spacetime' J.Math.Phys. 18 1399 (1977).
12. Budic, R. and Sachs, R.K. 'Causal Boundaries for General Relativistic Spacetimes' J.Math.Phys. 15 1302 (1974).
13. Zeeman, E.C. 'Causality Implies the Lorentz Group' J.Math.Phys. 5 490 (1964).
14. Lester, J.A. 'The Causal Automorphism of DeSitter and Einstein Cylinder Spacetime' J.Math.Phys. 25 113 (1984).
15. Vyas, U.D. and Akolia, G.M. 'Chronal Isomorphism' Gen.Rel.Grav. 16 1045 (1984).
16. Ishikawa, H. 'Properties of the Set of Non-Reflecting Points' Ge.Rel.Grav. 10 31 (1979).
17. Vyas, U.D. and Akolia, G.M. 'Causally Discontinuous Spacetimes' Gen.Rel.Grav. 18 309 (1986).
18. Tipler, F.J. 'Causality Violation in General Relativity' Annals of Physics 110 251 (1978).
19. Joshi, P.S. 'On Higher Order Causality Violations' Physics Letter 85A 319 (1981).

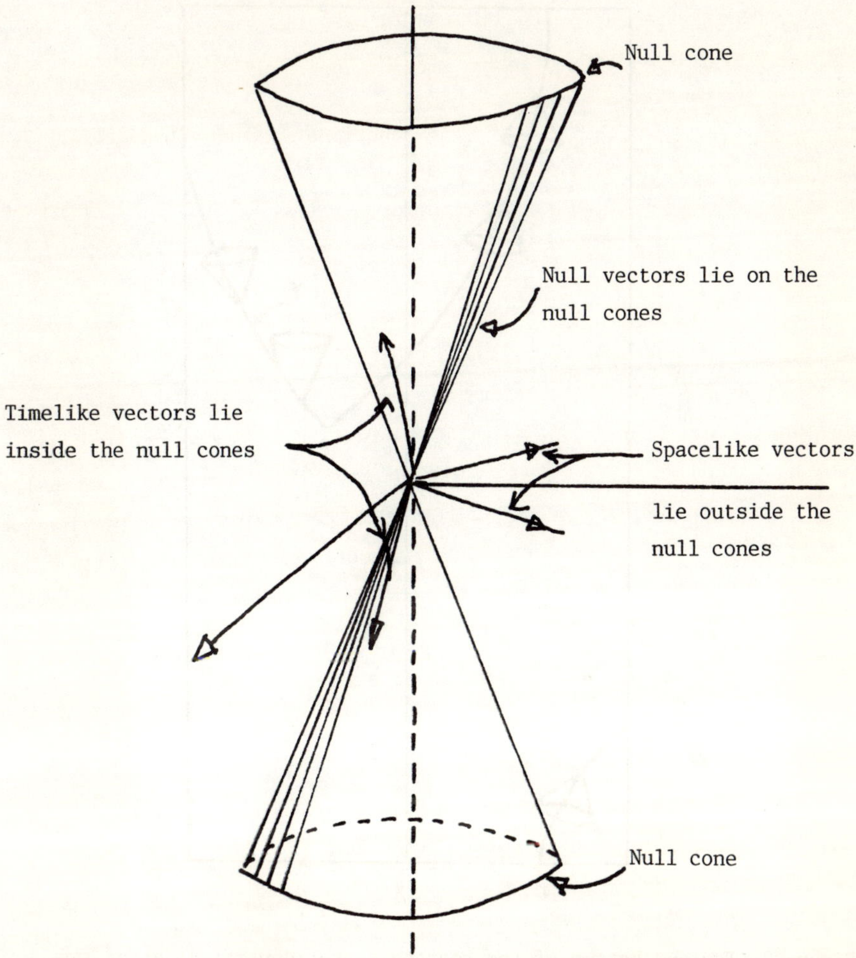

Figure 1: The division of vector space T_p into three mutually disjoint classes.

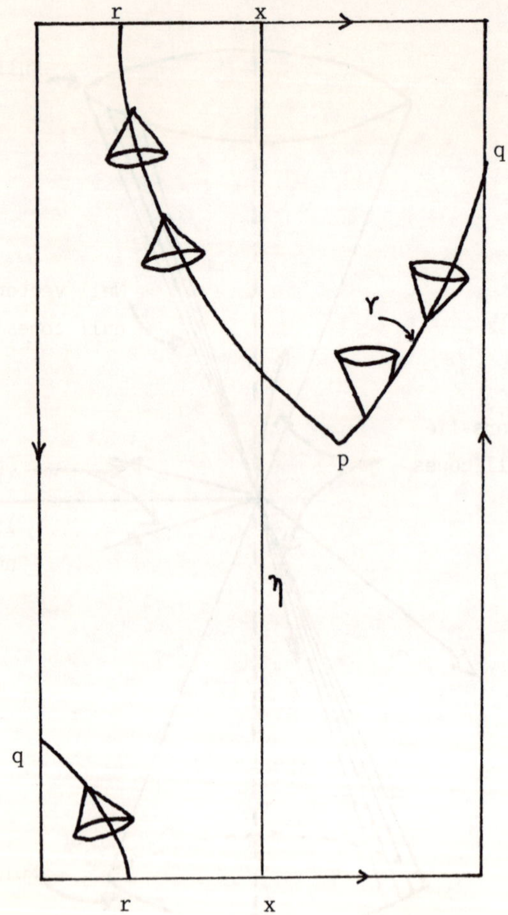

Figure 2: Top and bottom of the rectangle are identified; other two sides are identified with a twist. γ is a timelike curve and η is a null geodesic on it. F^+ and F^- both are constant on γ as well as on η .

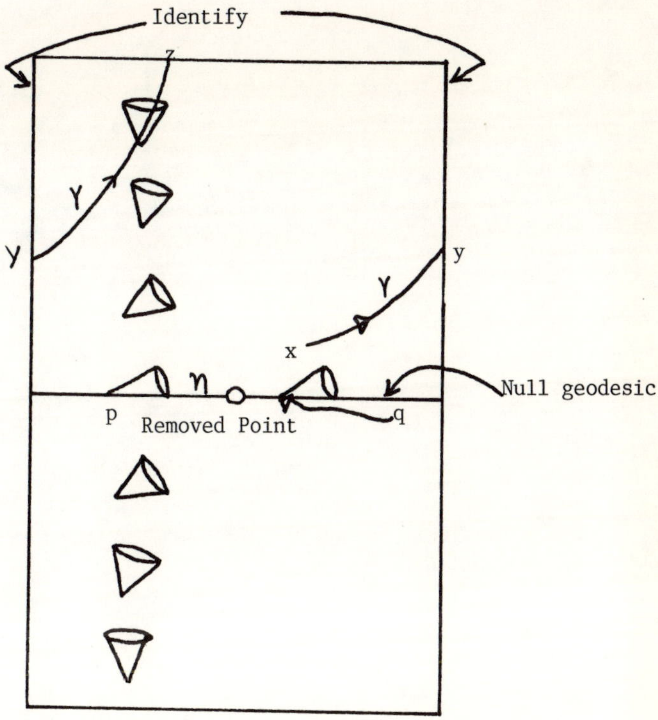

Figure 3: Two sides of a rectangle are identified to obtain a cylinder called Einstein's cylinder. The light cones at bottom are vertical and then gradually they bend and ultimately their null direction coincide with the null geodesic η . Again light cones tip back and obtain original position at top. One point is removed from η . There are no closed non-spacelike curves but for $p \neq q$ we have $I^+(p) = I^+(q)$.

γ is a timelike curve x-y-z and η is a null geodesic which is not closed.